COMMUNITY OPERATIONAL RESEARCH

Contemporary Systems Thinking

Series Editor: Robert L. Flood
Monash University
Australia

COMMUNITY OPERATIONAL RESEARCH
OR and Systems Thinking for Community Development
Gerald Midgley and Alejandro Ochoa-Arias

CRITICAL SYSTEMIC PRAXIS FOR SOCIAL AND ENVIRONMENTAL JUSTICE
Participatory Policy Design and Governance for a Global Age
Janet McIntyre-Mills

DESIGNING SOCIAL SYSTEMS IN A CHANGING WORLD
Bela H. Banathy

GUIDED EVOLUTION OF SOCIETY
A Systems View
Bela H. Banathy

METADECISIONS
Rehabilitating Epistemology
John P. van Gigch

PROCESSES AND BOUNDARIES OF THE MIND
Extending the Limit Line
Yair Neuman

SELF-PRODUCING SYSTEMS
Implications and Applications of Autopoiesis
John Mingers

SOCIOPOLITICAL ECOLOGY
Human Systems and Ecological Fields
Frederick L. Bates

SYSTEMIC INTERVENTION
Philosophy, Methodology, and Practice
Gerald Midgley

COMMUNITY OPERATIONAL RESEARCH

OR and Systems Thinking for Community Development

Edited by

Gerald Midgley
Institute of Environmental Science and Research (ESR Ltd)
Christchurch, New Zealand
and
Centre for Systems Studies
Business School University of Hull
Hull, United Kingdom

and

Alejandro Ochoa-Arias
Universidad de Los Andes
Marida, Venezuela

Kluwer Academic / Plenum Publishers
New York, Boston, Dordrecht, London, Moscow

Library of Congress Cataloging-in-Publication Data

Community operational research: OR and systems thinking for community development/edited by Gerald Midgley and Alejandro Ochoa-Arias.
p. cm. — (Contemporary systems thinking)
Includes bibliographical references and index.
ISBN 0-306-48335-1
1. Community development. 2. Operations research. 3. Systems analysis. I. Midgley, Gerald. II. Ochoa-Arias, Alejandro. III. Series.

HN49.C6C673 2004
307.1'4—dc22

2003070329

ISBN: 0-306-48335-1

233 Spring Street, New York, New York 10013

http://www.wkap.nl

10 9 8 7 6 5 4 3 2 1

A C.I.P. record for this book is available from the Library of Congress

Printed in the United States of America

Preface

In this book we have sought to consolidate the previously published literature on Community Operational Research ('Community OR' for short), and at the same time introduce new writings that push out the boundaries of theory, methodology and practice. We hope the balance will be of value to you, whether you have a long-standing commitment to Community OR, or are coming to this subject for the first time.

Following an introductory chapter, the book is presented in three sections:

Section One offers 'Early Contributions and Later Reflections on Theory, Methodology and Practice'. Here, the nature of Community OR, its institutional development, and people's motivations for engaging in it are all explored. Some significant theoretical and methodological issues are also a focus.

Section Two covers 'Local Action for Community Development'. This is concerned with how people have translated the theoretical insights of Community OR into practice, and how practice has informed theory. Since the inception of Community OR, the vast majority of projects have dealt with complex, localised community issues using participative methods (mostly in the developed world). This section of the book presents a variety of methodological ideas evolved for Community OR practice, and illustrates them with examples of projects with community groups, voluntary organisations and welfare services.

Section Three is entitled 'Dealing Locally with Global Issues'. The authors represented in this section seek to extend the practice of Community OR in two new directions. The first takes Community OR into developing countries, and the second takes it into the arena of environmental management. In both these areas, there are significant opportunities for Community OR practitioners to make real contributions to human welfare and environmental sustainability.

We hope that you enjoy the chapters to follow, and feel inspired (like the authors who have contributed to this book) to participate in both Community OR practice and the unfolding debate surrounding it.

Gerald Midgley
Alejandro E. Ochoa-Arias

Contents

1

An Introduction to Community Operational Research

Gerald Midgley and Alejandro E. Ochoa-Arias

1.1 Introduction

It is common for introductory chapters of edited books to start by defining their subject matter. However, in this case, it is really the task of the whole book, not this paragraph, to communicate the meaning of Community Operational Research ('Community OR' for short). We say this because we believe that healthy research networks involve a diversity of participants with both common and divergent interests, and the Community OR network is no exception. For us to specify right at the start of this book what Community OR is all about would almost certainly mean marginalising other people's concerns. We therefore prefer to portray Community OR, not as a finite set of ideas or activities, but as a label used by a variety of people engaged in debate and on-going learning about their own and other people's community development practices.

Having said this, there are two things that all participants in our research network have in common (as far as we are aware). First and most

Accepted for publication September 2003.

Gerald Midgley • Institute of Environmental Science and Research (ESR) Ltd., Christchurch Science Centre, 27 Creyke Road, PO Box 29-181, Christchurch, New Zealand. Centre for Systems Studies, Business School, University of Hull, Hull, HU6 7RX, UK.
Alejandro E. Ochoa-Arias • Centro de Investigaciones en Sistemologia Interpretiva, Faculty of Engineering, University of Los Andes, Mérida, Venezuela.

Community Operational Research: OR and Systems Thinking for Community Development, edited by G. Midgley and A.E. Ochoa-Arias, Kluwer Academic/Plenum Publishers, New York, 2004

obvious is a desire to make a contribution to change in communities. There are disparate motivations for this: some have political, religious and/or professional reasons for getting involved, while others simply have a desire to do something 'good' or 'useful' in their local neighbourhoods (Wong and Mingers, 1994). The second thing that Community OR practitioners seem to have in common is a concern with the design of methodologies, processes of engagement, methods and techniques. Of course, many Community OR practitioners (but not all) are also interested in the thorny political and theoretical issues surrounding the concept of 'community', and what it should mean for intervention. However, we suggest that it is the particular focus on *methodology and methods* (usually grounded in theories of community, organisation, social systems, complexity, power and/or participation) that marks the distinctive contribution of the Community OR network to community development. This is not to say that Community OR practitioners advocate the 'new Methodism' that Reason and Bradbury (2000) criticise (reducing the complexities of practice to the implementation of methods alone). Nevertheless, Community OR practitioners value the creation and use of methods as a *contribution* to the wider theoretically informed practice of community development. Some people write up their innovations for publication, while others choose to engage in practice and discuss methodological ideas without writing about them.

Over the thirteen years or so that we have both been reading and writing about Community OR, we have witnessed many interesting and fruitful debates. However, we have become very aware that the literature in our field is not easily accessible to people coming to it from outside our research network. This is largely due to the origins of Community OR in the disciplines of operational research and management science (see later in this chapter for details). Many of the key papers on Community OR have been published in places like the *Journal of the Operational Research Society*, *Omega*, *Systems Practice* and *Dragon*—arguably not the most obvious journals for people to consult for information on methods for community development. Therefore, unless a reader already knows the term 'Community OR' and searches the electronic databases using it, s/he is unlikely to come across our work by chance. This is one of the reasons we decided to edit the book you are now reading: we saw a need to bring together a variety of ideas about Community OR in one place so that they could be communicated to a wider audience.

Two previous books on Community OR have been published before this one (Ritchie *et al*, 1994; Bowen, 1995), and both make valuable contributions to the literature by presenting accounts of practice in a relatively jargon-free manner. However, these books were primarily aimed at community development practitioners and students coming to Community

OR for the first time. They therefore tended to avoid some of the more 'difficult' theoretical and methodological issues that have occupied the attention of our research network over the years (we should stress that this is not a criticism: it was the appropriate thing for the editors to do given their target audiences). Also, they were marketed solely in the UK. In contrast, this book is deliberately being aimed at an international readership, and it seeks to consolidate the literature for a wider audience that includes academics as well as practitioners and students.

However, consolidating the literature and presenting it to a wider readership is not our only motivation for editing this book. Another is the desire to introduce some new ideas into Community OR and thereby further develop its theory and practice. We have therefore tried to strike a balance between, on the one hand, reprinting some influential papers for the benefit of the next generation of readers and, on the other, extending the boundaries of the Community OR debate. We leave it for you to judge whether we have been successful in this regard.

In the remainder of this introductory chapter we do three things. First, we give a brief overview of the origins and early development of Community OR. Then we explain the sub-title of this book, *OR and Systems Thinking for Community Development*. It is no accident that the book is being published in a series on 'Contemporary Systems Thinking'. Although we recognise that the link between Community OR and systems thinking is not universally accepted (see Chapter 9, by John Friend, for an argument against it), we believe there are nevertheless good reasons for Community OR practitioners to draw upon systems approaches. Finally, we end with a detailed review of the structure and contents of this book, explaining what, in our view, each chapter contributes to the on-going debate about Community OR.

1.2 The Origins and Early Development of Community OR

Community OR is a child of the wider operational research (OR) movement. To understand how and why Community OR came into being, it is necessary to begin the story with the birth of OR in the 2nd World War. Many histories of OR have been written, but two that we have found particularly useful are Trefethen (1954) and Mar Molinero (1992). The account below is partly based on these writings, and partly on the oral history that is often recounted at OR meetings and conferences.

In both the UK and USA, scientists from a wide variety of disciplinary backgrounds came together during the 2nd World War to support the planning of military operations. They mainly applied mathematical modelling techniques to find optimal solutions to complex problems. Churchman (1987) provides a typical example: he describes how he applied mathematics to the problem of determining an optimal quality control procedure in the manufacture of ammunitions.

After the war some OR practitioners continued to work in defence, but most went into civilian positions in the public and private sectors. In the UK, they had a large part to play in the development of the newly nationalised industries—most notably, coal and steel (see, for example, Jones, 1992). Right from the early days, many people entered the OR profession because they saw it as a means to generate social improvement—to do something useful in society. Indeed, Rosenhead (1986) points out that a lot of the early OR practitioners who joined the profession in the 1940s and 1950s had strong and overtly socialist ideals. However, in our view, three major factors combined to alter or suppress this political orientation in post-war years.

First, OR came to be seen as largely synonymous with 'management science'. As Rosenhead (1986) points out, a great deal of management science supports the aims of the managers of industry within the confines of the political status quo, regardless of the consequences for other stakeholders. There are, of course, exceptions to this generalisation (see, for example, the work of the 'critical management science' movement as represented by Mingers, 1992a). However, by the 1970s, most OR was embedded in industrial organisations and simply offered a problem solving service to management.

The second significant factor that affected the political consciousness of practitioners was the increasing focus of a new generation of OR academics on teaching the rigours of mathematical modelling. This focus was reinforced by a prevalent belief in the 1950s and 1960s that nothing of generalisable value can be written about the human processes involved in conducting OR projects because each project is unique, so mathematical methods and techniques have to be the primary concern of the practitioner. Even Stansfield (1981), one of the founders of OR in the 2nd World War who was critical of the increasing focus on mathematical techniques, accepted the consensus that not much of value could be said about processes of engagement with human beings. Over the years, several high-profile writers (e.g., Churchman, 1970; Ackoff, 1979a,b; and Checkland, 1985), plus just about everybody writing about Community OR, stood out against this view, yet most efforts in educating the next generation of OR practitioners were still channelled into teaching applied mathematics. Only a small minority of modules on University degree programmes offered alternative methodologies. The result was that most of the next generation of OR practitioners saw themselves as technical experts rather than as agents of social improvement. In our view, many new practitioners left University with a good training in analytical methods and techniques, but an inadequate understanding of the human systems they needed to engage with. Most significantly, the degree courses that many students took offered little or no opportunity to reflect on the social and political roles that it is possible for professional OR practitioners to assume.

As we see it, the third factor affecting people's understandings of what OR was all about was the massive closure of industrial OR departments in the 1970s and 1980s. This was partly due to the over-concentration of many OR groups on solving tactical problems using mathematical modelling techniques (strategic planning was often placed outside their remit), and partly because of major changes in the business, economic and political environments of organisations employing operational researchers (Fildes and Ranyard, 1997). Of course, many of the activities of OR departments were later reinvented using new labels (quality management, knowledge management, etc.), but this did not avert a major crisis in OR and a massive decline in the number of people describing themselves as 'operational researchers'. Many struggling OR groups were forced to justify their work to their employers in terms of the contributions they made to profitability, and this was hard for them to do because of the difficulty of disentangling their contributions from those made by other parts of their organisations. The crisis in OR therefore intensified the pressure for its practitioners to portray themselves as simply offering a technical problem solving service to managers—any wider ideals had to be set to one side.

It is perhaps ironic that this crisis in the 1970s and 1980s not only had the effect of making the mainstream OR movement more business-focused than ever, but also led to the birth of Community OR. With the near collapse of its traditional base in industry, the Operational Research Society in the UK (and to a lesser extent its counterparts in other countries) cast around for new areas in which OR could be applied. Jonathan Rosenhead made an important intervention at this time: when he was President of the Operational Research Society in 1986, he realised that the timing was perfect for launching a new initiative, taking OR into the arena of community development. Rosenhead created a synergy between two different forces, both pushing for change: the Operational Research Society, which wanted to find new areas of application for the expertise of its members, and a significant minority of OR practitioners who still saw OR as a means to generate social improvement. Importantly, this latter group brought together several older practitioners with relatively high profiles in the OR community and a new wave of younger people (including us) who were influenced by action research, systems thinking, applied social science and community development. All these people were united in wanting to create an OR practice that was more participative, personally reflective and socially aware than the practices usually promoted using the label of OR. The irony is that the birth of Community OR was a reaction against the mainstream with its primarily managerial and mathematical foci, yet simultaneously was a result of the action of the Society that was *representing* this mainstream in looking for new 'markets' for OR.

Although the term 'Community OR' was first coined in 1986, we should acknowledge that a good deal of work applying OR methods to community development had already been done prior to that—it's just that no research network with specific interests in community issues had formed. In the United States, OR practitioners had been working with community groups since the late 1960s (e.g., Ackoff, 1970) and in the UK since the mid-1970s (e.g., Noad and King, 1977; Trist and Burgess, 1978; Jones and Eden, 1981). However, actually creating the label 'Community OR' facilitated a significant expansion of the number of community-based interventions.

In 1986, funding from the Operational Research Society was secured to establish a research centre (the Community OR Unit) at Northern College in the UK, and this opened in 1988 (it has since moved to Lincoln University). In 1987, the Community OR Network (a study group of the Operational Research Society) was formed (Mason, 1987). Within months, 300 members of the Society had joined, making it one of their largest study groups. Also in 1987, the Centre for Community OR was opened at the University of Hull (again with financial support from the Operational Research Society). This was later merged into the Centre for Systems Studies, also at Hull, resulting in a significant investment of new resources and a consequent expansion of Community OR activity. Since 1987, there have been Community OR streams at a number of UK and international conferences, and two conferences have been dedicated solely to Community OR. More detailed histories of the early institutional development of the Community OR movement can be found elsewhere (e.g., Carter *et al*, 1987; Parry and Mingers, 1991; Mar Molinero, 1992; Ritchie, 1994; Ritchie *et al*, 1994).

In the 1990s, after this initial period of growth and development, the number of people in the UK engaged in Community OR stabilised. We estimate that, today, the Community OR network in the UK comprises about 10% of the OR movement as a whole. However, in our view, the most significant aspect of Community OR in the 1990s, and in the first couple of years of the 21st Century, was its gradual spread to other countries—most notably in developing regions of the world. We can see striking parallels between community issues in the North and development issues in the South, and we believe they both demand a similar methodological focus. The extension of Community OR into the arena of development will be discussed in more detail later in this introductory chapter.

As we look to the future, we see the potential for a new period of growth in Community OR. Most importantly, neighbourhood regeneration and social inclusion seem to be central to the political agenda of the UK government, and also the agendas of many other centre-left governments around the world. In Venezuela, self-organising groups of citizens are forming in response to attempts by conservative forces to bring down the democratically elected government. If we can build on the political support currently being given to community development work; capitalise on the major funding opportunities becoming available in the West; and effectively communicate our practice to both academics and practitioners, we will be in a good position to demonstrate the value of Community OR. This should help to raise our public profile, bringing in more work and eventually more participants to our research community.

1.3 *Systems Thinking*

However, having made this claim that there is considerable scope in the current political climate for the further growth and development of Community OR, we want to surface an issue that we believe is central to the question of how our practice should evolve. We argue that Community OR can gain something significant from systems thinking.

Now, before explaining what we mean by this, we want to acknowledge that some people may regard it as a controversial statement—and the controversy stretches back to a time before the birth of Community OR. The traditions of systems thinking and OR have long been intertwined, and a number of authors have been highly influential in both research communities (e.g., Churchman, 1970; Ackoff, 1979a,b; Checkland, 1981; Beer, 1984; Jackson, 1991). However, there have been many debates over the years about whether OR is an applied wing of the systems movement, or whether systems thinking is an aspect of OR, making the relationship between the two communities rather problematic (Keys, 1991). As this is a matter of intellectual territory which people stake their careers on, any argument suggesting that either OR or systems thinking is a sub-set of the other will immediately be seen as imperialist and illegitimate by those whose ideas are threatened with subsumption. Our own view is that this kind of imperialism is both fruitless and unnecessary: OR practitioners and systems thinkers constitute two 'communities of practice' (to borrow a term from Wenger, 1998) with an overlap in membership. As we see it, there are significant differences between the systems and OR communities that make the preservation of their separate identities important, and help explain the levels of hostility exhibited in the face of imperialism. For instance, while the vast majority of people in the systems community have an interest in systems philosophy and theory, many are uninterested in linking their theories into interventions (see M'Pherson, 1974, for an analysis of the different foci of systems research that we suggest still holds true today). Asserting that the systems community is a sub-set of the OR community, which is primarily concerned with intervention, is therefore an anathema to these people. Likewise, there are a significant number of OR practitioners who are strongly committed to theories and methodologies that come from outside the systems tradition. A good example is Kelly's (1955) personal construct theory that was originally proposed within the discipline of psychology, and is now used to underpin the OR practices of cognitive mapping (Eden, 1988) and journey making (Eden and Ackermann, 1998). Another example is 'negotiated project engagement' (see Chapter 9 of this book) which informs Friend and Hickling's (1997) strategic choice approach. It is therefore unsurprising that authors who do not see themselves as systems thinkers resent OR being described as a form of systems practice.

However, once we accept that OR practitioners and systems thinkers constitute two inter-linked communities of practice which both have a right to preserve their identities and differences, it is possible to think about how the communities can learn from (rather than subsume) one another. The differences between them have led to the elaboration in both communities of extraordinarily rich tapestries of ideas, so both have something unique and valuable to offer. It is our contention that Community OR practitioners, who are already familiar with operational research, can gain something significant from tapping into the ideas that have been developed by those systems thinkers who do have a concern with methodologies and methods for intervention. Let us explain.

One thing that we believe is strikingly evident when you read the case studies of Community OR practice in this book (and also in other writings on community OR, such as Ritchie *et al*, 1994) is that a key issue is *comprehensiveness* in planning. That is, all Community OR authors seem to recognise that, when social change is being planned, significant interactions and potential side-effects of intervention should (as far as possible) be accounted for, and possible impacts on people, their communities and environments should be properly considered as part of the planning process. As far as we can see, this concern with comprehensiveness is equally important for community OR practitioners who have never engaged with systems thinking as it is for systems thinkers themselves. However, because systems thinkers (starting with Bogdanov, 1910–1913) have been researching the meaning of 'comprehensiveness' and its implications for planning and management for almost 100 years, we argue that there is a great deal that Community OR practitioners can learn from their work.

Again, there may be some controversy here. John Friend, a leading practitioner of Community OR, has explicitly argued that the notion of comprehensiveness is problematic (Friend and Hickling, 1997). Therefore, it is unlikely that he (or other people who use his strategic choice approach) will agree with our claim that comprehensiveness is a key issue unless we can provide further argumentation. It is our contention that two significantly different understandings of comprehensiveness have been developed in the systems community, and that one (proposed by Churchman, 1970; Ulrich, 1983; and Midgley, 2000) is similar in many ways to the idea of 'selectivity' that Friend and Hickling wish to use instead of 'comprehensiveness'. Superficially, the words 'selectivity' and 'comprehensiveness' sound like opposites, but we argue that (given Churchman's and Ulrich's view of comprehensiveness) they are not. Below, we explain the two different ideas about comprehensiveness that have been proposed by systems thinkers. We then return to the work of Friend and Hickling (1997) to demonstrate the similarity between this and their concept of 'selectivity'. Finally, we make reference to 'boundary critique' (a systems theory of comprehensiveness) that we believe can provide significant added value to Community OR practitioners (the details of this theory are not reproduced here as they can be found in Chapter 10, by Boyd *et al*, in this book).

Let us start by asking, 'what does it mean when we say that we have conducted a *systemic* analysis?' One answer is that, if something can be described as 'systemic', it is (as far as possible) comprehensively understood. This is the answer that was first proposed by general systems theorists (e.g., Von Bertalanffy, 1956; Boulding, 1956) whose primary interest was in transcending the limitations imposed on science by seemingly arbitrary disciplinary boundaries. General systems theory (GST) offers a language that describes the generic properties of all systems (e.g., cells, organs, organisms, families, organisations, communities, ecosystems, planets, solar systems and galaxies). In GST, a systems view (understanding the properties of systems *in general* and analysing *particular* systems with reference to these) is said to be the most comprehensive view that it is possible to attain.

GST was highly popular in the mid-20th century and still has strong proponents, especially in the USA (e.g., Troncale, 1985). A similar view of comprehensiveness can also be found in the earlier work of

Bogdanov (1910–1913) in Eastern Europe. According to Bogdanov, effective management requires a theory of organisation based on the idea that organisations are systems that share common characteristics with other systems found in the natural world. By exploring the common nature of *all* systems, we can generate more comprehensive understandings of *particular types* of system (e.g., organisations).

In Western Europe, however, these ideas had less influence on the development of systems thinking than it did in Eastern Europe and the USA. It was another strand of systems thinking that was influential in Western Europe, and this embodied quite a different understanding of what it means to be 'systemic'. The key author in this alternative tradition was C. West Churchman (1968a,b, 1970, 1971, 1979), whose fundamental ideas have been taken in a variety of different directions (e.g., by Ackoff, 1981; Checkland, 1981; Mason and Mitroff, 1981; Ulrich, 1983; Fuenmayor, 1991a–c; López-Garay, 1991; Midgley, 1992). For Churchman, to be systemic means to reflect on the boundaries of inclusion and exclusion in analyses.

Prior to the work of Churchman, many people (most notably general systems theorists) assumed that the boundaries of a system are 'given' by the structure of reality. In contrast, Churchman made it clear that boundaries are social or personal constructs defining the limits of the knowledge that is to be taken as pertinent in an analysis. There is also another important element of Churchman's understanding of being 'systemic'. When it comes to *social* systems, pushing out the boundaries of analysis may also involve pushing out the boundaries of who may legitimately be considered a decision maker (Churchman, 1970). Thus, the activity of reflecting on and setting boundaries defines both the knowledge to be considered pertinent *and* the people who generate that knowledge (and who also have a stake in the results of any attempts to improve the system). This means that there are no 'experts' in Churchman's systems approach, at least in the traditional sense of expertise where all relevant knowledge is seen as emanating from just one group or class of people: wide-spread stakeholder involvement is required, sweeping in a variety of relevant perspectives. Far from signalling 'comprehensive understanding', Churchman's view is that the systems idea highlights the bounded nature of all understandings, and hence the inevitable *lack* of comprehensiveness. The key to systems thinking is to sweep in sufficient information (from various different viewpoints) to paint a rich and complex picture, but without compromising the meaning of an analysis by over-inclusion, thereby paralysing action. Also see Ulrich (1983) for some important developments of this idea.

The above two views on comprehensiveness appear to be opposites. GST assumes that systems exist in the real world and that our models of them represent (as near as possible) comprehensive knowledge, while Churchman's and Ulrich's idea is that the boundaries of systems are personal or social constructs which highlight the inevitable *lack* of comprehensiveness. What GST and these authors have in common, however, is the focus on comprehensiveness as an *ideal*. Even though Churchman focuses on the *lack* of comprehensiveness apparent when a system is defined, he does so to highlight the need to generate further insights by 'sweeping' more information into our understanding of a situation. While *full* understanding is unattainable, *greater* understanding than we currently have at any particular moment is usually possible—although Churchman stresses that it's important not to compromise intelligibility through over-inclusion, and Ulrich observes that the purposes of any analysis will necessarily impose limits on the sweep-in process.

In the tradition of Churchman (1970) and Ulrich (1983) we suggest that, once we acknowledge that no view of the world can ever be comprehensive, the boundary concept becomes crucial. Where exactly boundaries are constructed, and what values guide the construction, will determine how issues are seen and what actions will be taken. It is this insight, elaborated by a variety of authors over 35 years, that we believe is important for Community OR.

So, let us return to the work of Friend and Hickling (1997) who argue that striving to be comprehensive in analyses is problematic because, in 'difficult and complex planning problems the norms of linearity, objectivity, certainty and comprehensiveness keep on breaking down' (p.22). If one defines comprehensiveness as conformity to the saying 'don't do things by halves' (Friend and Hickling, 1997, p.21), then we couldn't agree more. However, if we follow Churchman and subsequent writers in the systems domain, we need to recognise a crucial paradox. By viewing the pursuit of comprehensiveness as *dealing with its inevitable absence*, and by making this explicit in the form of boundary judgements that can be explored and critiqued, we are likely to be *more* comprehensive than if we simply take our boundary judgements for granted. It is our contention that this is actually quite similar to Friend and Hickling's (1997) prescription, 'learn to work with selectivity' (p.22). Being selective essentially means, to use systems terminology, making boundary judgements. However, selectivity also gives rise to uncertainty and the potential for new and surprising problems because there is always the possibility that what we do will interact with things that are outside the boundaries of our analysis or understanding. If we have understood Friend and Hickling correctly, this is why they emphasise *learning to work* with selectivity. Although they do not explicitly define this phrase, we interpret it to mean remaining aware of the limits to knowledge that come with selectivity, and exploring these limits periodically during interventions (the basis of this interpretation is Friend and Hickling's emphasis on improving knowledge of potential interactions between phenomena). This is precisely what Ulrich (1996) and Midgley *et al* (1998) mean when they use the term 'boundary critique' in relation to systems practice: being critical of taken for granted assumptions about who and what is relevant to an analysis that may hide important impacts of our interventions.

We can now clarify that, when we say that all Community OR practitioners seem to show an interest in the issue of comprehensiveness, we mean this in the sense given to the term by Churchman (1970), Ulrich (1983) and Midgley (2000). We hope that, given our argument above that this sense is different to the negative one ascribed to it by Friend and Hickling (1997), most people will agree with us that the pursuit of comprehensiveness (or rather, exploring our inevitable *lack* of comprehensiveness) is vitally important to Community OR practice. It is because the systems community has developed a whole network of theoretical and methodological ideas about what it means to explore the boundaries of our knowledge during intervention that we believe systems thinkers can indeed contribute something significant to Community OR practice. One such idea is the theory of boundary critique (mentioned above), and this is explained in more detail (with a practical example of its application in a Community OR project) in Chapter 10. Ulrich (1983) and Midgley (2000) also provide further information for people wanting to read more.

1.4 The Structure of this Book and the Contributions of the Individual Chapters

Having discussed the history of Community OR and the added value that systems thinking can offer, we now move on to explain how we have structured this book. We have divided it into three sections, as follows:

The first section, entitled 'Early Contributions and Later Reflections on Theory, Methodology and Practice', reprints some key works from the 1980s when Community OR was being founded, and also presents a more recent contribution raising new issues. All but one of the chapters in this section offer the reader theoretical and methodological reflections without any practical illustrations. Most of the early papers on Community OR took this form. The initial impetus to launch Community OR came from an awareness of political and methodological problems with mainstream OR (as discussed earlier), and most of the well-known examples of practice came later. It was really not until the 1990s that a significant number of case studies began to appear in the literature (e.g., Thunhurst and Ritchie, 1992; Thunhurst *et al*, 1992; Gregory and Jackson, 1992a,b; Ritchie *et al*, 1994; Taket and White, 1994).

Section Two of this book, called 'Local Action for Community Development', is concerned with how people have translated the theoretical insights of Community OR into practice, and how practice has informed theory. Since the inception of Community OR, the vast majority of projects have dealt with complex, localised community issues using participative methods (mostly in the developed world). This section of the book presents a variety of methodological ideas evolved for Community OR practice, and illustrates them with examples of projects with community groups, voluntary organisations and welfare services.

Section Three, 'Dealing Locally with Global Issues', presents a set of chapters seeking to extend the practice of Community OR in two directions. The first direction takes Community OR into developing countries. We believe it is no accident that a number of prominent Community OR practitioners also engage with development issues (e.g., Rosenhead, 1993; White, 1994; Friend, 1998; Ochoa-Arias, 1998, 2000; Córdoba and Midgley, 2003). Not only does development work involve dealing with issues that can make a significant difference to the lives of some of the world's poorest people, but there are striking similarities between many of the participative methods used in Community OR and those employed in development. For just one example of a development approach that bears comparison with some writings on Community OR, see the work of Chambers (1983, 1997, 2002). Like Rosenhead (1993), we believe that a dialogue between development workers and Community OR practitioners on methodology and methods, as well as the nature and potential purposes of development itself, would be very fruitful.

The other extension to Community OR practice advocated in Section Three takes us into the arena of environmental management. Until relatively recently, most Community OR practitioners have only concerned themselves with social issues (as distinct from ecological ones). However, since the UN Conference on Environment and Development (commonly known as the 'Rio Earth Summit') in 1992, there has been a wide spread recognition of the importance of local community participation in formulating policies and practices for environmental sustainability. The success of these policies depends on how strongly local people are committed to them: if they remain uncommitted, they are unlikely to change their behaviours. Arguably, the best way to generate commitment to new policies and practices is to ensure that they are developed participatively by those who will have to act on them (Ackoff, 1981; Checkland, 1981). However, given that policy makers generally have very little experience of facilitating participation (their usual mode of community involvement being consultation on pre-prepared plans), they may need some assistance. As Community OR practitioners have been developing methodologies and methods for enhancing community participation for some years, it seems to us that we are ideally placed to provide this assistance. In doing so, we will need to pay particular attention to the interpenetration of ecological, social and personal issues (Midgley, 1992); and potentially, the opportunity is there to provide models of community participation that are transferable to many different domains of governance.

1.4.1 *Early Contributions and Later Reflections on Theory, Methodology and Practice*

Having described the three sections of this book, let us now examine what we see as the main contributions of the individual chapters. We start with the first section, which mainly contains general discussions of Community OR and other theoretical and methodological reflections.

1.4.1.1 Community Operational Research: Its Context and its Future

Section One starts off with a chapter, originally published in *Omega* in 1991, by Richard Parry and John Mingers entitled, 'Community Operational Research: its Context and its Future'. We have chosen this as the lead chapter primarily because it gives a good introduction to the institutional history and development of the field. In addition, it asks a series of challenging questions about the nature of, and future possible directions for, Community OR. We believe that many of these questions are still relevant today, such as 'Is Community OR different from traditional OR?' and 'Is Community OR often just community work?'

One of the most important messages of the chapter is that practitioners gain credibility from how they are talked about in the various volunteer, activist and service provider networks that operate both formally and informally in the vast majority of local communities. Without local credibility, it is difficult to get access to and successfully complete Community OR projects. This means that, unlike many traditional disciplinary researchers who depend for their success solely on *academic* achievement (publishing in journals and presenting papers at conferences), Community OR practitioners are equally (if not more) dependent on good grass-roots networking and organising activities within their localities. Without the latter, Community OR practitioners would have nothing to discuss in academic circles.

1.4.1.2 Community Operational Research: Purposes, Theory and Practice

The second chapter is Mike Jackson's 'Community Operational Research: Purposes, Theory and Practice'. This was originally published in 1987 in a special issue of *Dragon* on the work of the Centre for Community OR at the University of Hull (Jackson, 1987a). We have chosen to include this for two reasons. First, it provides a very clear critique of the dominant paradigm in 'mainstream' OR; and second, when it originally appeared in print, it was the first paper in the literature (as far as we are aware) to make an explicit link between Community OR and systems thinking.

The chapter starts by contrasting 'impoverished' and 'enhanced' OR. The former reduces OR to a tool kit of mathematical techniques which its practitioners use to solve narrowly defined problems without any reference to wider issues of social improvement. In contrast, 'enhanced OR' is explicit about its social role; is systemic in its orientation; and embraces a wide range of methodological ideas. Its practitioners use participative methods for facilitating debate alongside the more usual quantitative techniques applied by operational researchers and social scientists. Jackson argues that Community OR will only be successful if its practitioners embrace this 'enhanced OR': community contexts are often highly complex with multiple actors holding different views on problematic situations, and 'impoverished OR' is simply inadequate in the face of this complexity.

The chapter then moves on to discuss a variety of political and theoretical issues surrounding what it might mean to develop a thriving practice of (enhanced) Community OR, focusing in particular on the relevance to practitioners of the critical social theory of Habermas (1972) (see the chapter for details). While many 'mainstream' OR practitioners shun social theory, claiming it is largely irrelevant to OR practice (e.g., Ormerod, 1996), Jackson's desire to embrace theory mirrors the belief of many people involved in Community OR that theoretical reflections can cast new light on taken for granted practices.

Jackson then ends by advocating a framework for classifying systems and OR methodologies according to the different assumptions they make about problem contexts. This framework, originally proposed by Jackson and Keys (1984) and elaborated by Jackson (1987b), allows the practitioner to choose the 'most appropriate' approach for the context at hand. Although a number of writers have critiqued Jackson's actual framework (e.g., Gregory, 1990, 1992; Midgley, 1990, 1996, 2000; Mansell, 1991; Mingers, 1992a,b, 1993; Tsoukas, 1992; Jones, 1993; Dutt, 1994; Sutton, 1995), the basic message about the value of methodological pluralism to Community OR is, we believe, now widely accepted. Being able to draw upon diverse methods, originally born in different paradigms, makes the practice of Community OR much more flexible and responsive to stakeholder concerns than adherence to just one or two methods drawn from a single paradigm.

1.4.1.3 Visions of Community for Community OR

The final chapter in Section One reprints a paper of our own, first published in *Omega* in 1999, called 'Visions of Community for Community OR'. We originally wrote this because of our concern that the meaning of the word 'community' was not receiving sufficient attention, yet it is a key word that differentiates our practice from 'mainstream' OR. There are authors who argue that 'community' is merely a feel-good term, referring to a non-existent utopia of harmony and belonging, so its use should be regarded with scepticism (e.g., Bauman, 2001). However, our reading of the literature on the politics of community indicated to us that the word takes on distinctly different meanings in different political discourses (for instance, the ideal of community pursued by liberals is quite different to the one that Marxists or communitarians want to embrace). It quickly became evident to us that, without some reflection on alternative visions of community and their implications for community OR, many researchers will simply operate by default with whatever concept of community is culturally dominant. For most, this will be the understanding of community that is implicit in the Western liberal/capitalist political tradition. For some this may not be a problem, but for others who want to pursue alternative political agendas it certainly will be. We therefore set out to make the different visions of community that are present in a variety of political discourses more visible, and asked what kind of Community OR practice would support each one. Our analysis shows that a number of choices are open to practitioners, and indeed the range of choices is not limited by the literature we have reviewed: as people's political understandings evolve, there is always scope for the development of *new* visions of community to inform practice.

1.4.2 Local Action for Community Development

We now move on to discuss the second section of this book, where each of the chapters illustrates theoretical and methodological ideas with case studies of practice.

1.4.2.1 A Black Ghetto's Research on a University

The opening chapter is a reprint of a classic paper from Russell Ackoff, first published in 1970 in *Operations Research*. The title is 'A Black Ghetto's Research on a University', and we believe most people will be happy to acknowledge that this represents a significant contribution to our field, even though it originally appeared well before the term 'Community OR' was first coined.

Ackoff's paper is regarded as a classic because, in 1970, it provided a substantial challenge to the common (but implicit) assumption in the OR literature that the practitioner should always play the role of the expert who solves other people's problems. Ackoff was concerned that many write-ups of OR projects portray the practitioner as an active agent, and his or her clients as passive recipients of this agency. In talking in his title about 'a black ghetto's research on a University' (rather than the other way around), the challenge he makes to conventional wisdom is made clear right from the start. The project that Ackoff and his colleagues set up with community leaders in Mantua, Philadelphia, was funded from charitable sources, and enabled three people from the local community to be employed by the University of Philadelphia as researchers. It was these three, working with the community leaders and local people, who developed plans for investment in new industries to benefit the ghetto. The OR practitioners did not set the boundaries of how their 'expertise' would be used: they received requests for aid from the community researchers and found ways to provide what was needed. Some of these requests were for services that would not be perceived as 'research' using any conventional definition of the term, but were absolutely necessary to the success of the project. OR became part of a wider remit of community activism, driven by the local researchers who were acknowledged to be the primary experts in their own community's affairs.

After describing the project and some of its outcomes, which were regarded as highly significant by people in the ghetto, Ackoff goes on to draw out the wider implications of his work for both OR and public sector management. The chapter makes clear the value of working in a responsive mode rather than trying to impose ones own 'expertise' on others, and also advocates a strongly participative model of public sector accountability that is just as relevant today as it was when Ackoff first proposed it.

We should also acknowledge that Ackoff is a well-known systems thinker as well as being a major contributor to the OR literature (see, for example, Ackoff, 1981). Although 'A Black Ghetto's Research on a University' is not *explicitly* about systems thinking, the influence of the systems discourse is plainly visible. In our view this is especially so in the way Ackoff describes how the community leaders viewed their work: he claims that they were much more sensitive than most public sector managers to how changes would fit with wider systems (within and beyond the local neighbourhood). It is very interesting to note that here Ackoff is not claiming expertise in systems thinking for himself alone. Consistent with the message of the chapter that OR practitioners should not be seen as having a monopoly on expertise, he draws attention to the systems thinking of others rather than his own.

1.4.2.2 Housing in the Dearne Valley: Doing Community OR with the Thurnscoe Tenants Housing Co-operative

Moving on to the second chapter in this section, we have a contribution from Charles Ritchie entitled 'Housing in the Dearne Valley: Doing Community OR with the Thurnscoe Tenants Housing Co-operative'. This is a write-up of an intervention that is very well known in the UK research community, as it was the flagship project of the Community Operational Research Unit (based at Northern College, Barnsley) in its early years. Some details of the first phases of the work have been published previously (Thunhurst and Ritchie, 1992; Thunhurst *et al*, 1992), but this chapter completes the story by taking it to the conclusion of the intervention in 1995.

One interesting feature of this project is how Ritchie and his colleagues used both 'soft' (debate-orientated, problem structuring) approaches and 'hard' (quantitative, modelling) methods in parallel. Given that the need for methodological pluralism is widely accepted by Community OR practitioners, this in itself is not remarkable. However, what is worthy of note is the *manner* in which 'hard' techniques (like financial modelling) were employed, preserving the participatory ethos of the project as a whole. While it is relatively easy to enable participation when using methods for structuring debate, financial modelling requires particular skills that are not necessarily going to be wide spread in the community. It is interesting to read about how the researchers built a financial model themselves, but instead of handing down recommendations (as many traditional OR consultants would have done), they involved the tenants in experimenting with the parameters of the model. This way the tenants could get some understanding of the assumptions built into it, and could therefore make their own informed decisions.

One of the lessons we believe can usefully be derived from this intervention is that it is worth thinking carefully about how barriers to participation can be overcome when there is a need to use specialist methods. These methods should not necessarily be thrown away just because people in the community have not (yet) got the skills to use them. The easy path for Ritchie and his colleagues would have been to ignore the financial planning aspect of the project. However, this would have been irresponsible, not to say unethical: poor financial decision making could have destroyed the housing co-operative. They therefore took the harder path of taking on the financial modelling themselves, at the same time thinking through how the tenants could engage with the model so that it didn't become disempowering.

After presenting the intervention itself, Ritchie ends his chapter with some reflections on a range of significant issues raised for Community OR. One of these is a suggestion that Community OR is not as different from traditional OR (when the latter is practised most successfully) as some people might think. Here, the words in brackets are significant: Ritchie's suggestion is that *all* good OR practice requires a strong focus on client involvement and methodological flexibility. This is in accord with Jackson's distinction (discussed in Chapter 3) between 'impoverished' and 'enhanced' OR, the latter being the only really adequate practice.

1.4.2.3 Looking Forward with a Council for Voluntary Service

Following on from Ritchie's contribution, the third chapter in this section on local action for community development, authored by Andy Marlow and Jim Bryant, is entitled 'Looking Forward with a Council for Voluntary Service'. Councils for Voluntary Service (CVSs) can be found throughout the UK, and their role is to provide advice and support of various kinds (help with bids for charitable funding, venues in which to hold meetings, etc.) to voluntary organisations and community groups. The chapter reports on the authors' strategic planning work with the CVS, and then proceeds to reflect on how this project typifies Community OR engagements. Marlow and Bryant discuss the often-encountered need to deal with situations characterised by multi-stakeholder interactions. They also reflect on the particular features of voluntary organisations that differ from those found in the public or private sectors. Dealing with these features creates a unique challenge for Community OR practitioners.

One particular contribution of this chapter that we believe is worthy of special note is that it explains how Marlow and Bryant simultaneously addressed strategic planning issues (the explicit purpose of the intervention) and tensions in the relationships between various people in the CVS who were championing different viewpoints. Dealing with the latter was not an explicit purpose of the intervention identified by the CVS, but the tensions they were experiencing nevertheless significantly impacted on the strategic planning (and indeed, day-to-day management), so could not be ignored without jeopardising the success of the work. In our experience (e.g., Midgley *et al*, 1998; Midgley, 2000), complex situations like this are commonly encountered in Community OR. It is quite usual to have to account for power relations and emotional dynamics within a group; between groups (or stakeholders); and between the research team and other participants *at the same time* as pursuing the explicit purposes of the intervention (most often expressed as a task to be undertaken). In the literature, this has been described in various ways: for example, as undertaking 'multi-layered' interventions (Weil, 1998), or as simultaneously addressing a variety of interrelated questions about the situation through the creative design of methods (Midgley, 1997, 2000). In their chapter, Marlow and Bryant not only demonstrate awareness of how 'process' (relationship) and 'content' (task) issues interact, but they also show how, in their intervention, this shaped their choices of methods.

Before moving on, it is worth making one final point about this chapter in light of our earlier comments on the value of systems thinking to Community OR. Marlow and Bryant's intervention provides a good example of how practitioners can draw upon and mix systems ideas and methods (e.g., parts of Checkland's, 1981, soft systems methodology; and Flood and Jackson's, 1991, total systems intervention) with OR methods from other sources (e.g., Eden's, 1989, strategic options development and analysis).

1.4.2.4 Facilitation as Fair Intervention

Next we have 'Facilitation as Fair Intervention' by Wendy Gregory and Norma Romm. The practice of facilitating workshops is an essential part of most Community OR projects. Gregory and Romm present their own ideas about facilitation and illustrate these with material from an intervention commissioned by the Trent Regional Health Authority in the UK, designed to investigate quality standards in services for people who are blind or partially sighted.

In our view, this chapter makes at least two significant contributions to the literature on Community OR. The first is that it challenges some common, taken-for-granted assumptions about facilitation. One is that facilitators should remain 'neutral' about whatever it is that group participants are discussing. Gregory and Romm not only argue that neutrality is impossible, they also suggest that contributions from facilitators can be positively beneficial—as long as the facilitators are careful to open up their ideas to critique through processes of dialogue rather than imposing them on others. Over the course of the chapter, the authors introduce their theory of how facilitators can participate actively in discussions while still remaining accountable to the other group participants for their inputs.

The other significant contribution made by this chapter is that it provides case study evidence for the value of one particular social theory (Habermas's, 1976, 1984a,b, theory of communicative action) to Community OR practice. Earlier, when discussing Mike Jackson's chapter (Chapter 3), we mentioned the tendency amongst many Community OR writers to draw upon social and political theories to inform methodology. Habermas has been a particular favourite for many people, including Jackson (Chapter 3) and Ulrich (1983) amongst others. Here, Gregory and Romm draw on one particular aspect of Habermas's theory (his understanding of rational argumentation) to inform their theory of accountability. Habermas (1976) suggests that any sentence intended for communication carries four implicit 'validity claims': that it is intelligible; that its propositional content is true; that it is being uttered appropriately (i.e., the speaker is doing the right thing in making the utterance); and that it is meant sincerely. Gregory and Romm argue that people in dialogue can be facilitated so they become more aware of the claims made in their own and other people's statements, and can therefore challenge the viewpoints of others more effectively than they would otherwise be able to do. In establishing dialogue on quality between stakeholders in their project, the participants were explicitly encouraged by Gregory and Romm to question the intelligibility, truth, rightness and sincerity of the opinions being expressed—not in a hostile manner, but in a spirit of open and critical inquiry. Importantly, the facilitators opened *their own* interventions to this kind of questioning as well. This served two purposes: it made the facilitators accountable to the group for their inputs, and it served to demonstrate (model) the process of critical questioning.

From the case study evidence presented by Gregory and Romm, it would appear that this mode of facilitation enabled the participants to reveal and critique assumptions in the dialogue that might otherwise have gone unquestioned. It therefore offers some promise for other practitioners interested in establishing critically questioning dialogues around topics of mutual interest, although we also suggest that the theory of 'rational argumentation' might be enhanced by the inclusion of ideas about emotional dynamics too (see Taket, 2002, for a useful discussion of these).

1.4.2.5 Perspectives of Engagement in Community Operational Research

The next chapter, 'Perspectives of Engagement in Community Operational Research' by John Friend, offers some reflections on two different theories about what it means to engage in Community OR. These are the 'systems intervention' perspective and 'negotiated project engagement'. Friend is critical of the systems tradition because, in his view, it takes as its starting point the idea that an organisation is an 'open system'. An open system is an interacting set of elements exchanging matter and energy with its environment, with properties that can only be attributable to the workings of the whole system, not any of its elements in isolation (e.g., Von Bertalanffy, 1950; Marchal, 1975; Emery, 1993; Emmeche *et al*, 1997). While acknowledging that some systems approaches have a good appreciation of complexity, Friend nevertheless argues that, by taking the existence of systems with a stable identity over time as their starting point, systems thinkers fail to sufficiently appreciate the transient nature of some important relationships. The 'negotiated project engagement' approach, in contrast, takes a particular problem or difficult decision as its starting point, and then builds transitory alliances between stakeholders to address this. The relationships involved in these alliances are not stable enough for the set of stakeholders to be classified as a system (in the 'open system' sense of the term), and will most likely dissolve when the problem is dealt with, or the difficult decision under scrutiny is made. In Friend's view, negotiated project engagement provides a more accurate interpretation of what actually happens in Community OR projects than systems thinking.

In pursuing this argument over the course of his chapter, Friend does two particularly useful things. First, he traces the origins of the negotiated project engagement approach back to a group of practitioners (including Friend himself) who worked at the Tavistock Institute in the mid-1960s. These institutional origins are important because they signal the existence of a coherent alternative to the systems thinking of Emery and Trist (1965, 1981). In the eyes of many systems practitioners, the name of the Tavistock Institute is virtually synonymous with the 'socio-technical systems approach' that Emery, Trist and others developed (e.g., Trist and Bamforth, 1951; Trist *et al*, 1963; Emery and Thorsrud, 1969, 1976; Emery, 1993). By revealing the alternative discourse of negotiated project engagement, Friend makes visible a debate that has been largely neglected by systems thinkers (if not by Community OR practitioners), and is still very relevant today.

A second useful contribution that Friend makes in his chapter is to provide case study evidence from two Community OR projects to support his claim that negotiated project engagement provides a more accurate representation of Community OR projects than the open systems perspective. It is because of the inclusion of these case studies that we decided to place Friend's chapter in this section on local action for community development, rather than the earlier section on theoretical and methodological reflections. In our opinion, because of these practical examples, Friend has produced a stronger case against the systems approach than the one (in Friend and Hickling, 1997) that we discussed earlier in this introductory chapter. We leave it to you to read Friend's chapter; our earlier case for systems thinking; and Chapter 10 by Boyd *et al* (which, in our view, presents a systems theory with more promise for Community OR than the open systems perspective), and make up your own mind who is right. It would be an abuse of our editorial role to counter the arguments in Friend's chapter in this book, but suffice it to say that we believe that clarifying our own position in the light of Friend's arguments *is* both possible and desirable. However, it will mean explaining how our own systems approach differs significantly from the open systems perspective that Friend criticises. We have long been aware of this difference, but have never had any reason to articulate it publicly. We would therefore like to thank Friend for giving us a very good reason indeed! As Churchman said way back in 1979, people should not accept theories unless they can be justified in debate with their strongest 'enemies', and Friend is one of the most open minded and constructive 'enemies' of the systems approach that we have had the pleasure to engage with.

1.4.2.6 Systemic Intervention for Community OR: Developing Services with Young People (Under 16) Living on the Streets

Moving on, we come to 'Systemic Intervention for Community OR' by Alan Boyd, Mandy Brown and Gerald Midgley. This discusses the relevance of the authors' systemic intervention methodology for Community OR, focusing on the concepts of boundary critique (reflecting on the inclusion/exclusion/marginalisation of people and issues, and exploring value judgements as part of intervention) and methodological pluralism (mixing methods drawn from a diversity of methodological origins). After the main methodological ideas are presented, Boyd *et al* narrate a case study where the authors worked with young people (under 16) living on the streets, and a variety of voluntary and statutory agencies, to support the design of new services to meet the young people's needs. At various points during the narration, the authors show how their thinking about boundary critique and methodological pluralism informed their decision making about methods and processes of intervention.

This project is the latest in a series of case studies that have been used by researchers at the Centre for Systems Studies, University of Hull (UK), to test out systems ideas and reflect on their strengths and weaknesses in order to further develop systemic intervention methodology.

For more details of this research programme (and other case studies), see Midgley (2000).

We believe that the main contribution of this chapter is to demonstrate the added value of the synergy between boundary critique and methodological pluralism. Boundary critique provides a way of thinking systemically about issues of inclusion and exclusion given the impossibility of genuinely comprehensive understanding, and it provides a means to conceptualise the marginalisation of people and issues in complex social situations. However, unless it is related to a strong discourse on methodology and methods, there is a risk of generating interesting sociological insights but no action in response to them (Flood and Jackson, 1991). In contrast, methodological pluralism brings flexibility and responsiveness to stakeholder concerns, making the move to action easier for pluralists than for those who only champion a limited set of methods drawn from a single paradigm (see Jackson, 1987c, for one of many arguments for this view). However, methodological pluralism carries its own dangers. If the choice of methods is based on a superficial diagnosis of the problematic situation (failing to take account of alternative viewpoints and power relations), then the action that is taken may be problematic in itself (Midgley, 1996, 2000; Mingers, 1997; Flood, 1999; Han, 2000). It may simply play into the hands of those with the capacity to exclude or marginalise other people (and the issues that concern them), generating unanticipated side effects, social discord or even perceptions of alienation, helplessness and oppression. The answer to these problems, according to Boyd *et al*, is to create a *synergy* of boundary critique and methodological pluralism: each addresses the weaknesses of the other. Boundary critique is made more action-orientated by the use of multiple methods, and the diagnosis of problematic situations is deepened by the reflections on values and processes of marginalisation that boundary critique brings to it.

1.4.2.7 Playing with PANDA: The CybOrg and the Rhizome

The next chapter, 'Playing with PANDA' by Ann Taket and Leroy White, briefly discusses their approach to the 'participatory appraisal of needs and development of action' (PANDA for short), and then goes on to explore how the authors use metaphors in their practice. Two particular metaphors, the 'CybOrg' and the 'rhizome', are given as examples, and their use is illustrated through a couple of detailed case studies of Community OR projects.

As we see it, there are two significant contributions to Community OR in this chapter: Taket and White's *general* argument for using metaphors; and their analysis of the *particular* implications of the CybOrg and rhizome metaphors. We will discuss their general argument first. A number of writers on management (most notably Morgan, 1986) have observed that particular metaphors are very often implicit in both theories of organisation and the day-to-day speech of managers. For example, we may talk in terms of organisations 'breaking down' (a machine metaphor), 'learning' (a brain metaphor) or 'trapping us' (a prison metaphor). Because using each metaphor involves a different 'way of seeing', it enables particular insights into organisational functioning to be grasped. Morgan's (1986) argument is that most people are unaware of the metaphors they use and tend to employ just one or two on a regular basis. They therefore have only a limited range of insights at their disposal. By using a wider range of metaphors more explicitly, people can begin to explore the limitations of their usual ways of thinking and sweep in a richer variety of insights.

Taket and White accept the logic of this, but add a couple of interesting new dimensions to the argument. Because many Community OR projects involve the participation of multiple organisations, they stress the value of exploring metaphors within and across organisational boundaries without (i) limiting the exploration to a pre-designed set of metaphors, or (ii) trying to establish a consensus on metaphors between organisations. They are critical of authors (like Flood and Jackson, 1991) who offer only a limited number of metaphors for exploration, tying these to recommendations for 'appropriate' action (e.g., in Flood and Jackson's case, choosing a particular systems methodology). Taket and White argue for a greater degree of flexibility and creativity in the exploration process, and point out that when multiple organisations are involved it is unlikely that the same set of metaphors will be relevant across the board. Their argument against seeking a consensus between organisations on 'appropriate' metaphors is also linked to this point. Multi-agency partnerships in Community OR projects are usually entered into on a voluntary basis, so it is important for the researcher to adopt an approach that maintains and/or builds an atmosphere of mutual respect. If organisations are pushed into agreeing a 'common' metaphor, this will almost certainly be of more value to some people than to others, possibly leading to feelings of alienation from both the process and the participants for whom the supposedly common metaphor works. This in turn may lead to resentment and/or mistrust, undermining the mutual respect that is necessary for the voluntary partnership to work. In our view, Taket and White make an important point here that is often neglected when there is a desire for multi-agency commitment to a single strategy or plan: pressure to ignore, minimise or reduce differences between agency viewpoints may actually undermine the realisation of this desire. We are not saying that multi-agency commitment to a single plan is impossible to facilitate in a Community OR project (see Gregory and Midgley, 2000, for an example of this working), but a respect for the differences between participants is crucial to success.

The second major contribution of Taket and White's chapter (from our perspective) is in their discussion of the CybOrg and rhizome metaphors. The word 'CybOrg' is a shortened version of 'cybernetic organism', and the implication that the organisation is being viewed as part human and part machine is significant. Taket and White borrow the CybOrg metaphor from Haraway (1985), acknowledging earlier contributions from the discourse of cybernetics (e.g., Wiener, 1948; Beer, 1966) and socio-technical systems thinking (Emery and Trist, 1981) along the way. The key to the CybOrg metaphor, according to Taket and White, is that it points to 'leaky boundaries', particularly between the human and animal (the word 'organism' blurs this distinction); the organism and the machine; and the physical and non-physical (organisms have minds which penetrate the inorganic world). The CybOrg metaphor can therefore not only be used to explore cybernetic feedback processes maintaining organisational viability, and the interface between human and technical systems, but raising awareness of leaky boundaries as an issue can help people make challenges to *any* boundaries that are presented as absolute. As should have been clear from our own earlier exploration of the boundary concept, boundaries signal cut-offs in our understandings of interactions (also see Ulrich, 1983), so when they are presented as absolute markers of closure this should make us suspicious of what is hidden from our understanding.

The rhizome metaphor, borrowed from Deleuze and Guattari (1980), is also interesting. A rhizome is the root mass of a plant; essentially, a complex tangle of interconnections. However, as Taket and White make clear, 'an attempt to picture a rhizome is almost a contradiction in terms of the level of connectivity involved, not all of which is equally (or simultaneously) visible'. Indeed, the level of connectivity is so high that it signals the near impossibility of predicting influences, and the complexity is added to as the rhizome continuously grows *new* connections, but without any central planning. While the CybOrg metaphor points to the non-absolute status of boundaries, the idea of the rhizome makes any kind of useful boundary difficult to contemplate. If human situations are like rhizomes, then attempts to define boundaries are inevitably undermined by the connections crossing them. Even when it appears that a meaningful boundary has been established at one point in time, the continuing growth of new connections at diverse sites across the network will eventually problematise this.

Not only does the rhizome metaphor create a challenge to the neatness of the boundary concept, but (as Taket and White explain) it also casts doubt on the utility of hierarchy theories. Hierarchy theories view the world in terms of nested systems, where each system is both a whole in itself and a part of a larger system (e.g., Boulding, 1956; Miller, 1965, 1978; Von Bertalanffy, 1968; Salthe, 1989). The problem with hierarchy theories, if the world is seen as a highly complex web of interconnections, is that elements at different hierarchical levels are connected in so many ways that it undermines the very idea of defining discrete levels (see also Capra, 1996).

It should be clear that the CybOrg and rhizome metaphors offer the Community OR practitioner two 'ways of seeing' that emphasise connectivity; dynamism; boundary crossing; the transgression of easy categories; multiple, local sites of action; and continual change. Taket and White's case studies, presented after the explication of the two metaphors, serve to demonstrate the practical value of these emphases.

Having explored the contributions of the chapter itself, we believe it is also worth looking at these in light of the previous two chapters (by Friend and Boyd *et al*). *All three* chapters problematise the idea of certainty around boundaries; stability in relationships; and what we would describe as the 'structural-functionalist' systems paradigm, exemplified by general systems theory (e.g., Boulding, 1956; Von Bertalanffy, 1956; Troncale, 1985) and the open systems perspective (e.g., Emery and Trist, 1981; Emery, 1993). However, while their critiques have some similarities, the methodological languages the various authors develop to frame the experience of engaging in Community OR are quite different, especially in relation to their inclusion or exclusion of systems thinking. Friend distances himself from systems thinking; Boyd *et al* adopt a reconstructed language of boundaries that is more sensitive to issues of connectivity; and Taket and White are willing to use the language of systems at some moments and undermine it at others (whatever appears most useful at the time). It seems to us that the debate has only just started. The authors have identified a major issue here, and we suggest that dialogue between them (also including others with similar concerns) may give rise to some significant new developments of wider value to Community OR practitioners (not to say systems thinkers and applied researchers more generally). We look forward to continuing research.

1.4.3 Dealing Locally with Global Issues

Having completed our review of the section on 'local action for community development', we now move on to examine the three chapters pointing to new possibilities for widening the boundaries of Community OR. The first focuses on the use of Community OR in developing countries, and the other two discuss environmental issues (and their interface with social concerns).

1.4.3.1 An Interpretive Systemic Exploration of Community Action in Venezuela

'An Interpretive Systemic Exploration of Community Action in Venezuela', by Alejandro Ochoa-Arias, presents the author's work, conducted over a twelve year period, with a Venezuelan 'neighbourhood association' (a local community association that acts as both a public service watchdog and a campaign group). The focus is on how political inquiry can, and in Ochoa-Arias's opinion should, be an integral part of Community OR practice.

In addition to presenting his argument for the value of political inquiry, in our view Ochoa-Arias also makes two other contributions. First, there is the *content* of the political inquiries conducted with the neighbourhood association. These focus primarily on the concept of 'participation', and they reveal different meanings given to the term within different political discourses. Most importantly, the chapter shows how the neighbourhood association came to the realisation that most participation in Venezuela is orientated to building capacity in local communities so they can manage their own affairs (including taking responsibility for services that were at one time publicly owned). However, this is not the sort of participation that is necessary for participatory democracy, which also requires debate on the 'common good' that can transcend the narrow boundaries of particular local communities and interest groups. By revealing the different meanings given to participation within different political discourses, Ochoa-Arias has not only helped open up new possibilities for action by the neighbourhood association he is working with, but also raises an issue that we believe is relevant to all Community OR practitioners wherever they are located. There is a strong emphasis on participation in most Community OR work (see, for example, Ritchie *et al*, 1994), so the realisation that there are different forms of participation supporting different political ends should give us pause for thought. We suggest that reflections on the types of participation being promoted by Community OR projects is worthwhile if we do not want to unwittingly promote political trends that we, and the groups we work with, would normally be opposed to.

The other contribution of this chapter (as we see it) is Ochoa-Arias's discussion of the methodology he used with the neighbourhood association to develop the political inquiries. This methodology is called interpretive systemology (e.g., Fuenmayor, 1991a–c; Fuenmayor *et al*, 1991;

López-Garay, 1991; Ochoa-Arias, 1996), and it seeks to unveil different possible interpretations of a problematic phenomenon in order to inform participative debate. Ochoa-Arias ends his chapter by recommending this to Community OR practitioners wishing to enhance the role of political inquiry in their projects.

One final point is worthy of note with regard to this chapter. In our view, it demonstrates that the strong focus on politics that was present in the work of Rosenhead (1986, 1987) when he first launched the Community OR initiative is still alive and well. While some of the philosophical and theoretical assumptions underlying Ochoa-Arias's work differ from Rosenhead's (1987) explicitly 'materialist' perspective, the idea that Community OR should embrace political consciousness raising is held in common by both authors.

1.4.3.2 Community and Environmental OR: Towards a New Agenda

This chapter, by Gerald Midgley and Martin Reynolds, summarises the process of engagement and results of a project (Midgley and Reynolds, 2001) designed to support OR practitioners and their stakeholders in reflecting on how OR can be enhanced to deal more effectively with environmental management issues. Following completion of the original project, the authors reflected on the substantial similarities between Community OR and the 'enhanced Environmental OR' advocated by the practitioners who participated in their research. The results of these reflections have been written up in their chapter, 'Community and Environmental OR', advocating a strategic alliance between practitioners in these two areas.

Midgley and Reynolds identify three reasons for pursuing this alliance. First, to forge a critical mass of activists who can build a stronger movement for sustainable community development than currently exists. Second, to give due recognition to the importance of environmental issues for local communities, and ultimately the whole human species. Third, to encourage learning amongst OR practitioners working in the area of environmental management about methodologies and methods for participative community development (the province of Community OR).

As we see it, the contribution that this chapter makes to Community OR is to highlight the potentially problematic separation of community and environmental issues (and the different sets of methods that have been developed for addressing them). This separation is evident in the practice of many practitioners who focus on *either* community *or* environmental issues, rarely linking the two. However, if we embrace a systemic perspective, where the boundaries of projects are negotiated with participants in relation to whatever problematic situation is being dealt with (see Chapter 10), the work can, in principle, stray into any territory (Midgley, 2001). Maintaining a strict demarcation between 'community' and 'environment' therefore introduces an anti-systemic constraint on the practice of Community OR: in the language of Taket and White (Chapter 11), it represents a failure to recognise the 'leaky boundaries' between the two categories.

1.4.3.3 Adaptive Methodology for Ecosystem Sustainability and Health (AMESH): An Introduction

The final chapter in this book (by David Waltner-Toews, James Kay, Tamsyn Murray, and Cynthia Neudoerffer) presents these authors' Adaptive Methodology for Ecosystem Sustainability and Health (AMESH). While Chapter 13, by Midgley and Reynolds, offers a theoretical rationale for blurring the boundaries of Environmental and Community OR, and explores some of the methodological developments that are needed, this chapter presents the work of some Community OR writers whose practice is already crossing these boundaries.

Waltner-Toews *et al* begin their explanation of AMESH with a discussion of the complex systems paradigm in which their work is based. These authors are certainly not the first in the OR community to advocate a systems perspective for sustainable community development (see, for example, Grossmann and Watt, 1992; Dzidonu and Foster, 1993; Clayton and Radcliffe, 1996; Straussfogel and Becker, 1996; Hendrickson and Tuttle, 1997; Makowski, 2000; Nakamori, 2000; Ravetz, 2000). However, in our view, they demonstrate a level of methodological sophistication that is, in some respects, ahead of much of the field. We make this judgement based on the review of the Environmental OR literature conducted by Midgley and Reynolds (2001), who point out that most authors either specialise in the use of quantitative techniques or the facilitation of debate, but only a small number bring these together. Waltner-Toews *et al*, like most other Community OR practitioners, are committed to both the inclusion of multiple viewpoints and methodological pluralism. However, when complex environmental issues are involved, maintaining an inclusive orientation can be difficult: very often specialist analytical methods are needed to research ecosystem health, and making these accessible to ordinary members of local communities is a major challenge. In our view, one of the significant contributions of this chapter (as with Chapter 6 by Ritchie) is to explain how the use of quantitative approaches can be embedded within a wider participative methodology. Waltner-Toews *et al* put the dialogic exploration of ecosystem parameters first, so that any quantitative analysis undertaken by researchers separately from the community (if it is needed) is already firmly based on assumptions that have been explored and understood by local people.

Finally, we wish to point out that the application of AMESH is illustrated through examples drawn from the authors' interventions in Peru, Kenya and Nepal. This chapter is therefore not only relevant to environmental management, but also to the theme of extending Community OR into developing countries. Indeed, it neatly ties together a number of the emerging trends in Community OR practice that have been explored in this book. It uses a variety of systems ideas; blurs the boundaries between environmental and community issues; and demonstrates the applicability of AMESH to development contexts. It is therefore a fitting chapter for us to close on.

1.5 Concluding Comments

In this book we have sought to consolidate the previously published literature on Community OR, and at the same time introduce new writings that push out the boundaries of theory, methodology and practice. We hope the balance will be of value to you, whether you have a long-standing commitment to Community OR, or are coming to the literature on this subject for the first time. Most of all we hope that you enjoy the chapters to follow, and feel inspired (like the authors who have contributed to this book) to participate in both Community OR practice and the unfolding debate surrounding it.

References

Ackoff, R.L. (1970). A black ghetto's research on a university. *Operations Research*, **18**, 761–771.

Ackoff, R.L. (1979a). The future of operational research is past. *Journal of the Operational Research Society*, **30**, 93–104.

Ackoff, R.L. (1979b). Resurrecting the future of operational research. *Journal of the Operational Research Society*, **30**, 189–199.

Ackoff, R.L. (1981). *Creating the Corporate Future*. Wiley, New York.

Bauman, Z. (2001). *Community: Seeking Safety in an Insecure World*. Polity Press, Cambridge.

Beer, S. (1966). *Decision and Control: The Meaning of Operational Research and Management Cybernetics*. Wiley, Chichester.

Beer, S. (1984). The viable system model: Its provenance, development, methodology and pathology. *Journal of the Operational Research Society*, **35**, 7–25.

Bogdanov, A.A. (1913–1917). *Bogdanov's Tektology*. 1996 edn. Dudley, P. (ed.). Centre for Systems Studies Press, Hull.

Boulding, K. (1956). General systems theory—The skeleton of science. *Management Science*, **2**, 197–208.
Bowen, K. (1995). *In at the Deep End: MSc Student Projects in Community Operational Research*. Community Operational Research Unit, Barnsley.
Capra, F. (1996). *The Web of Life: A New Synthesis of Mind and Matter*. HarperCollins, London.
Carter, P., Jackson, M.C., Jackson, N.V., and Keys, P. (1987). Announcing a 'Centre for Community Operational Research' at Hull University. *Dragon*, **2**(2), 5–18.
Chambers, R. (1983). *Rural Development: Putting the Last First*. Prentice Hall, Harlow.
Chambers, R. (1997). *Whose Reality Counts? Putting the First Last*. Intermediate Technology Development Group, London.
Chambers, R. (2002). *Participatory Workshops: A Sourcebook of 21 Sets of Ideas and Activities*. Earthscan, London.
Checkland, P. (1981). *Systems Thinking, Systems Practice*. Wiley, Chichester.
Checkland, P. (1985). From optimizing to learning: A development of systems thinking for the 1990s. *Journal of the Operational Research Society*, **36**, 757–767.
Churchman, C.W. (1968a). *Challenge to Reason*. McGraw-Hill, New York.
Churchman, C.W. (1968b). *The Systems Approach*. Dell, New York.
Churchman, C.W. (1970). Operations research as a profession. *Management Science*, **17**, B37–B53.
Churchman, C.W. (1971). *The Design of Inquiring Systems*. Basic Books, New York.
Churchman, C.W. (1979). *The Systems Approach and its Enemies*. Basic Books, New York.
Churchman, C.W. (1987). Systems profile: Discoveries in an exploration into systems thinking. *Systems Research*, **4**, 139–146.
Clayton, A.M.H. and Radcliffe, N.J. (1996). *Sustainability: A Systems Approach*. Earthscan, London.
Córdoba, J.R. and Midgley, G. (2003). Addressing organizational and societal concerns: An application of critical systems thinking to information systems planning in Colombia. In: *Critical Reflections on Information Systems: A Systemic Approach*. Cano, J.J. (ed.). Idea Group, Hershey.
Deleuze, G. and Guattari, F. (1980). *A Thousand Plateaus: Capitalism and Schizophrenia*. Massumi, B. (trans.), 1988 edn. Athlone, London.
Dutt, P.K. (1994). Problem contexts—A consultant's perspective. *Systems Practice*, **7**, 539–550.
Dzidonu, C.K. and Foster, F.G. (1993). Prolegomena to operational research modelling of the global environment development problems. *Journal of the Operational Research Society*, **44**, 321–331.
Eden, C. (1988). Cognitive mapping: A review. *European Journal of Operational Research*, **36**, 1–13.
Eden, C. (1989). Using cognitive mapping for strategic options development and analysis (SODA). In: *Rational Analysis for a Problematic World: Problem Structuring Methods for Complexity, Uncertainty and Conflict*. Rosenhead, J. (ed.). Wiley, Chichester.
Eden, C. and Ackermann, F. (1998). *Strategy Making: The Journey of Strategic Management*. Sage, London.
Emery, F. (1993). Characteristics of socio-technical systems. In: *The Social Engagement of Social Science, vol. II: The Socio-Technical Perspective*. Trist, E. and Murray, H. (eds.). University of Pennsylvania Press, Philadelphia.
Emery, F.E. and Thorsrud, E. (1969). *Form and Content in Industrial Democracy*. Tavistock, London.
Emery, F.E. and Thorsrud, E. (1976). *Democracy at Work*. Nijhoff, Leiden.
Emery, F.E. and Trist, E.L. (1965). The causal texture of organizational environments. *Human Relations*, **18**, 21–32.
Emery F.E. and Trist E.L. (1981). Sociotechnical systems. In: *Systems Behaviour*, 3rd edn. Open Systems Group (ed.). Harper and Row, London.

Emmeche, C., Køppe, S., and Stjernfelt, F. (1997). Explaining emergence: Towards an ontology of levels. *Journal for General Philosophy of Science*, **28**, 83–119.

Fildes, R. and Ranyard, J.C. (1997). Success and survival of operational research groups—A review. *Journal of the Operational Research Society*, **48**, 336–360.

Flood, R.L. (1999). *Rethinking the Fifth Discipline: Learning Within the Unknowable*. Routledge, London.

Flood, R.L. and Jackson, M.C. (1991). *Creative Problem Solving: Total Systems Intervention*. Wiley, Chichester.

Friend, J. (1998). Managing development projects and programmes: Fresh perspectives towards an action research agenda. University of Lincolnshire & Humberside Working Paper #21.

Friend, J. and Hickling, A. (1997). *Planning Under Pressure*, 2nd edn. Butterworth-Heinemann, Oxford.

Fuenmayor, R.L. (1991a). The roots of reductionism: A counter-ontoepistemology for a systems approach. *Systems Practice*, **4**, 419–448.

Fuenmayor, R.L. (1991b). The self-referential structure of an everyday living situation: A phenomenological ontology for interpretive systemology. *Systems Practice*, **4**, 449–472.

Fuenmayor, R.L. (1991c). Truth and openness: An epistemology for interpretive systemology. *Systems Practice*, **4**, 473–490.

Fuenmayor, R.L., Bonucci, M., and López-Garay, H. (1991). An interpretive-systemic study of the University of Los Andes. *Systems Practice*, **4**, 507–525.

Gregory, A.J. and Jackson, M.C. (1992a). Evaluating organizations: A systems and contingency approach. *Systems Practice*, **5**, 37–60.

Gregory, A.J. and Jackson, M.C. (1992b). Evaluation methodologies: A system for use. *Journal of the Operational Research Society*, **43**, 19–28.

Gregory, W.J. (1990). Critical systems thinking and LST: How 'liberating' are contemporary critical and liberating systems approaches? In: *Toward a Just Society for Future Generations, vol. I: Systems Design*. Banathy, B.H. and Banathy, B.A. (eds.). International Society for the Systems Sciences, Pomona, CA.

Gregory, W.J. (1992). Critical systems thinking and pluralism: A new constellation. PhD Thesis, City University, London.

Gregory, W.J. and Midgley, G. (2000). Planning for disaster: Developing a multi-agency counselling service. *Journal of the Operational Research Society*, **51**, 278–290.

Grossmann, W.D. and Watt, K.E.F. (1992). Viability and sustainability of civilizations, corporations, institutions and environmental systems. *Systems Research*, **9**, 3–41.

Habermas, J. (1972). *Knowledge and Human Interests*. Heinemann, London.

Habermas, J. (1976). *Communication and the Evolution of Society*. English edn., 1979. Heinemann, London.

Habermas, J. (1984a). *The Theory of Communicative Action, vol. I: Reason and the Rationalisation of Society*. Polity Press, Cambridge.

Habermas, J. (1984b). *The Theory of Communicative Action, vol. II: The Critique of Functionalist Reason*. Polity Press, Cambridge.

Han, J. (2000). A framework of value-focused systems thinking. *Systemic Practice and Action Research*, **13**, 97–109.

Haraway, D.J. (1985). A manifesto for cyborgs: Science, technology and socialist feminism in the 1980s. *Socialist Review*, **80**, 65–108.

Hendrickson, L.U. and Tuttle, D.B. (1997). Dynamic management of the environmental enterprise: A qualitative analysis. *Journal of Organizational Change Management*, **10**, 363–382.

Jackson, M.C. (1987a). Community operational research: Purposes, theory and practice. *Dragon*, **2**(2), 47–73.

Jackson, M.C. (1987b). New directions in management science. In: *New Directions in Management Science*. Jackson, M.C. and Keys, P. (eds). Gower, Aldershot.
Jackson, M.C. (1987c). Present positions and future prospects in management science. *Omega*, **15**, 455–466.
Jackson, M.C. (1991). *Systems Methodology for the Management Sciences*. Plenum, New York.
Jackson, M.C. and Keys, P. (1984). Towards a system of systems methodologies. *Journal of the Operational Research Society*, **35**, 473–486.
Jones, G.C. (1993). OR practice, systems methodologies, and the need to do better. *Journal of the Operational Research Society*, **44**, 845–848.
Jones, H.G. (1992). Early OR in the steel company of Wales. *Journal of the Operational Research Society*, **43**, 563–567.
Jones, S. and Eden, C. (1981). O.R. in the community. *Journal of the Operational Research Society*, **32**, 335–345.
Kelly, G.A. (1955). *The Psychology of Personal Constructs*, vol. I: *A Theory of Personality*. W.W. Norton, New York.
Keys, P. (1991). *Operational Research and Systems: The Systemic Nature of Operational Research*. Plenum, New York.
López-Garay, H. (1991). An interpretive-systemic study of the Regional Planning Corporation of Los Andes in Venezuela. *Systems Practice*, **4**, 491–505.
Makowski, M. (2000). Knowledge integration in model-based decision support. In: *Proceedings of an International Symposium on Knowledge and Systems Sciences: Challenges to Complexity (KSS 2000)*. Ishikawa, Japan, September 25–27.
Mansell, G. (1991). Methodology choice in a coercive context. *Systems Practice*, **4**, 37–46.
Marchal, J.H. (1975). On the concept of a system. *Philosophy of Science*, **42**, 448–468.
Mar Molinero, C. (1992). Operational research: From war to community. *Socio-Economic Planning Sciences*, **26**, 203–212.
Mason, N. (1987). The community operational research (OR) network. *Dragon*, **2**(2), 135–141.
Mason, R.O. and Mitroff, I.I. (1981). *Challenging Strategic Planning Assumptions*. Wiley, New York.
Midgley, G. (1990). Creative methodology design. *Systemist*, **12**, 108–113.
Midgley, G. (1992). The sacred and profane in critical systems thinking. *Systems Practice*, **5**, 5–16.
Midgley, G. (1996). What is this thing called critical systems thinking? In: *Critical Systems Thinking: Current Research and Practice*. Flood, R.L. and Romm, N.R.A. (eds.). Plenum, New York.
Midgley, G. (1997). Developing the methodology of TSI: From the oblique use of methods to creative design. *Systems Practice*, **10**, 305–319.
Midgley, G. (2000). *Systemic Intervention: Philosophy, Methodology, and Practice*. Kluwer/Plenum, New York.
Midgley, G. (2001). Rethinking the unity of science. *International Journal of General Systems*, **30**, 379–409.
Midgley, G., Munlo, I., and Brown, M. (1998). The theory and practice of boundary critique: Developing housing services for older people. *Journal of the Operational Research Society*, **49**, 467–478.
Midgley, G. and Ochoa-Arias, A.E. (1999). Visions of community for community OR. *Omega*, **27**, 259–274.
Midgley, G. and Reynolds, M. (2001). *Operational Research and Environmental Management: A New Agenda*. Operational Research Society, Birmingham.
Miller, J.G. (1965). Living systems: Basic concepts. *Behavioral Science*, **10**, 193–237.
Miller, J.G. (1978). *Living Systems*. McGraw-Hill, New York.

Mingers, J.C. (1992a). Recent developments in critical management science. *Journal of the Operational Research Society*, **43**, 1–10.
Mingers, J.C. (1992b). Technical, practical and critical OR—Past, present and future? In: *Critical Management Studies*. Alvesson, M. and Willmott, H. (eds.). Sage, London.
Mingers, J.C. (1993). The system of systems methodologies—A reply to Schecter. *Journal of the Operational Research Society*, **44**, 206–208.
Mingers, J.C. (1997). Towards critical pluralism. In: *Multimethodology: The Theory and Practice of Combining Management Science Methodologies*. Mingers, J. and Gill, A. (eds.). Wiley, Chichester.
Morgan, G. (1986). *Images of Organisation*. Sage, Beverly Hills, CA.
M'Pherson, P.K. (1974). A perspective on systems science and systems philosophy. *Futures*, **6**, 219–239.
Nakamori, Y. (2000). Knowledge management system toward sustainable society. In: *Proceedings of an International Symposium on Knowledge and Systems Sciences: Challenges to Complexity (KSS 2000)*. Ishikawa, Japan, September 25–27.
Noad, N. and King, L. (1977). Area co-ordination: Some examples compared. *Linkage*, **2**, 10–13.
Ochoa-Arias, A.E. (1996). An interpretive-systemic study of community organizations in Venezuela. PhD Thesis, University of Hull.
Ochoa-Arias, A.E. (1998). An interpretive-systemic framework for the study of community organizations. *Systemic Practice and Action Research*, **11**, 543–562.
Ochoa-Arias, A.E. (2000). Community organizations in Venezuela: Toward the disintegration of the modern state and the emergence of a 'community organized' society? *Systemic Practice and Action Research*, **13**, 165–187.
Ormerod, R. (1996). New methods for old. *Journal of the Operational Research Society*, **47**, 1317–1318.
Parry, R. and Mingers, J. (1991). Community operational research: Its context and its future. *Omega*, **19**, 577–586.
Ravetz, J. (2000). *City Region 2020: Integrated Planning for a Sustainable Environment*. Earthscan, London.
Reason, P. and Bradbury, H. (2000). *The Handbook of Action Research: Participative Inquiry and Practice*. Sage, London.
Ritchie, C. (1994). Community OR—Five years of organised activities and beyond. *International Transactions in Operational Research*, **1**, 41–49.
Ritchie, C., Taket, A., and Bryant, J. (eds.) (1994). *Community Works: 26 Case Studies Showing Community Operational Research in Action*. Pavic Press, Sheffield.
Rosenhead, J. (1986). Custom and practice. *Journal of the Operational Research Society*, **37**, 335–343.
Rosenhead, J. (1987). From management science to workers' science. In: *New Directions in Management Science*. Jackson, M.C. and Keys, P. (eds.). Gower, Aldershot.
Rosenhead, J. (1993). Enabling analysis: Across the developmental divide. *Systems Practice*, **6**, 117–138.
Salthe, S.N. (1989). Self-organization of/in hierarchically structured systems. *Systems Research*, **6**, 199–208.
Stansfield, R.G. (1981). Operational research: Dying to live? Paper presented (and text of talk distributed) at the Operational Research Society Annual Conference held at the University of Sussex, September 8–11.
Straussfogel, D. and Becker, M.L. (1996). An evolutionary systems approach to policy intervention for achieving ecologically sustainable societies. *Systems Practice*, **9**, 441–468.
Sutton, D.C. (1995). The enterprise design framework meets the system of systems methodologies. *Systems Practice*, **8**, 409–439.

Taket, A. (2002). Facilitation: Some contributions to theorising the practice of operational research. *Journal of the Operational Research Society*, **53**, 126–136.

Taket, A. and White, L. (1994). Doing community operational research with multicultural groups. *Omega*, **22**, 579–588.

Taket, A. and White, L. (2000). *Partnership and Participation: Decision-Making in the Multi-Agency Setting*. Wiley, Chichester.

Thunhurst, C. and Ritchie, C. (1992). Housing in the Dearne Valley: Doing community OR with the Thurnscoe Tenants Housing Cooperative. Part 2—An evaluation. *Journal of the Operational Research Society*, **43**, 677–690.

Thunhurst, C., Ritchie, C., Friend, J., and Booker, P. (1992). Housing in the Dearne Valley: Doing community OR with the Thurnscoe Tenants Housing Cooperative. Part 1—The involvement of the Community OR unit. *Journal of the Operational Research Society*, **43**, 81–94.

Trefethen, F.N. (1954). A history of operational research. In: *Operations Research for Management*, vol. 1. McCloskey, J.F. and Trefethen, F.N. (eds.). John Hopkins University Press, Baltimore.

Trist, E.L. and Bamforth, K.W. (1951). Some social and psychological consequences of the longwall method of coal-getting. *Human Relations*, **4**, 3–38.

Trist, E.L. and Burgess, S. (1978). Multiple deprivation: A human and economic approach. *Linkage*, **3**, 8–9.

Trist, E.L., Higgin, G.W., Murray, H., and Pollock, A.B. (1963). *Organizational Choice*. Tavistock, London.

Troncale, L.R. (1985). The future of general systems research: Obstacles, potentials, case studies. *Systems Research*, **2**, 43–84.

Tsoukas, H. (1992). Panoptic reason and the search for totality: A critical assessment of the critical systems perspective. *Human Relations*, **45**, 637–657.

Ulrich, W. (1983). *Critical Heuristics of Social Planning: A New Approach to Practical Philosophy*. Haupt, Berne.

Ulrich, W. (1996). *Critical Systems Thinking for Citizens: A Research Proposal*. Centre for Systems Studies Research Memorandum #10. Centre for Systems Studies, University of Hull, Hull.

Von Bertalanffy, L. (1950). The theory of open systems in physics and biology. *Science*, **111**, 23–29.

Von Bertalanffy, L. (1956). General system theory. *General Systems*, **1**, 1–10.

Von Bertalanffy, L. (1968). *General Systems Theory*. Penguin, London.

Weil, S. (1998). Rhetorics and realities in public service organizations: Systemic practice and organizational learning as critically reflexive action research (CRAR). *Systemic Practice and Action Research*, **11**, 37–62.

Wenger, E. (1998). *Communities of Practice: Learning, Meaning, and Identity*. Cambridge University Press, Cambridge.

White, L.A. (1994). Development options for a rural community in Belize. *International Transactions in Operational Research*, **1**, 453–462.

Wiener, N. (1948). *Cybernetics: Control and Communication in the Animal and Machine*. Wiley, New York.

Wong, N. and Mingers, J. (1994). The nature of community OR. *Journal of the Operational Research Society*, **45**, 245–254.

I

Early Contributions and Later Reflections on Theory, Methodology and Practice

2

Community Operational Research: Its Context and its Future

Richard Parry and John Mingers

2.1 Introduction

It is three years since the Community OR initiative was launched [*this chapter was originally published as a journal paper in 1991—eds.*]. A Community OR Unit has been set-up at Northern College, a Centre for Community OR is operating at Hull University, and a Network of interested members of the OR community has been established. However, there remain many questions about what operational research can offer to community groups and how relationships should be developed. This chapter aims to set the Community OR initiative in context, to trace its historical antecedents, and to discuss a number of issues which need to be addressed to help the initiative to succeed. The ideas and conclusions are based, in part, on a survey of community groups in the West Midlands (Parry and Mingers, 1989) and interviews both with members of community groups and with OR practitioners.

The chapter outlines the origins of a social role for OR and the development of the Community OR initiative. Following a number of

Originally published 1991.

Richard Parry • British Airports Authority, UK.
John Mingers • Canterbury Business School, University of Kent, Canterbury, Kent, UK.
Community Operational Research: OR and Systems Thinking for Community Development, edited by G. Midgley and A.E. Ochoa-Arias, Kluwer Academic/Plenum Publishers, New York, 2004

examples of recent projects the chapter goes on to discuss important issues and problems. This draws upon relevant lessons from a similar involvement in the community of other professions such as planning and law. Finally, recommendations are made for the continuing success of the Community OR initiative.

2.2 *History: OR and Social Responsibility*

Although OR seems intimately wedded to the needs of industrial and public companies, there has always been a concern that OR should be used as much for social good as for private profit. This can be traced back to the founding fathers such as Blackett and Waddington, and Community OR can be seen as the most well organised manifestation of this concern. In tracing this history, we do not wish to suggest that there is a single coherent development, rather that many different views have been at play. Some of these are discussed later in the chapter.

2.2.1 *The Early Days of OR*

The origins of the social role of Operational Research can be traced back as far as the origins of OR itself in the 1930s. Many of the principal figures in the development of OR, such as Blackett and Waddington, were members of the radical science movement. Bernal, Blackett and Waterson-Watt were all Presidents of the British Association of Scientific Workers which aimed:

> to secure the wider application of science and scientific method for the welfare of society. (Bernal, 1939)

In 1939 Bernal published the influential text, *The Social Function of Science* in which he spoke of the importance of science making a meaningful contribution to society:

> Science, conscious of its purpose, can in the long run become a major force in social change. (Bernal, 1939)

Blackett, arguably the father of OR, wrote:

> Socialism will want all the science it can get ... Scientists have ... to make up their minds on which side they stand. (Blackett, 1938)

2.2.2 *The Post-War Period*

The social role of science was further discussed at a post-war conference on *Science and Human Welfare* (Dick and Evans, 1946). The adoption of OR by the nationalised industries must have seemed compatible with this aim since the process of rebuilding the economy assumed prime importance in both economic and social terms. Later, as Cook (1973) points out and Rosenhead (1989) documents, the social function and responsibilities of OR were neglected as these goals diverged. This is reflected in a review of the application of OR to general community services in the USA in 1964 by Flagle.

In America, Ackoff and colleagues at the University of Philadelphia pioneered a relationship with Mantua, a poor city ghetto with a predominantly black population (Ackoff, 1970; Ackoff, 1974a). The relationship developed under the control of the community and drew on a wide range of the University's skills and facilities. People from a number of departments were involved at any one time, often providing quite modest assistance such as transport. This work re-activated an interest in OR's possible contribution to the wider social environment. Ackoff also spoke of the responsibilities of OR practitioners to all those affected by their decisions and, particularly relevant to the forthcoming Community OR initiative, the importance of OR workers giving time and effort where it was needed, for no financial reward, if OR was to take on the mantle of a 'profession' (Ackoff, 1974b).

In Britain, the Institute of OR (IOR) was formed in 1963 with the aim of carrying OR into social and public affairs. It was attached to the Tavistock Institute of Human Relations and later changed its name to the Centre of Organizational Operational Research (COOR). It carried out considerable work into inter-organisational relations (Stringer, 1967) and the provision of community services. Much of this research was with local government (Friend and Jessop, 1977), but of special interest, from a Community OR perspective, is work with voluntary community organisations (Noad and King, 1977; Trist and Burgess, 1978; Stern, 1980). A well-produced newsletter called *Linkage* was issued between 1977 and 1983.

2.2.3 *More Recent Experiences—the 70s and 80s*

A strong advocate of community-based OR was Professor Steve Cook who argued the need for science to benefit society and for OR to act in the interests of social well-being and humanity (Cook, 1973). He stressed the importance of participation by people in the decisions that affected them, and was the first person to explicitly discuss the ideas which would ultimately take shape as Community OR:

> Clients should be drawn not only from the power structure and from the formal business ... hierarchies. (Cook, 1973)

He identified as the groups relevant to this new approach,

> trade unions, consumer groups, ... citizens' groups, residents' associations, charitable bodies such as Oxfam, Shelter. (Cook, 1973)

Cook was also instrumental in promoting discussion around these ideas, notably in the form of the 1975 symposium on *Scientific Method and Social Progress* (Farmer, 1975) which focused around such issues as the measurement of the quality of life, and the evaluation of social group's objectives. Furthermore, influenced by the success of Ackoff's work, Cook formed links with the black community in Birmingham on the same model. This led to the Handsworth Employment Scheme and a series of workshops (Luck, 1984).

Also during the 70s, within the OR Society, the OR and Social Organisation study group discussed such issues as 'is OR doing society more harm than good?', and an occasional newsletter, with the title 'OR?gasm', was produced by a group called ?OR?. One of the chief aims of the latter group was the foundation of an Institute for Critical OR (OR for the People, 1974) committed to the development of new approaches and new partnerships to work towards alleviating the '... severe disparity of power between the "planners" and the "planned against"...' (Thunhurst, 1973), and to explore '... OR work for groups which have no "management" in the conventional sense' (Rosenhead, 1978). This failed to make much headway at the time. Why more success should occur in the 80s is unclear—perhaps OR was more uncertain, or perhaps Thatcherism actually stimulated more of an awareness of poverty and inequality.

At the same time the nature of OR and its role in strengthening the grip of capitalism was scrutinised in articles such as 'Who Does OR Operate For?' (Thunhurst, 1973; Hales, 1974; Chesterton *et al*, 1975; Rosenhead, 1976, 1986a; Sagasti, 1976) Some years later these ideas were carried through into a 'materialist' analysis of operational research (Rosenhead and Thunhurst, 1982) in which the harmful effects of OR's transition to 'management science' were described and a call made for balance by, for example, helping 'grass-roots organizations' with their own 'counter-planning' (Rosenhead, 1987b).

In a practical vein, several local projects were undertaken, notably with residents of North Islington as part of a campaign to improve the local environment. Working parties were set up to examine the potential contribution of OR to Community Health Councils, the finance of public housing and trade unions, though only the latter made real progress. A seminar series was developed, revolving primarily around the unions' concern about aspects of the new information technology (Rosenhead, 1978). More considerable work with trade unions took place in Scandinavia where OR workers were able to help with the development of information systems (Nygaard, 1977) in order to "... try to help them increase their power to control their own affairs" (Dando and Eden, 1977).

The potential for Community OR is also illustrated by the involvement of Norman Dombey (a physicist) with an action group opposed to the building of a new shopping centre (Dombey, 1975). He was able to challenge the model of the centre's likely effects which was used at the planning inquiry, developing a more realistic model to show that the environmental impact would be far greater than predicted.

Moving to the 1980s, significant work has been carried out by Eden and colleagues at Bath University with a number of national charities (Eden and Smithin, 1979; Sims and Smithin, 1982; Smithin and Sims 1982), a non-profit employment agency (Eden and Simpson, 1989), and with a project to help the unemployed black youth of Bath (Jones and Eden, 1981). In each case the technique of 'cognitive mapping' (Eden *et al*, 1983) was used to help the members of the organisation to think about their problems more clearly.

2.3 *The Establishment of Community OR*

In 1985 these various strands came together as the Community OR initiative with the election to the Presidency of the OR Society of Jonathan Rosenhead, a critic of OR's failure to deal with important issues. He had a mandate to '... find a more significant social role for OR...' (Rosenhead, 1985) and '... to dispel the notion that Operational Research is necessarily wedded to a single social interest and perspective' (Rosenhead, 1987a). The Community OR initiative was thereby formally established.

By the end of 1986, a decision had been made to establish a Community OR Unit at Northern College, an adult education college near Barnsley supported by local authorities and trade unions. What distinguishes the College is its contact with a '... wide variety of community groups ...' (Rosenhead, 1987a) and its emphasis on group development amongst the local 'grass-roots' community. The Unit was opened in September 1988 with the first full-time Community OR worker starting some months later. A newsletter called *Acorn* started in 1989.

During 1987 a further OR Society initiative was the establishment, as a study group, of the Community OR Network. This was an informal network of people interested in Community OR but with a committee to co-ordinate work done in the community sector by OR practitioners to '... match willing OR workers with receptive client groups, and with existing projects requiring additional help' (Mason, 1987a).

The third strand of OR's new trinity was provided by the decision to establish a Centre for Community OR at Hull University building upon the work of Jackson and Keys within their MA course, and drawing on their research interests (Carter *et al*, 1987). In what follows, the 'Unit' will refer to Northern College, and the 'Centre' to Hull.

2.3.1 Examples of Community OR

Very little has been published on Community OR, and as a result it is unclear how much has been achieved. Many articles examining general 'social' issues such as those by Bennett *et al* on football hooliganism (Bennett *et al*, 1980), and even Cook's thoughtful analysis of the feasibility of a social wage (Cook, 1984) are about *ideas* and not *action;* their focus is in marked contrast to the thrust of Community OR, to help groups deal with their problems.

In fact, the bulk of reported Community OR work has been carried out at the Hull University Centre through using student projects. These include using Soft Systems Methodology to help a Co-Operative Development Agency with its marketing (Chung and Jackson, 1987), and to help a Development Agency to scrutinise its objectives. A combination of methods, including Assumption Surfacing (Mason and Mitroff, 1981), have been used to help a Council for Voluntary Service to design an evaluation process (Clemson and Jackson, 1988).

Network projects include the development of strategies for fund-raising from the private sector for a Volunteer Bureau, advice on survey/interview work to discover the potential users of a Community Centre, and helping a bicycle pressure group to develop its strategy using a combination of visual decision-making aids (Cropper, 1988). The Unit has made contact and worked with a range of organisations including Thurnscoe Tenants Housing Cooperative (Friend, 1989; Ritchie, 1989), The National Childbirth Trust (Moullin, 1989) and the centre for Alternative Technology (Cobb, 1990).

Outside the boundary of OR, but worthy of note, is work done by Libby Cooper at the Community Research Advisory Centre (CRAC) in North London where advice is provided to local community groups on how to organise surveys. This work has resulted in both a catalogue of successful interventions and an increasing demand for help from the community.

2.3.2 Issues for Community OR

Now that Community OR has been firmly established, it is important that continual learning takes place. This should benefit both the wider OR community and those practitioners who wish to be involved in Community OR. The former in terms of an enrichment of OR practice and theory, and the latter through helping to create the most effective Community OR provision. What follows is an early attempt to raise and address important questions in both these areas. It is based on interviews with many people involved in Community OR, and on a survey of community organisations in the West Midlands area (Parry and Mingers, 1989). The survey examined the problems, needs, and attitudes to OR of community/voluntary organizations, using a questionnaire and follow-up interviews, to determine whether and how OR would be helpful to them. It elicited 53 responses (from 200) from a wide range of groups.

Several interrelated questions have emerged:

- Who is Community OR for? What target groups should it be aimed at?
- What has been learnt from Community OR practice about the characteristics of community groups and the problems of community OR?
- What therefore is the nature of Community OR? Is it different from traditional OR? What distinguishes it from other services and consultants available to community organisations?
- How is it best to organise the provision of Community OR—particularly the Network?

2.4 *Who are the Target Groups of Community OR?*

This is a difficult question to answer definitively, firstly because of the great diversity of possible types of organisation and secondly because there are different views within OR as to the purpose of the initiative. The general problem of defining the voluntary sector is discussed by Ware (1989) and Butler and Wilson (1990). Different purposes will lead to different definitions of the target groups and so will be discussed first. Rosenhead (1987a) illustrates this diversity of view:

> The idea behind the Unit is that it should give extensive experience of how formal problem-structuring approaches can assist non-hierarchical organizations, disposing of few resources, but attempting to represent the interests of their members ... we see the Unit as extending the range of OR's potential clients ... we shall be expanding the domain of rational argument, tackling a new and exciting range of unstructured problems, and contributing to making our society a better one to live in.

This quote articulates three main views (which are not mutually exclusive) of the motivation for Community OR and thus for defining client groups:

(i) That it is organisations with particular characteristics that are important—i.e. those with little money (which would therefore be unable to afford to pay for OR), that are participatively organised and that represent people's interests.
(ii) That Community OR can enhance traditional OR by providing a fertile group of unusual problems and new types of clients.
(iii) That OR should be socially committed, aiming to improve society.

The second view leads to the widest range of client groups as it would include any organisation that is not a traditional OR client. Traditional types of OR clients are profit-making organisations both publicly and privately owned (medium and large); local and central government; and non-profit state organisations such as the NHS and education. These are fairly easily defined, but what is left, i.e. non-traditional groups, is very heterogeneous and includes charities, voluntary organisations, community organisations, workers co-operatives, trade unions and small businesses. There are a vast number of such organisations covering a tremendous spectrum in terms of resources and size.

The other two views would exclude certain groups. Some organisations clearly have extensive resources, for example trade unions and the larger charities (Oxfam's income is over £30 million). These could afford to pay for OR in the normal way. They are also relatively well organised and quite large employers. Do they need to be the recipients of a special initiative? The view that Community OR is essentially motivated by social concerns would exclude small businesses, co-ops (if they were essentially for profit) and possibly trade unions (as they are essentially for the benefit of their members only).

In practice, completed projects have covered a broad range of groups, including action groups, national charities, and small-scale local initiatives. The Centre and the Unit target very different groups, the Centre working largely with professional 'community' organisations such as Councils for Voluntary Service, and the Unit focussing on the 'community groups themselves' rather than these professional intermediaries. In this sense the two initiatives complement each other. Very little work seems to have been done with self-help groups and other small essentially 'people'-based organizations. The survey suggested that there could be a hostile attitude amongst trade unions to OR workers from commercial environments having access to their operations for reasons of confidentiality.

2.5 *What has been Learnt from Community OR Practice?*

Despite the relatively small number, the projects so far undertaken reveal much about community organisations, and raise a number of questions about the nature of community OR.

2.5.1 *The Nature of Community Organisations*

What characterises the voluntary/non-profit type of organisation (which are clear target groups on any definition) at a local level? They are generally small and staffed largely by full-time or part-time volunteers with some paid workers. The volunteers often have no particular skills or training in management or organisation. The main feature of such organisations is that they tend to be overwhelmed by lack of time and money. Maintaining the organisation at all is a significant problem and most work is by way of response to environmental pressures and problems. There is little chance of planning for the future or assessing the overall performance of the organisation. The other main characteristic is that the style of management, or non-management, is very different from that of more traditional organisations. There is little formal hierarchy and decision making is participative and consensual. This often reflects the distinctive value systems of the workers who tend to view formal structures and outside consultants with suspicion if not hostility. The general problems of using 'management techniques' in voluntary/alternative organisations is well illustrated by Landry *et al* (1985).

2.5.2 *Problems in Undertaking Community OR*

A number of particular problems have been encountered, although it can be argued that these may occur in any OR project—a point that will be taken up in the next section.

The first difficulty, and possibly the most important, is the actual presentation of what OR is or can do—a difficult enough task in conventional circumstances. Yet unless this is done well the result will be an inappropriate or non-existent project. The actual emphasis that should be placed on the notion of 'operational research' has been much discussed. Thunhurst (1987) expresses caution about the introduction of OR: 'The words "operational research" may never enter into the engagement.' This attitude reflects the view that groups will be antagonistic to the use of jargon. However, if the word 'operational research' is excluded then under what banner will OR be presented? Sims (1981) and Jackson (1988) talk of 'undefining' OR, and while this is necessary up to a point, it does present the danger of leaving the community clients with no real idea of what is being offered. It might be possible to promote the role of an OR worker as a 'facilitator', although OR is obviously more than this, and this may represent a specific skill that not all OR workers can boast. The very different groups that make up the community sector have different attitudes to professionals and their jargon; a Council for Voluntary Service (CVS) is likely to be more comfortable with it than a self-help group. Is a single definition or explanation of OR appropriate to them all?

Second, there are a number of factors concerned with the on-going relationship such as maintaining credibility, knowing the area, and disengaging from the project. Gaining credibility can be difficult because of the nature of community organisations. Generally there will be no prior credibility accruing from being part of the organisation or from a known consultancy, and there is no management hierarchy to compel cooperation; you will be judged at least in part by your values and commitment. To some extent it may be necessary to work as an ordinary member of the organisation ('making jam' in Eden's phrase) to provide some proof of commitment before an explicit OR contribution can be made. However, whilst it is necessary to perform a range of tasks that may not all be genuine OR, it is important to maintain the distinction between the analyst and the organisation. Too general an involvement is dangerous since once immersed in the organisation it may be difficult to withdraw at a later stage.

Third, specifying the contract or extent of commitment, and some associated success measure is made difficult by the analyst generally being unpaid and part-time. If the work does not appear to be successful the community group will quickly decide not to pursue it, but if they find it useful the OR analyst may find the demands made become too great. Sims and Smithin (1982) identify money as the 'great legitimator', and claim that a voluntary service is valued less highly than that which has been paid for. It is certainly necessary to establish some sort of measure of the project's progress such as 'milestones' to review the partnership or contracts to set expectations of what the study aims to achieve. There is a need to define the OR contribution to clarify the analyst's role and to make clear the point at which the study reaches completion.

2.6 *What is the Nature of Community OR?*

A number of questions arise—what exactly has OR got to offer community organisations? Is community OR different from traditional OR? Are the problems tackled and difficulties faced different, or effectively more intense, or just the same? Is much of what has actually been done not OR at all, merely general community work? Does it matter if it is? How, if at all, is it different from what other consultants can offer? These again are difficult questions which reach to the core of OR itself.

2.6.1 *Is Community OR Different from Traditional OR?*

The first debate concerns the basic nature of community OR and whether it is different from traditional OR. The arguments for it being different are that the organisational context for projects, as described in the previous section, is radically different from that faced by the in-company OR group or mainline consultancy; that this leads to greater preponderance of soft and ill-defined problems; and that a major task for community OR should be the transference of skills to the community thus requiring appropriate simple and robust methods. Rosenhead speaks of the need for 'alternative' methodologies which reflect the participative management that characterise the community sector (Rosenhead, 1981). Jackson (1988) has employed his grid of 'problem-contexts' to suggest that the 'soft' techniques, including the methodologies of Checkland and Eden, are suitable for use in the community sector. However, the work done by Checkland (Checkland, 1987), Eden and his colleagues, and Jackson and Keys may well have been influenced by a desire to test out favoured techniques in new contexts.

Traditional OR methods have also been used in Community OR, notably in the work of Mar Molinero (1988) and Conway (Harman, 1988) into school closures. The survey found a need for 'hard' analysis/advice equal to that for 'soft' OR. Straightforward analytical skills, particularly in the statistical and survey area, have been the basis of the work of the Community Research Advisory Centre (CRAC) which has been very successful. The survey also found that small-scale voluntary groups were probably more interested in an advisory service they could contact when needed. However, it also identified a number of situations where groups were interested in the contribution that OR could make to strategic issues.

2.6.2 *Is Community OR often just Community Work?*

A second issue is whether it is possible to define a boundary between a veritable OR contribution and a more general sort of help? Is community OR really just community work (Lally, 1989; Mingers, 1989)? A number of projects have been criticised for not being 'real' OR but simply involving common-sense or general management skills, such as drafting reports, helping define objectives, designing questionnaires. Community OR, it is argued, must be based on the particular skills of OR people in the quantitative and modelling areas. This fundamental issue depends on what is seen as the nature of OR—is it specifically related to quantification and measurement, or is it a more wide-ranging set of skills? Moreover, given the particular difficulties of gaining cooperation and credibility, undertaking basic work within a community organisation may be a necessary prerequisite for more specific OR projects. Too much concern with OR has its own dangers. A conflict may be created between what the group needs and what the OR worker, wanting to make a definite OR contribution, delivers.

2.6.3 Does Community OR Differ from other Consultancy Services?

A third question is how does community OR differ from other services available? This may determine both the importance and success of the initiative. If the skills that Community OR offers to the community are not unique then the same contribution could be made by others and so the input of OR is not crucial. There are certainly a number of community consultants and national schemes such as Technical Aid and the National Council for Voluntary Service's new Organizational Problem Advice Link which aim to help the community sector. There are also local schemes such as CRAC (already cited) and the growing number of 'Science Shops' on the European model. Some of these services may cover different areas from Community OR, with Technical Aid more concerned with the built-up environment and Science Shops with technology, but there is some definite overlap.

However, the survey revealed that only one community group was aware of a similar form of voluntary help. Others were vaguely aware of consultancy support which could offer similar skills but felt that either (in most cases) they could not afford such a service, or that a relationship with a voluntary OR person would be fundamentally different, allowing a more investigative relationship. What matters is surely the knowledge of groups themselves; if there is notionally some similar service but no groups know of it then there would still be a role for Community OR. If these other services have some overlap with Community OR, it may in any case be productive to seek to form partnerships with them to extend the collective service.

2.7 The Structure of Community OR Provision

What has been learnt about the way that Community OR should be organised?

2.7.1 *The Community OR Network*

As has been noted earlier, the Centre and the Unit are complementary, targeting different groups within the community sector. The Network's status is more ambiguous, being a co-ordinator and no more than the sum of its members' activities. There are many possible roles for the Network to adopt. One possibility is for the central committee to generate projects, notifying members about them so that they can get involved. This is clearly unrealistic without a full-time organiser. Another possibility is to liase with national organisations to prepare the ground for members to follow-up locally. This is dependent on significant co-ordination between the Network and its membership, and national and local organisations, as well as the availability of Network members. Hence it is unlikely to be a productive strategy. Any lesser role, such as a discussion group, would surely represent a failure to fulfil its original ambitions.

The problem with the Network trying to operate as a single whole is that it is a national organisation (or at least tries to be, against the significant London bias in its membership) and the community groups are extremely local in their activities. This offends against Beer's ideas about a system matching the variety of the system it is seeking to control (Beer, 1985). Although talk of 'control' is inappropriate, the basic idea still holds true: that if the Network is to work with local groups it too must work through *local* networks of interested practitioners to integrate in the way that the Centre has done in the Humberside area. This has proved effective in the case of Planning Aid (Cowan, 1981), where the regional branches of both the Planning Institute and the Planning Association stay closely in touch.

2.7.2 *Using Students for Projects*

OR practitioners, working on Community OR in their spare time, can make only a limited commitment. Is using students to sustain the initiative viable? It has proved successful at Hull where students have been invaluable in providing the momentum for the Centre's work through MA projects. It is also the model on which Aston University maintained the contacts established by Steve Cook (Checkland, 1987). However there are problems: students will have to cope with the variable pace at which community groups operate, often frustratingly slow and sporadic with the frantic pace of a sudden crisis in which a student might get swept aside. The danger of performing actual OR at the cost of the group's needs will be acute in the case of students needing to produce work relevant to their courses. Careful supervision is important since work has to be continued to maintain contacts throughout those periods when students are unavailable.

As well as sustaining the initiative, it might be in the interests of students themselves to carry out Community OR projects. Work done for community groups will be taken seriously, students will have to cope with organisational politics (a key feature of their potential future employment) and they will be in a position where it will be possible to achieve meaningful change. This is not necessarily the case for projects in business.

2.7.3 *Finance*

The OR Society has made a financial commitment for three years, and there is a definite need to locate finance for the future. The possibility of some of the work being 'income-generating' has been mooted (Thunhurst, 1987) and hopes have been expressed that Community OR can become at least partially 'economically self-sufficient' (Mason, 1987b). However, most community groups would not be able to pay for OR, since they simply do not have the resources. If ideas of income-generation are to be pursued it would probably raise the spectre of means-testing to determine the ability to pay (Harman, 1986), even if only a nominal fee. This will damage the unique facility that Community OR has to offer. A voluntary partnership would open up far greater possibilities; without financial pressure a truly valuable relationship could be formed, striking at the heart of the group's problems.

2.8 *Recommendations for the Future*

The discussion in the previous section identifies a number of areas where change is appropriate. This section makes some explicit recommendations. These focus largely on the Network, partly because it seems to be the area which is experiencing the most problems, and partly because the success of Community OR may depend ultimately on its success since it is the aspect of the initiative directly involved with the OR community.

To operate effectively in the local environments in which community groups work the Network should concentrate on developing *local* networks. These can then develop according to local circumstances, with the Network centre acting to co-ordinate the local networks. The regional groups of the OR Society might also become more directly involved.

At the same time these networks should integrate with the local networks by forming partnerships with Councils for Voluntary Service, Citizens' Advice Bureaux, local councils or any other influential group in the local environment. This will provide the credibility needed to establish Community OR. Initial projects should be undertaken with those organisations which are most professional and influential locally in order to gain the greatest credibility.

For Community OR to learn collectively from its experience a simple mechanism for reporting is needed, such as a project form that could be completed, detailing the work done and reflections on it. This will also provide up-to-date examples to publicise Community OR. The reports could be kept centrally, providing information on previous experience should similar projects arise.

Planning Aid proves another observation:

> The most effective and significant … work at present is being done by people who are committed to the cause for which they are providing help. (Dick and Evans, 1946).

In order to get started, OR workers should look for groups which share their interests.

A policy on how to pitch the publicity for Community OR should be developed; greatest success will probably be had by presenting OR in different ways to the different types of group. The most productive strategy will be to use examples of successful projects relevant to the different groups so that they can identify similar situations to their own, using 'operational research' as a brand label. The volume of publicity must be treated carefully; not enough and the groups that come for help may not be those that the scheme is aimed at, whereas too much may provide the flexibility to choose which groups to help but will mean that not every one can be helped. These groups may be discouraged from approaching for help in the future.

2.9 *Conclusion*

This chapter has looked at the background to Community OR, and has focused on a number of issues that must be faced. Recommendations for *how* they might be faced are provided. To summarise, the main factor determining the success of Community OR is the interest expressed by the community sector in it, which depends upon its *credibility*. This requires integration with local 'voluntary' networks, which can only be achieved by Community OR becoming known in the community in much the same way as Law Centres. Hence the Community OR Network must organise locally. The creation of local networks will also give OR workers greater confidence to embark on projects with support close at hand. The success of both Planning Aid and Law Centres indicates that all types of group, including those at the very grass-roots, are interested in professional support if it is seen as useful. As the survey research has demonstrated, it is possible to find community groups eager to benefit from the skills OR offers.

References

Ackoff, R.L. (1970). A black ghetto's research on a university. *Operations Research*, **18**, 761–771.

Ackoff, R.L. (1974a). *Redesigning the Future*. Wiley-Interscience, New York.

Ackoff, R.L. (1974b). The social responsibility of Operational Research. *Operational Research Quarterly*, **25**, 361–371.

Beer, S. (1985). *Diagnosing the System for Organisations*. Wiley, Chichester.

Bennett, P.G., Dando, M.R., and Sharp, R.G. (1980). Using hypergames to model difficult social issues: An approach to the case of soccer hooliganism. *Journal of the Operational Research Society*, **31**, 621–635.

Bernal, J.D. (1939). *The Social Function of Science*. Rouledge & Kegan Paul, London, pp. 410.

Blackett, P.M.S. (1938). *The frustration of science*. In: *The Social Function of Science*. Bernal, J.D. (1939), pp. 396. Routledge, London.

Butler, R. and Wilson, D. (1990). *Managing Voluntary and Non-profit Organizations*. Routledge, London.

Carter, P., Jackson, M.C., Jackson, N.V., and Key, P. (1987). Announcing a centre for Community OR at Hull University. *Dragon* **2**, 5–18.

Checkland, P.B. (1987). The application of systems thinking in a real-world problem situations; the emergence of Soft Systems Methodology. In: *New Directions in Management Science*. Jackson, M.C. and Keys, P. (eds.). Gower, Aldershot.

Chesterton, K., Goodsman, R., Rosenhead, J., and Thunhurst, C. (1975). A comment on 'The social responsibility of Operational Research'. *Operational Research Quarterly*, **26**, 91–95.

Chung, C.E. and Jackson, M.C. (1987). Soft systems methodology: A case study for Winterton Co-Operative Developoment Agency. Working Paper 16, Department of Management Systems and Sciences, University of Hull, UK.

Clemson, M. and Jackson, M.C. (1988). Evaluating organisations with multiple goals. *OR Insight*, **2**, 2–5.

Cobb, M. (1990). Acorn: *The Newsletter of the Community OR Unit* 3. Northern College, Sheffield.

Cook, S.L. (1973). Operational Research, social well-being and the zero growth concept. *Omega*, **1**, 647–667.

Cook, S.L. (1984). Can a social wage solve unemployment? In: *The Writings of Steve Cook*, pp. 52–65. Bowen, K., Cook, A., and Luck, M. (eds.). The OR Society, Birmingham.

Cowan, R. (1981). *Providing Planning Aid*. Town and Country Planning Association, London.

Cropper, S.A. (1988). Bricycles, bricycles. Presented at Young OR Conference, Coventry, UK.

Dando, M. and Eden, C. (1977). Reflections on Operational Research: A report from the Euro II Congress. *Omega*, **5**, 225–270.

Dick, W.E. and Evans, I.B.N. (1946). *Science and Human Welfare*. Temple Fortune Press, London.

Dombey, N. (1975). Experts at planning inquiries. *The Planner*, **61**, 299–301.

Eden, C. and Smithin, T. (1979). Operational gaming in action research. *European Journal of Operational Research*, **3**, 450–458.

Eden, C., Jones, S., and Sims, D. (1983). *Messing About in Problems*. Pergamon Press, Oxford.

Eden, D. and Simpson, P. (1989). SODA and cognitive mapping in practice. In: *Rational Analysis for a Problematic World*. Rosenhead, J. (ed.). Wiley, London.

Farmer, J. (1975). Scientific method and social progress. *Operational Research Society Newsletter*, **4**(7).

Flagle, C.D. (1964). Operations Research in community services. In: *Proceedings in Operations Research, vol. 2*. Hertz and Edison (eds.). Wiley, New York.

Friend, J.W. (1989). *Acorn: The Newsletter of the Community OR Unit 1*. Northern College, Sheffield.

Friend, J.W. and Jessop, W.N. (1977). *Local Government and Strategic Choice*, 2nd edn. Pergamon Press, Oxford.

Hales, M. (1974). Management science and the second industrial revolution. *Journal of Radiation Science*, **1**, 5–8.

Harman, D. (1986). Letter. *Operational Research Society Newsletter*, **16**(1).

Harman, D. (1988). Report on conference. *Process of OR Study Group—Methodology in Community OR Conference 1987*. Health Service Management Centre, University of Birmingham.

Jackson, M.C. (1988). Some methodologies for community Operational Research. *Journal of Operational Research Society*, **39**, 715–724.

Jones, S. and Eden, C. (1981). OR in the community. *Journal of Operational Research Society*, **32**, 335–345.

Lally, R. (1989). Letter. *Operational Research Newsletter*, **18**, 12.

Landry, C., Morley, D., Southwood, R., and Wright, P. (1985). *What a way to Run a Railroad: An Analysis of Radical Failure*. Comedia, London.

Luck, G.M. (1984). Working with inner-city community organisations. In: *The Writings of Steve Cook*, pp. 77–78. Bowen, K., Cook, A., and Luck, M. (eds.). The OR Society, Birmingham.

Mar Molinero, C. (1988). Schools in Southampton: A quantitative approach to school location, closure and staffing. *Journal of Operational Research Society*, **39**, 339–350.

Mason, N. (1987a). Community OR Network to be launched in May. *Operational Research Society Newsletter*, **16**(8).

Mason, N. (1987b). The Community Operational Research (OR) network. *Dragon*, **2**, 135–141.

Mason, R.O. and Mitroff, I.I. (1981). *Challenging Strategic Assumptions*. Wiley, New York.

Mingers, J. (1989). Letter. *Operational Research Newsletter*, **19**, 1.

Moullin, M. (1989). *Acorn: The Newsletter of the Community OR Unit* **2**. Northern College, Sheffield.

Noad, N. and King, L. (1977). Area co-ordination: Some examples compared. *Linkage*, **2**, 10–13.

Nygaard, K. (1977). Co-operation with the trade unions: A challenge to OR workers. Presented at Euro II Congress, Stockholm.

OR for the People (1974). *OR?gasm*, **2**.

Parry, R. and Mingers, J. (1989). *Contacting Community Clients*. SIBS, University of Warwick, UK.

Policy of the British Association of Scientific Workers (1939). In: *The Social Function of Science*. p. 461. Bernal, J.D. (ed.). Routledge & Keegan Paul, London.

Ritchie, C. (1989). *Acorn: The Newsletter of the Community OR Unit* 2. Northern College, Sheffield.

Rosenhead, J. (1976). Some further comments on 'The social responsibility of Operational Research'. *Journal of Operational Research Quarterly*, **27**, 266–272.

Rosenhead, J. (1978). Letter. *Operational Research Society Newsletter*, **8**(3).

Rosenhead, J. (1981). OR and trade unions: Report of the OR society working party. *Operational Research Society Newsletter*, **10**(6).

Rosenhead, J. (1985). *Manifesto for Election as OR Society President.*
Rosenhead, J. (1986a). Community OR and conflict research: A comment. *Journal of Operational Research Society*, **37**, 1199–1201.
Rosenhead, J. (1986b). Custom and practice—Inaugural Presidential address. *Journal of Operational Research Society*, **37**, 335–343.
Rosenhead, J. (1987a). From the President. *Operational Research Society Newsletter*, **16**(5).
Rosenhead, J. (1987b). From management science to workers' science. In: *New Directions in Management Science*. Jackson, M.C. and Keys, P. (eds.). Gower, Aldershot.
Rosenhead, J. (1989). Operational Research at the cross-roads: Cecil Gordon and the development of post-war OR. *Journal of Operational Research Society*, **40**, 3–28.
Rosenhead, J. and Thunhurst, C. (1982). A materialist analysis of Operation Research. *Journal of Operational Research Society*, **33**, 111–122.
Sagasti, F.R. (1976). Thoughts on the use and abuse of OR/MS in the planning and management of development—Can OR/MS help in the planning and management of revolutions? *Operational Research Quarterly*, **27**, 937–947.
Sims, D. (1981). The practice of OR: Operationalising Ackoff. *Management Decision*, **19**, 39–47.
Sims, D. and Smithin, T. (1982). Voluntary OR. *Journal of Operational Research Society*, **33**, 21–28.
Smithin, T. and Sims, D. (1982). Ubi caritas? Modelling beliefs about charities. *European Journal of Operational Research*, **10**, 237–243.
Stern, E. (1980). Community organisations in changing settings. *Linkage*, **5**, 6–7.
Stringer, J. (1967). OR for multi-organisations. *Operational Research Quarterly*, **18**, 105–120.
Thunhurst, C. (1973). Who does OR operate for? Presented at OR Society Conference, Torquay, November.
Thunhurst, C. (1987). Doing OR with the community. *Dragon*, **2**, 143–153.
Trist, E. and Burgess, S. (1978). Multiple deprivation: A human and economic approach. *Linkage*, **3**, 8–9.
Ware, A. (ed.) (1989). *Charities and Government*. Manchester University Press, Manchester.

3

Community Operational Research: Purposes, Theory and Practice

Michael C. Jackson

The usual impression of operational research (OR) held by social scientists is of a discipline which seeks to use quantitative techniques to solve tactical problems in pursuit of goals specified by management in large organisations. This view of OR, and especially as it is not a wholly unfair characterisation of much contemporary practice, should make us pause in considering the prospects for 'community OR'. For OR of this type (let us call it 'impoverished OR') would appear, for many reasons, to be unsuitable for the different context of community and co-operative organisations. It is as well to be aware, therefore, that OR began not as a mathematical but as an interdisciplinary science, and that the creation of interdisciplinary teams was seen as one of the most important elements of OR practice in the early textbooks (Churchman *et al*, 1957; Ackoff and Sasieni, 1968). We should also note that these same textbooks emphasise that OR is a 'systems' approach, aiming to be relevant to strategic as well as tactical problems, and appreciate that many of the pioneers of OR were socialist scientists who believed that OR should be used for public rather than sectional interests, and that only under socialism could science realise its full potential for increasing human well-being (Rosenhead, 1987).

Originally published 1987.

Michael C. Jackson • Centre for Systems Studies, Business School, University of Hull, Hull, HU6 7RX, UK.

Community Operational Research: OR and Systems Thinking for Community Development, edited by G. Midgley and A.E. Ochoa-Arias, Kluwer Academic/Plenum Publishers, New York, 2004

Given the aspirations of early OR theorists and practitioners, the way the discipline and profession actually developed, and was employed, came to many as a major disappointment. In Churchman's (1979a) opinion, the original intention of a holistic, interdisciplinary, experimental science addressed to problems of social systems was betrayed, as OR degenerated in the 1960s into little more than mathematical modelling. Rosenhead (1987), taking a more materialist line, rues the way OR was transformed into *management* science as its techniques and methods were appropriated and harnessed for use in the service of capitalism. The standard form of OR from the 1960s to the present has been the 'impoverished'.

Some within OR, however, did keep faith with the original intention. Cook (1973), for example, lamented that its current methodology had led OR to move from the position of 'science helping society' to that of 'science helping the establishment', and sought to develop new methods with clients outside the normal power structure, while others remained true, but felt obliged to change the disciplinary banner under which they worked. Churchman's educational programme in 'social system design' at Berkeley was an attempt to keep the original OR dream alive under a different label (Churchman, 1979a). Ackoff (1979, 1981a) continued to develop an approach that could deal with ill-structured and strategic issues, giving the name 'social systems sciences' (S3) to the educational and consulting activities he initiated at the University of Pennsylvania, while Beer (1967) developed management cybernetics to deal with the significant problems arising in economies, brains and companies, and retained OR for lesser control problems.

We can, therefore, reasonably talk in terms of an 'enhanced OR' based on the spirit of the early pioneers of OR, and drawing upon the work of those, either in OR or on the fringe of the subject, who have continued to develop the discipline according to original intention. This kind of OR is an interdisciplinary science employing rational methods to alleviate ill-structured and strategic (as well as tactical) problems arising in social systems, for the benefit of society. On the basis of 'enhanced OR' the idea of community OR begins to look much more plausible. Furthermore, a long tradition of thought upon which community OR can build is identified. Of particular significance, from this point of view, must be Ackoff's work with leaders of the black ghetto of Mantua (Ackoff, 1970, 1974a), work carried out under Cook's guidance with inner-city community organisations (Luck, 1984), Beer's project with the Allende Government in Chile (Beer, 1981), and various projects undertaken from Bath University with charitable and community groups (Jones and Eden, 1981; Sims and Smithin, 1982).

The recent surge in community OR activity in the UK—including the Centre at Hull, the Community OR Unit at Northern College and the national Community OR Network—can, however, be more immediately traced to the 'Community OR Initiative' launched by the Operational Research Society and inspired by its current President, Jonathan Rosenhead [*this chapter was originally published as a journal paper in 1987—eds.*]. In his Presidential address to the Society, he set out the route he believes OR must follow if it is to carve out for itself a significant role in society and if future progress in the discipline is to be facilitated (Rosenhead, 1986). This involves expanding the range of OR's clients beyond the managements of large organisations, and, on the basis of the challenges arising in assisting 'alternative' clients, developing available theories and methodologies to make them more appropriate to new problem situations. Rosenhead's preferred vehicle for travelling this route turned cut to be 'community OR' and the announcement of the Operational Research Society initiative followed.

Two assumptions underpin Rosenhead's conception of community OR. The first is that OR as an interdisciplinary, systems based subject with a modelling orientation can be of great assistance to a range of alternative clients. The second is that, in trying to help such clients, the OR approach can be enriched in a way that will increase its capacity to aid decision-making in society at large. The following will hopefully lend support to these assumptions by considering, in turn, the purposes that community OR can serve, a possible theoretical framework to underpin the activity and some examples of community OR practice.

3.1 Purposes

During the course of inviting proposals from institutions to house a 'Community OR Unit', the Steering Group for the Operational Research Society initiative stated (Steering Group, 1986) the aims of community OR to be:

(a) To extend awareness of OR to new sections of the community, thus broadening the range of clients.
(b) To demonstrate the relevance of OR to a wider range of problem situations.
(c) To enrich OR methodology and revitalise intellectual life through involvement in novel types of problem.
(d) To contribute to improving the quality of discussion and decision-making in society at large.

It is also possible to interpret community OR as having a more critical and radical dimension. This interpretation will be followed here and can be summarised under two more headings:

(e) To help redress the resource imbalance that exists under capitalism by assisting those underprivileged in this respect.
(f) To develop decision-aiding and problem-solving methods appropriate to a more democratic and socialist milieu.

The implications of these possible purposes will now be explained as a prelude, in the following sections, to suggesting how they might be realised.

(a) The intention to broaden the range of OR's clients implies that the present set of customers is seriously limited. Rosenhead (1986) argues that the customers of OR have been:

> Almost exclusively ... the managements of formally established and legally entrenched organisations disposing of substantial resources (capital, equipment, buildings, supplies), including the labour power of their employees.

Following Rosenhead (1986) and Cook (1973) and also applying our own minds, the list of excluded groups will number, among others, Patients Associations, Community Health Councils, trade unions, consumer groups, political parties, charitable bodies, citizen groups, residents associations, Councils for Voluntary Service, voluntary organisations and workers' co-operatives. As Rosenhead (1986) suggests, some of these have been ignored because they lack the funds to pay OR consultants, while others, such as trade unions, have been neglected because of mutual suspicions regarding aims and intentions and because of fears of alienating the traditional clientele of managers. In order to extend the range of organisations served, OR consultants will have to accept the enhanced version of OR, be prepared to sympathise with the more varied concerns exhibited by 'alternative' clients, and be 'willing to provide services at reduced cost to needy clients—as long ago suggested by Ackoff (1974b). To extend awareness of the usefulness of OR to new groups will entail publishing relevant work done through the various networks that link non-traditional clients (e.g. the 'National Network of Local Co-operative Development Agencies' and the 'National Council of Voluntary Organisations'). It will also help if methods and techniques suitable for community OR are taught on degree courses and as a specialist provision to alternative clients.

(b) As has already been suggested, to demonstrate the relevance of OR to a wider range of problem situations will require the abandonment of 'impoverished OR' and a commitment to the creation of an 'enhanced

OR', drawing upon the spirit of the original pioneers of the discipline and upon the work of those who, whether calling themselves operational researchers or not, have developed OR or 'systems' according to that spirit. The kind of impoverished OR that became the norm in the 1960s, and which still remains in the ascendancy, is unsuitable for the great majority of problems found in the community context. In large part this is because, as Rosenhead (1986) argues, it evolved and was fashioned according to the needs of large bureaucratic organisations, with a tendency towards centralisation, an emphasis on controlling their members' activities and an interest in deskilling their workers. Traditional OR methodology reflects the need for a hierarchically organised decision-making system which can dictate the goals to be pursued and can ensure implementation of recommended procedures using autocratic control devices. In these circumstances human elements could be regarded as passive, and quantitative and optimising techniques employed. In doing community OR there will be fewer situations where a clearly defined goal can be agreed or can be enforced through the managerial hierarchy; and many more situations where debate and consensus building will be necessary before action can be taken. Wide involvement of personnel in decision-making is necessary if satisfactory results are to be obtained and implementation achieved. Fortunately, during the 1970s and the 1980s a member of 'softer' approaches were developed in OR (Eden *et al*, 1983) and systems thinking (Checkland, 1981; Ackoff, 1981b; Mason and Mitroff, 1981) which encourage and facilitate participation and debate. An enhanced OR can utilise these methods in assisting alternative clients to come to terms with the 'messes' (Ackoff, 1981a) or 'wicked problems' (Rittel and Webber, 1981) with which they are constantly confronted.

(c) Just as the OR methods wrought in the service of large corporations took on a specific form tuned to the needs of these enterprises, so the OR fashioned in the community context will take on characteristics which, hopefully, make it adept in these circumstances. Thus the experience of doing OR for alternative clients should enrich OR methodology and revitalise the intellectual life of the discipline and profession. The organisations which community OR will serve usually lack the resources of traditional clients, may be wedded to democratic decision-making procedures and may lack a clear managerial hierarchy which can delimit preferences and ensure the implementation of recommended changes. The usual props supporting the success of impoverished OR will therefore be missing and practitioners will be forced in a different direction if their work is to prove useful. Rosenhead (1987) has listed some of the characteristics that an alternative OR might have to take on, stressing the extent to which these are polar opposites of its usual features. Differences of interest will have to be accepted and methods designed which facilitate negotiation between different groups. Analysis must be conducted so as to be transparent to all concerned and should assist judgement rather than replace it. Human beings will have to be treated as active subjects, and self-control rather than external control of their activities encouraged. Optimisation must be dropped and a greater degree of flexibility and openness in the face of turbulent environments accepted. A number of soft OR and soft systems thinking approaches already demonstrate some of these characteristics and can provide a good foundation for the further enrichment of OR methodology.

(d) The tools and techniques developed to aid alternative clients should, of course, contribute to increasing the quality of discussion and decision-making in the sectors of society in which these clients operate. This, however, is not the only benefit to emerge for society. It can be argued that in many other sections of society too, methods capable of alleviating 'messes' and dissolving 'wicked problems' will be welcomed. For example, strategic problems in organisations and social issues facing government are of a type with those confronting community OR's clients—ill-structured, involving many 'stakeholders' with multiple perceptions and embedded in an uncertain environment. The work conducted by Ackoff and his colleagues, with leaders of the Mantua black ghetto, illustrates the point (Ackoff, 1970, 1974a). The project revealed the need for continuous, adaptive, participative planning in the situation faced by the ghetto leaders—a situation characterised by uncertainty, lack of hierarchy and the need for active intervention. The idea that planning should be continuous, adaptive and participative later became the cornerstone of the 'interactive planning' methodology recommended by Ackoff (1981b) to corporate executives facing turbulent environments. So, in exposing itself to alternative clients and learning from the experience, OR may become more relevant to the strategic and social concerns from which it has been effectively excluded because impoverished OR is only deemed suitable for resolving tactical questions.

(e) It has been argued by Ackoff and Churchman that OR has a social responsibility to serve all the stakeholders of the systems in which it intervenes and all segments of society. For Ackoff (1974b) OR practitioners, if they are to deserve the title 'professional', must ensure that the interests of those not participating in a decision but affected by it are considered. And they should make provision for all those who might benefit from OR but cannot afford it. Ackoff (1974a, 1981b) has given detailed consideration to how planning can be made more participative, involving more of the interested parties, and how resources to assist with planning can be made available to the more underprivileged of these groups.

Churchman (1970), similarly, argues that the primary responsibility of OR as a profession must be to serve all the 'customers' of the systems it is supposed to be benefiting, and not just the powerful decision-makers. Both Ackoff and Churchman seem to believe, however, that the resource imbalances which are admitted to exist under capitalism, and which prevent full participation by those underprivileged in this respect, can be sidestepped by good OR practice which persuades the powerful that they too have something to gain by admitting the powerless into decision-making (Ackoff, 1982; Jackson, 1982a,b). While this is worth a try, self-respecting community OR practitioners will have, on occasion, to take sides. They will come across conflicts of interest that cannot be simply resolved through debate with the powerful. To challenge resource imbalances, the community OR practitioner may be called upon to help develop methods of struggle against the advantaged group or groups.

(f) Rosenhead (1986) has argued that there is nothing inherently capitalist about OR and that

> ... despite the market/control bias of the dominant methodology which it has actually accreted ... it could be argued that OR prefigures a planning mechanism for a society whose impetus does not come from the dynamic of capital accumulation.

Community OR workers of a radical bent will be hoping to show the truth of this and will certainly content themselves with the opinion that the tools and techniques perfected with alternative clients are those which will be most useful in some democratic, non-exploitative, socialist society of the future. Various methods already developed in the tradition of 'enhanced OR' show potential in this respect. Checkland (1981) has suggested that his 'soft systems methodology' can help enrich and structure debates approximating to the model of 'communicative competence' which Habermas (1970) believes must underpin any rational society. Beer's (1981) work in Chile illustrates how organisational cybernetics could be used to design a form of economic co-ordination that might prove more acceptable than the market mechanisms employed in capitalist countries and the centralised bureaucratic planning which, until recent experiments, dominated in East European communist states.

3.2 *Theory*

A consideration of the purposes outlined above suggests that the Centre for Community OR should engage in action research. The purposes embrace both a commitment to practical problem alleviation and to science. These two commitments can, to a large extent, be realised *together* through action research. Existing knowledge will be employed in the choice and use of appropriate OR methodologies and techniques to help clients for whom, at present, little OR gets done. This experience will then become the basis for learning to what extent and in what ways existing knowledge needs to be modified to provide better guidance and more effective practice in future.

Successful action research demands a set of integrated and clearly formulated theoretical hypotheses and a closely monitored programme of practical intervention. In this section of the paper we concentrate on the first of these requirements.

In order for any action research programme to realise its aims it needs to be built upon theoretical premises which can guide practice and against which the results of practice can be interpreted. The idea that it is possible to start with 'pure' observation on practice and, on the basis of this, construct theory is not defensible. All observation and all practical intervention is already and always premised upon some theoretical assumptions—whether made explicit or not. Science and technology best progress if theoretical and methodological assumptions are made explicit at the outset. This permits other scholars and practitioners to carry out 'falsification' tests and to learn and contribute to knowledge in a discipline. A body of work and literature can develop around the original contributions. The process of incrementally building up knowledge in the field of interest is set in motion.

We need some theoretical guidelines on which to base community OR practice. In order that our search is not limited by any pre-conceptions about the current nature of OR, let us begin by offering a very general account of organisations as 'systems'. This will allow us to develop a theoretical codification of the problems likely to arise within organisations for OR practitioners, from which it is a short step to specifying the nature of an enhanced OR suitable for steering community OR work.

Organisations are often characterised in management theory as open socio-technical systems (Kast and Rosenzweig, 1985). This means that they have a technical aspect and are systems which seek to efficiently pursue goals in often volatile environments. They also have a human and social aspect and depend for their viability and effectiveness on the establishment of shared meanings among the actors who make them up. Either social or technical characteristics may predominate in particular organisations or particular parts or levels of organisations, but the successful evolution of organisations would seem to depend upon sufficient attention being given to both.

Although it is not generally recognised, there are some interesting parallels between socio-technical systems thinking and the sociological theory of Habermas (1972, 1974, 1979). Habermas' work can be read as providing a deep understanding of the nature of organisations as open socio-technical systems. According to Habermas there are two fundamental conditions underpinning the socio-cultural form of life of the human species. These he calls 'work' and 'interaction'. Work enables human beings to achieve goals and to bring about material well-being through social labour. Its success depends upon achieving technical mastery over the natural and social environments. The importance of work to the human species leads us to have what Habermas calls a 'technical interest' in the prediction and control of natural and social systems. The technical interest is one of two anthropologically based cognitive interests which Habermas recognises the human species as possessing. The other is linked to 'interaction' and is called the 'practical interest'. Its concern is with expanding the possibilities of mutual understanding among all those involved in the reproduction of social life. Disagreement among different groups can be just as much a threat to the reproduction of the socio-cultural form of life as a failure to predict and control natural and social affairs.

While work and interaction have for Habermas (at least in his later work) pre-eminent anthropological status, the analysis of 'power' and the way it is exercised is equally essential, Habermas argues, for the understanding of all past and present social arrangements. The exercise of power in the social process can prevent the open and free discussion necessary for the success of interaction. Human beings, therefore, also have an 'emancipatory interest' in freeing themselves from the constraints imposed by power relations and in learning, through a process of genuine participatory democracy, involving discursive will-formation, to control their own destiny.

As organisations form the very basis of the socio-cultural life of the human species, it seems clear that they will be the primary centres of social labour, social interaction and the exercise of power. Human beings will have a technical, a practical and an emancipatory interest in their functioning.

It is now possible to use these insights to consider the types of problem situation likely to confront OR practitioners addressing social systems. The starting point is the two anthropologically based cognitive interests of the human species—the technical interest and the practical interest. The technical interest is concerned with the prediction and control of natural and social processes. It follows that problems arise for the human species if the *systems* with which they deal become difficult to predict and control. The practical interest is concerned with securing and expanding the possibilities of mutual understanding among all those involved in the reproduction of social life. It follows that problems also arise for the human species when there is a lack of genuine agreement among *'participants'* about the objectives they wish to see realised through organisational systems. Jackson and Keys (1984) and Jackson (1987) have constructed a grid of 'problem-contexts' based upon a classification of systems and a classification of participants. These classifications, which are now provided, reflect the way these two key dimensions can prove difficult for would-be OR analysts.

Systems can be classified in various ways. Very generally, however, we can talk about a continuum of system types. At one end of the continuum are systems which will be relatively simple and are characterised by having a small number of elements with few, or at least regular, interactions between them. Such systems are likely to be governed by well understood laws of behaviour, to be closed to the environment—in von Bertalanffy's (1969) sense—and to be unaffected by behavioural and social factors such as sub-systems pursuing their own goals. At the other end of the continuum of system types will be systems which are very complex and are characterised by having a large number of elements which are highly inter-related. Such systems are probabilistic, open to the environment, are subject to behavioural and social influences, and have purposeful parts. Ackoff (1974c) has used the terms 'machine age' and 'systems age' to refer to eras characterised by a dominant concern with these two different types of system. Applying Ackoff's terminology, reference will be made in what follows to 'mechanical' problem-contexts which contain relatively simple systems and 'systemic' problem-contexts which contain very complex systems, naturally manifesting more difficult problems.

The kind of relationship existing between participants is the other factor which can greatly affect the nature of problem-contexts. If the participants associated with a particular problem-context are in genuine agreement about objectives, then that problem-context will be easier to handle than if their objectives conflict. Where there is, in general, genuine agreement among the relevant participants, then the problem-context will be called 'unitary'. If the participants have to some extent differing objectives, but a genuine compromise can be reached upon which all agree (because their fundamental interests are not irreconcilable), then the problem-context will be called 'pluralist'. If, however, any consensus that exists is only achieved through the exercise of power and by domination (overt, or more or less concealed) of one or more groups of participants over others, then the problem-context will be called 'coercive'. This final distinction points to the additional emancipatory interest as identified by Habermas.

If the classifications of systems and participants are now combined to yield a six-celled matrix, problem-contexts can be seen to fall into the following categories: *mechanical-unitary, systemic-unitary, mechanical-pluralist, systemic-pluralist* and *systemic-coercive*. Each differs in a significant way from the others and its characteristics determine the nature of the problems found within it and the degree of difficulty posed for the OR analyst. Important differences between problem-contexts should be reflected in the type of theoretical guidance available and in the different kinds of problem-solving methodology that exist. This sets the standard for enhanced OR. If its ability to help alternative clients is not to be constrained then it should possess the facility to operate in each of the six problem-contexts identified. It is an exacting standard, but pessimism gives way to a degree of hope once the distance already traversed is recognised. Community OR can again take sustenance from the work of those outside the OR mainstream who, nevertheless, are closer to the original intention of the discipline. It is worth asking, therefore, to what extent the six problem-contexts are already served by existing OR, cybernetic and systems approaches. This examination of the present strength of enhanced OR can be studied in detail in other publications (Jackson and Keys, 1984; Jackson, 1987) and it is only necessary to make some brief points here.

Mechanical-unitary problem-contexts are well served by the kind of impoverished OR perfected during the 1950s, 1960s and 1970s and still dominating the discipline. The methodology of impoverished OR (and the same can be argued for systems analysis and systems engineering) is useful as long as the problem-solver can establish the objectives of the system in which the problem resides and represent that system in a quantitative model which will simulate its performance under different operational circumstances. These two conditions are only found together in mechanical-unitary contexts.

In *systemic-unitary* problem-contexts the Systems of concern are what Beer (1967) calls exceedingly complex, probabilistic systems. They are not easily represented in a quantitative model or optimised. There is, however, general agreement about the objectives to be pursued. The tools provided by what can be called a 'cybernetic' approach are most useful in these circumstances. This is taken to embrace the organisational cybernetics of Beer (1979, 1981), general system theory (van Gigch, 1978), socio-technical Systems thinking (Rice, 1958; Trist *et al*, 1963) and modern contingency theory (Lawrence and Lorsch, 1969; Galbraith, 1973). The emphasis is on the design of adaptive systems. It is claimed that organisations designed according to 'cybernetic' principles will be self-regulating and even self-organising in the face of environmental perturbations.

Mechanical-pluralist problem-contexts cause difficulties because of disagreement among the participants about the objectives to be pursued by the system of concern. Work by Churchman (1979b), Mason and Mitroff (1981) and Eden *et al* (1983) provides valuable suggestions about how such situations can be alleviated. The emphasis is upon dialectically developing 'world-views' until a temporary consensus is achieved. There is little attention paid to any structural principles that might underpin successful systems design.

Particularly suitable for *systemic-pluralist* problem-contexts is the soft systems thinking of Ackoff (1981b) and Checkland (1981). The soft systems approach can be used to tackle contexts in which there is a lack of agreement about objectives among the participants concerned. Ackoff and Checkland are also prepared to give advice on how to deal with difficulties stemming from the systemic character of the context. They offer cybernetic guidelines for the design of whatever systems we might wish to bring into existence.

Finally, *mechanical-coercive* and *systemic-coercive* problem-contexts should be discussed. If OR practitioners are simply content to serve the powerful in organisations, they can use one of the approaches already mentioned (impoverished OR, cybernetic, dialectical, soft) in these contexts, content that the exercise of coercion will help guarantee success, at least in the short term. Often the existence of power can make coercive situations resemble the unitary and OR workers can take advantage of this fact to progress their analyses. Impoverished OR approaches, for example, often 'work' not because they are the most suitable for the situation in which they are employed, but because they reinforce the position of the powerful, and implementation is therefore enforced. Community OR workers will, however, wish to bring about 'whole system improvement' and must, therefore, find some way of equalising the balance of power in the situation and getting rid of the structures of inequality which allow some groups to dominate others. Enhanced OR is only just beginning to admit the existence of coercive contexts. However, Ulrich's (1983) 'critical systems heuristics' offers a way of proceeding in mechanical-coercive situations, while Jackson (1985) has suggested an approach based upon Habermas' thought as being apt in systemic-coercive contexts.

If, therefore, we abandon the narrow scientism of impoverished OR, an enhanced OR can be identified, embracing approaches such as cybernetics and soft systems thinking, which has gone some way towards developing the necessary range of methods that will be required in doing OR work in the community situation. The classification of enhanced OR methodologies, outlined above, demonstrate this case. It can also be used to help further enrich the methods available to community OR. It insists on the continual development on the theoretical side of the work and hints at the areas for which theoretical support might be forthcoming; for example, once pluralism and coercion are encountered, the social sciences become the obvious source of insight, however underdeveloped they may be at present. Increased dialogue between operational researchers and social scientists is to be encouraged. The classification can also help on the practical side, giving cohesion to the work undertaken, guidance on the choice of appropriate tools and techniques and providing criteria for the choice of projects. It is to the practice of community OR that we now turn.

3.3 *Practice*

No doubt all of the problem-contexts identified in the preceding theoretical section will be discovered when doing community OR and, therefore, all of the different methods and techniques mentioned (and many more) will be appropriate in different circumstances in serving alternative clients. There will be times when the techniques of impoverished OR will be the best choice because a system can be optimised in pursuit of agreed upon goals. The optimal way to deliver services efficiently to the members of a trade union may differ not a jot from the optimal way of carrying out the same activity in a private business organisation. Nevertheless it is reasonable to expect a shift in the balance of problem-contexts found when serving non-traditional clients, away from the unitary or, at least, assumed-unitary type, towards the pluralist and coercive. There will be fewer situations where a clearly defined goal can be agreed or can be enforced through the managerial hierarchy. There will be more situations where debate and consensus building will be necessary before action can be taken. And more situations where the client needs to be supported in the face of powerful opposing interests.

All is not quite as simple as this, of course, because problem-contexts are not unambiguous features of the 'real-world'. To claim they are would be to fall prey to a completely indefensible form of positivism. Problem-contexts are formed from the perceptions and interactions of human beings as purposeful actors. To state that a problem-context is unitary (for example) implies no more than that the perceptions of participants—actors inside the system and relevant observers outside—are all in agreement regarding the purposes to be achieved and about the need to pursue those purposes efficiently. Once this is admitted, it must follow that perceptions and opinions about the character of particular problem-contexts are quite likely to differ. Because they are not 'objective' features of the 'real-world', problem-contexts rarely display their nature with certainty. The way any problem-context is classified, therefore, will depend very much on the person who observes it. In postulating a shift from unitary towards pluralist and coercive contexts, therefore, we are suggesting that community OR workers will, along with the alternative clients themselves, feel inclined to define more problem-contexts in this way.

With these comments in mind we shall attempt to show, in the remainder of this section, how the theory and the methodological procedures mentioned can be utilized in community OR practice. The examples which follow are meant to illustrate the sorts of problem that can arise in community OR and the kind of enhanced OR approach that can be adopted. They are drawn from projects undertaken by members at the Centre at Hull, but are not intended to be accurate reports of those projects. The examples are arranged in order according to whether the context was defined as unitary, pluralist or coercive.

The first example concerns a project with a small Entertainments Group (Jackson and Alabi, 1986). We can take this as a community organisation because of its commitment to the local area and strong orientation to conservation. The converted warehouses it uses for some of its activities are highly regarded by conservationists and have won five major awards. The organisation which had begun originally as one club had grown considerably in size and had plans for further expansion. The directors felt the need to move towards a more formal structure with proper planning, control and information handling procedures. At the same time they did not wish to lose the informal management style that created a particular atmosphere within the company. It was decided to treat the problem-context as systemic-unitary (there were other possibilities) and to try a cybernetic approach to the problem based upon Beer's (1953, 1979) work. Various models were produced and recommendations made for an essentially decentralised structure of decision-making with overall control being maintained through efficient information flows. The recommendations were geared to maintaining informality at the human level while formalising routine business processes.

The majority of problem-contexts encountered in dealing with small community and co-operative organisations are probably best defined as pluralist. Though there will be strong divergences of opinion on the exact aims to be pursued and the best means of achieving these, there is usually an underlying shared value system which can be tapped as the basis for a temporary consensus around a particular issue. There will often be, as well, some commitment towards democratic decision-making. Two examples might suffice to give an indication of this type of situation.

The first involved an assumption surfacing exercise using Mason and Mitroff's (1981) 'Strategic Assumption Surfacing and Testing' approach.

The aim was to give the development workers, in a Co-operative Development Agency, insights into their behaviour as a collective and to help them in decision-making in the future. The particular focus of the exercise was a disagreement in the Agency over the relative merit of 'top–down' as against 'bottom–up' co-operative development work. The 'top–down' approach, which involves identifying business opportunities and then recruiting individuals to form worker co-operatives in those fields, is usually viewed with great distrust in co-operative circles. The preferred approach is 'bottom–up', essentially encouraging and assisting groups already thinking about starting co-operatives in particular fields. Group formation, stakeholder analysis, assumption surfacing and synthesis sessions were staged in accordance with the approach, and the exercise helped clarify where differences of opinions lay and produced consensus on the need for experiments with a modified top–down approach. The second involved assisting the staff of a Council for Voluntary Service, and others, to establish some criteria for evaluating the performance of the Council. It was recognised that the established 'goal' and 'system' models of organisational effectiveness were inappropriate, given the multitude of different viewpoints possible about what success for the Council would entail. Using stakeholder analysis, part of the Mason and Mitroff methodology, it was possible to draw up a list of relevant stakeholder groups, to develop a richer appreciation of the various expectations that the different groups might have of the Council and to consider how these might be taken into account in conducting the evaluation.

The two examples discussed in the last paragraph were relatively short interventions as part of longer projects. They did not themselves involve any redesign of procedures and structures and can therefore be perceived as taking place within mechanical-pluralist contexts. The Co-operative Development Agency provides another example, this time of a problem-context defined to be systemic-pluralist. The Agency was disappointed with the number of co-operatives being set up, by the number of enquiries received from the public and by the number of people coming to the agency to ask for help. It was presumed that something was wrong with the marketing strategy and Checkland's soft systems methodology (1981) was employed to look into this activity. Various possibilities for improving marketing were considered in systems terms—planning, organising, and controlling marketing activities; mobilising potential support for marketing; using more appropriate techniques; a more aggressive 'top–down' strategy; and ways of improving the image of the Agency and the idea of workers' co-operatives. Recommendations were made asking for a more positive attitude to marketing, a division of labour to ensure marketing was not neglected (structural change) and regular marketing audits (procedural charge). The context facing a voluntary organisation, a Community Furniture venture was also considered to be systemic-pluralist. This organisation collects furniture from people who no longer need it and provides it, after refurbishment if necessary, to those in need. Demand for the services of the venture was decreasing and problems loomed associated with the possible end of funding. Checkland's methodology was again used to investigate advertising strategy and possible income generating activities, including selling surplus furniture.

Another project was in a distinctly coercive setting and, since the sources of power imbalance were relatively clear, might be regarded as being within a mechanical-coercive context. This involved the proposed development of a piece of land as a nuclear waste disposal site. Test drilling had already begun on the site. The resources available to the company concerned, and the Government, were clearly much greater than those of the local protest group and the local councils who were opposing the development. An attempt was made using Ulrich's (1983) 'critical systems heuristics' to clarify the value judgements embedded in different proposals for the site, with a view to securing more meaningful debate on the issue.

As a final example, it was decided that a particular business organisation could be most appropriately viewed as providing a systemic-coercive context. The owners of this enterprise liked to regard it as a family firm with a genuine concern for the workforce. However, this image was in danger of shattering in the face of increasing industrial unrest. Managers were eager to paper over the cracks that were emerging with appropriate human relations techniques. We were asked to see the industrial disputes as a result of a short term break-down in communications. It was much more insightful, in fact, to view the situation as systemic-coercive. The firm actually operated in a very traditional manner and employed autocratic control procedures. The coercion had at last come to the consciousness of the workforce. The question of consultant obligations came to the fore and a somewhat tense situation ensued.

The purpose of this section has been to give some examples of community OR work using enhanced OR, and to show how the theory outlined in the last section can both guide community OR practice, and assist reflection on it.

3.4 Conclusion

In this chapter the idea of community OR has been introduced and an attempt made to show how it links to a long tradition of enhanced OR work. The purposes of pursuing community OR were outlined and a framework provided which can assist in realising those purposes by linking theoretical and practical endeavours. Finally some examples of community OR projects were given, showing how these relate to the classificatory framework. There is every reason to expect that an enhanced OR can be of significant use to a wide range of non-traditional clients. There is every reason to believe that this activity will yield significant advances in enhanced OR theory and methodology. Community OR has a major contribution to make in helping to realise the full potential of OR as a discipline and profession.

Acknowledgement

I would like to thank the M.A. students who worked on projects referred to above—B.O. Alabi, E. Chung, F. Lajumoke, and S. Medjedoub.

References

Ackoff, R.L. (1970). A black Ghetto's research on a University. *Operations Research*, **18**(5).

Ackoff, R.L. (1974a). *Redesigning the Future*. New York, Wiley.

Ackoff, R.L. (1974b). The social responsibility of OR. *Operational Research Quarterly*, **25**(3).

Ackoff, R.L. (1974c). The systems revolution. *Long Range Planning*, **7**(6).

Ackoff, R.L. (1979). Resurrecting the future of OR. *Journal of the Operational Research Society*, **30**(3).

Ackoff, R.L. (1981a). The art and science of mess management. *Interfaces*, **11**(1).

Ackoff, R.L. (1981b). *Creating the Corporate Future*. New York, Wiley.

Ackoff, R.L. (1982). On the hard headedness and soft heartedness of M.C. Jackson. *Journal of Applied Systems Analysis*, **10**.

Ackoff, R.L. and Sasieni, M.W. (1968). *Fundamentals of OR*. New York, Wiley.

Beer, S. (1953). *Diagnosing the System for Organizations*. Chichester, Wiley.

Beer, S. (1967). *Cybernetics and Management*, 2nd edn. English University Press, London.

Beer, S. (1979). *The Heart of Enterprise*. Chichester, Wiley.

Beer, S. (1981). *Brain of the Firm*, 2nd edn. Chichester, Wiley.

Checkland, P.B. (1981). *Systems Thinking, System Practice*. Chichester, Wiley.

Churchman, C.W. (1970). OR as a profession. *Management Science*, **17**(2).

Churchman, C.W. (1979a). Paradise regained: A hope for the future of systems design education. In *Education in Systems Science*. B. A. Bayraktar *et al* (ed.). Taylor and Francis, London.

Churchman, C.W. (1979b). *The Systems Approach*. New York, Dell.

Churchman, C.W. *et al* (1957). *Introduction to OR*. New York, Wiley.

Cook, S.L. (1973). Operational research, social well-being and the zero growth concept. *Omega*, **1**(6).

Eden, C. *et al* (1983). *Messing About in Problems*. Oxford, Pergamon.

Galbraith, J. (1973). *Designing Complex Organization*. Addison-Wesley, Reading, MA.

Habermas, J. (1970). Towards a theory of communicative competence. *Inquiry*, **13**.

Habermas, J. (1972). *Knowledge and Human Interests*. Heineman, London.
Habermas, J. (1974). *Theory and Practice*. Heineman, London.
Habermas, J. (1979). *Communication and the Evolution of Society*. Heineman, London.
Jackson, M.C. (1982a). The nature of soft systems thinking: The work of Churchman, Ackoff and Checkland. *Journal of Applied Systems Analysis*, **9**.
Jackson, M.C. (1982b). The nature of soft systems thinking: Comments on the three replies. *Journal of Applied Systems Analysis*, **10**.
Jackson, M.C. (1985). Social systems theory and practice: The need for a critical approach. *International Journal of General Systems*, **10**(3).
Jackson, M.C. (1987). New directions in management science. In, *New Directions in Management Science*. Jackson, M. C. and Keys P. (eds.). Gower, Aldershot.
Jackson, M.C. and Alabi, B.O. (1986). Viable system All/A diagnosis for XY entertainments. *Working Paper*, 9, Department of Management Systems and Sciences, Hull University.
Jackson, M.C. and Keys, P. (1984). Towards a system of systems methodologies. *Journal of the Operational Research Society*, **35**(6).
Jones, S. and Eden, C. (1981). OR in the community. *Journal of the Operational Research Society*, **32**(5).
Kast, F.E. and Rosenzweig, J.R. (1985). *Organization and Management. A Systems and Contingency Approach*, 4th edn. McGraw Hill, New York.
Lawrence, P.R. and Lorsch, J.W. (1969). *Developing Organizations: Diagnosis and Action*. Addison-Wesley, Reading, MA.
Luck, M. (1984). Working with inner city community organizations. In *The Writings of Steve Cook*. Bowen K. *et al* (ed.). The OR Society, Birmingham.
Mason, R.O. and Mitroff, I.I. (1981). *Challenging Strategic Planning Assumptions*. Wiley, New York.
Rice, A.K. (1958). *Productivity and Social Organisation*. Tavistock, London.
Rittel, H.J.W. and Webber, M.M. (1981). Dilemmas in a general theory of planning. In *System Thinking, vol. 2*. Emery F. E. (ed.). Penguin, Harmondsworth.
Rosenhead, J. (1986). Custom and practice. *Journal of the Operational Research Society*, **37**(4).
Rosenhead, J. (1987). From management science to workers' science. In *New Directions in Management Science*. Jackson M. C. and Keys P. (eds.). Gower, Aldershot.
Sims, D. and Smithin, T. (1982). Voluntary OR. *Journal of the Operational Research Society*, **33**(1).
Steering Group for the 'Community OR Initiative' (1986). Community OR: Notes for those submitting proposals.
Trist, E.L. *et al* (1963). *Organizational Choice*. Tavistock, London.
Ulrich, W. (1983). *Critical Heuristics of Social Planning: A New Approach to Practical Philosophy*. Paul Haupt, Berne.
van Gigch, J.P. (1978). *Applied General Systems Theory*, 2nd edn. Harper and Row, New York.
von Bertalanffy, L. (1969). The theory of open systems in physics and biology. In *Systems Thinking*. Emery F. E. (ed.). Penguin, Marmondsworth.

4

Visions of Community for Community OR

Gerald Midgley and Alejandro E. Ochoa-Arias

4.1 Introduction

Operational researchers in the United States have been working with community groups since the late 1960s (Ackoff, 1970), and in the UK since the mid-1970s (Noad and King, 1977; Trist and Burgess, 1978)—some ten years before the term 'community OR' was first coined. However, the labelling of community OR as a specific sub-discipline of operational research facilitated a significant expansion of this kind of activity. In 1986, funding from the Operational Research Society was secured to base a Community OR Unit at Northern College in the UK, and this opened in 1988. In 1987, the Community OR Network (a study group of the Operational Research Society) was formed (with over 300 members), and the Centre for Community OR was opened at the University of Hull (later to be merged into the Centre for Systems Studies, resulting in a massive expansion of community OR activity at Hull). Since 1987, there have been community OR streams at many national and international conferences,

Originally published 1999.

Gerald Midgley • Institute of Environmental Science and Research (ESR) Ltd., Christchurch Science Centre, 27 Creyke Road, PO Box 29-181, Christchurch, New Zealand. Centre for Systems Studies, Business School, University of Hull, Hull, HU6 7RX, UK.
Alejandro E. Ochoa-Arias • Centro de Investigaciones en Sistemologia Interpretiva, Faculty of Engineering, University of Los Andes, Mérida, Venezuela.

Community Operational Research: OR and Systems Thinking for Community Development, edited by G. Midgley and A.E. Ochoa-Arias, Kluwer Academic/Plenum Publishers, New York, 2004

and two conferences have been dedicated solely to community OR. More detailed histories of the institutional development of the community OR movement can be found elsewhere (Parry and Mingers, 1991; Ritchie, 1994).

What is surprising, given the fact that community OR has been practiced using that label for over ten years, is that (as far as we are aware) there have been no in-depth analyses of what 'community' actually means. Some may say that it is obvious: it refers to the use of OR with community groups (the term 'community group' being used in the UK to denote a group of people who are organising or campaigning in their local community on a voluntary basis without significant funds). However, such an answer is not reflected in the diverse reports of community OR practice, which include work with the British National Health Service (Taket, 1994), voluntary organisations with paid staff (Gregory and Jackson, 1992a,b), local government bodies (Vahl, 1994; Midgley *et al*, 1996) and multi-agency groups (Barr and Vangen, 1994; Cohen and Midgley, 1994; Pindar, 1994; Midgley and Milne, 1995; Gregory and Midgley, 2000). If the twenty-six case studies presented by Ritchie *et al* (1994) are representative, then these kinds of projects account for well over half of all community OR activity.

Of course, it would be possible to claim that these projects are not really community OR, but to do so with any justification one would need a normative theory of community that makes clear why community OR should be one particular thing rather than another. Any claim that community OR should only be about, say, working with community groups, would then be open to challenge from others who disagree with this definition and support a different normative theory of community.

Ritchie *et al* (1994) are explicit that they do not wish to impose a restrictive definition on community OR, and argue that diversity is healthy. There is some sense to this, especially when community OR practitioners are trying to grow their discipline. However, there is a danger. Just because no normative theories of community are declared in reports of practice does not mean that they are not present: in deciding who to accept as clients, how to set boundaries for the inclusion and exclusion of issues and participants, and what methods to suggest, the community OR practitioner is necessarily operating with a normative understanding of community—even if s/he is unaware of this. The danger is that, without some reflection on alternative visions of community and their implications for community OR practice, many researchers will simply operate by default with whatever concept of community is culturally dominant. For most, this will be the understanding of community that is implicit in the Western liberal/capitalist political tradition. For those that are happy with this, there is no problem. However, for some people, an important reason to practice community OR is that they believe it has the potential to create space (however small) for critiques of this dominant order, and may also allow the design of human practices and organisations that signal possibilities for alternative political agendas.

It is fair to say that, while the institutional history of community OR is well documented (Parry and Mingers, 1991; Ritchie, 1994; Ritchie *et al*, 1994), its political history remains largely unwritten. That is, there is relatively little documentation about the political motivations people have had in building the institutions of community OR and engaging in its practice, except for some quite general comments about 'doing good' (Wong and Mingers, 1994). Yet many of us who are active in the academic community know that some people's motivations have included the desire to promote socialist revolution; to serve God; to develop participative democracy; to halt the advance of capitalist ideology; to reconstitute civil society; or several of these at once. There are only a handful of writers in OR who have publicly expressed their own political motivations, perhaps the best known being Rosenhead and Thunhurst (1982) and Rosenhead (1986, 1987) in their argument for the creation of a 'workers' science'—but these papers do not address community OR in any detail.

More mundane reasons for engaging in community OR that *are* well documented include broadening the client base of OR (Rosenhead, 1986); broadening and developing the set of methods available to OR practitioners (Rosenhead, 1986; Keys, 1987); promoting methodological pluralism, given that many community OR practitioners have felt the need to choose between methods, or mix them, to deal with novel situations (Jackson, 1988); the challenge of a new experience; practicing new techniques; trying out old techniques in a new environment; promoting community OR as a discipline; gaining a qualification or some other professional reason; and practicing in the locality in which one lives (the last six motivations are mentioned by Wong and Mingers, 1994). These are valid and important reasons for an interest in community OR, especially as some of them indicate the potential for it to inform wider methodological debates in operational research (Jackson, 1988), but we find it disturbing that they have been focused on largely to the exclusion of the political, moral and religious motivations of community OR practitioners (beyond simply 'doing good'). Reflection on these has as much, if not more, potential to transform our understanding of operational research.

It is because of this potential for transformation, and also because of the need (if this potential is to be realised) for community OR practitioners to become more aware of the visions of community that inform their work, that we are writing this paper. Its purpose is to open the space for a debate about normative views of community and their implications for community OR. As such, it does not make a detailed case for any one vision of community to inform our practice: rather, it sketches the landscape of community politics, highlighting some of the strengths and weaknesses of the different visions of community that have been reported in the literature. However, the paper also goes beyond this. While much of the literature on communitarianism (a term often used to describe political positions that emphasise the value of community cohesion) takes for granted that liberalism, and to a lesser extent Marxism, are in some sense 'enemies' of community, we start by seeking to reveal the understandings of community that are implicit in these two political traditions. We therefore follow Frazer and Lacey (1993) in refusing the simple opposition between communitarianism and other practices that writers often try to set up. Liberalism is the first to be discussed because this is the dominant cultural tradition in the West (and increasingly most of the rest of the world). We then move on to Marxism because this is the perspective that, at least in part, informs the analysis of Jonathan Rosenhead, who is widely regarded as the founder of community OR (as a formal discipline), and who has therefore had a substantial influence on its construction.

Following liberalism and Marxism, we discuss communitarianism. Since the death of State socialism in Eastern Europe, the distinction between 'liberalism' and 'communitarianism' has largely replaced the older opposition between 'capitalism' and 'socialism'. However, it is difficult to argue that communitarianism is a unitary political position, despite the common emphasis on the value of community cohesion (Booth Fowler, 1991), so we will try as best we can to capture at least some of its diversity.

In addressing the three political traditions in turn, we should acknowledge that the literature on each of them is vast, so the impression we will give cannot be anything other than superficial. However, it is sufficient to our purpose (opening a debate) to simply sketch some of the different positions within the three traditions, acknowledging (of course) that sketching implies a construction of narratives based on inevitably partial information. It is up to the interested reader to pursue particular aspects of the literature in more depth.

In examining the three traditions in turn, we will first provide general outlines before looking at their implications for the practice of community OR. Taken as a whole, we hope that this paper will help community OR practitioners make some initial judgements about which position(s) within the three traditions they may or may not wish to explore further, or support. Indeed, we will argue in our conclusion that the discussion may even raise possibilities for new developments beyond the positions we describe.

4.2 *Liberalism*

Let us start with liberalism. Liberalism is said to have its roots in the philosophical treatises of Locke (1689), Kant (1787) and Mill (1859) which, while in many ways substantially different, share a common assumption: that the individual, or subject, is an irreducible moral agent with the inherent potential for rational thought. For Rawls (1971), one of several contemporary liberal writers who have received wide-spread attention, individuals are in a sense 'pre-social': they come into the world as selves, or choosing subjects. When they participate in social interactions and use common understandings they do so as active agents with the capacity for rational choice. This is not to say that human beings are completely autonomous, but they do have partial autonomy within their communities (Caney, 1992).

This basic assumption about human nature directly informs the politics of liberalism. Because all individuals are capable of rational choice, and rationality is of fundamental value, it follows that all human lives have the same worth. Thus, equality is a central principle of liberalism: equality before God (if belief in God is accepted) and the law. There is, of course, some debate amongst liberals about what equality really means. According to Frazer and Lacey (1993), most stick with the formal definition just given, but there are a minority (e.g., Marshall, 1950) who argue that material equality is just as important as equality under the law because deprivation of material wealth restricts choice. Not surprisingly, there are others (Friedman, 1963; Nozick, 1974) who are vehemently opposed to such a view on the grounds that the redistribution of wealth that it implies is a fundamental violation of individual liberty.

The notion of equality under the law, for most liberals, means that it should be impartial and universal, in the sense that it should guarantee the freedoms of all and be equally applicable to all. For many liberals, the importance of the role of the law in defending individual freedoms leads to the concept of human rights: thus, rights to free speech, free association, freedom from torture, etc., come to be enshrined in the law.

While the importance of the law is central to the politics of liberalism, writers are more equivocal when it comes to the State. There are what we might term 'welfare liberals' (like Rawls, 1971), who believe that the State should take some responsibility for ensuring that individuals are at least minimally provided for (preferably by helping people to help themselves), and there are others (we could call them 'radical liberals', like Nozick, 1974) for whom any State intervention beyond ensuring the protection of individual liberties should be resisted. Important to most liberals, however, is a strong distinction between the public and private spheres, with the role of the State confined largely, if not wholly, to the former (Simpson, 1994; Tomasi, 1994). Thus, Rorty (1989) argues that individuals should be free to make their own ethical choices, and the role of the State is to create the environment in which this freedom may be protected and maintained. The danger of rampant individualism, where people do as they like regardless of the effects of their behaviour on others, can be averted, according to Rorty, by the power of literature to produce empathy and thereby engender human solidarity.

Of course, a strong, mutually supportive relationship has developed in the West (and increasingly elsewhere) between liberalism and capitalism. Liberalism provides a vision of the individual with a capacity for free choice, and capitalism provides an economic practice in which the individual's capacity to choose may be exercised in the domains of production and consumption.

With all this emphasis on the individual, it would appear that consideration of 'community' is entirely absent. Indeed, in the works of some writers, this is so. However, there is a small handful of authors who have set out to propose specifically liberal understandings of community (rather than simply touch upon the subject as part of a larger thesis), and one of the most comprehensive visions has been produced by Dworkin (1989). The vital feature of community in the liberal tradition is that it can be voluntarily chosen. While people are obviously born into existing communities, they have the capacity for self-reflection and therefore also have a significant degree of choice about their roles within a community, or indeed which communities out of those available they wish to affiliate to—unless this choice is circumscribed by coercive power. A consequence of this focus on choice is that the good liberal community should be pluralist in character: while the shared understandings that grant community cohesion are vital (Caney, 1992), a liberal community should still make available a range of role and life-style options, and be sufficiently open to allow individuals to enter, move around the community, and leave.

Having briefly reviewed liberalism and the understanding of community it embraces, we can now ask the question—what kind of community OR practice would support this tradition? To answer this, we first have to differentiate between welfare and radical liberalism. Each of these will be dealt with in turn.

4.2.1 Welfare Liberalism and Community OR

In the case of welfare liberalism, community OR may play several roles. A significant aspect of the Welfare State is the public provision of services to those who are deemed to be disadvantaged in society, or in particular need. In recent years, a significant issue for service providers and their paymasters has been service quality (Gregory and Walsh, 1993). Community OR practitioners may support service providers in improving quality by working with their managers, workers and/or clients to identify problems and design solutions. They may also support the promotion of user involvement in the planning and management of services in order to enable their customers to have a more influential voice in decision-making. A great deal of this activity is already being conducted under the banner of community OR, especially in the UK (e.g., Gregory and Walsh, 1993; Barr and Vangen, 1994; Gregory *et al*, 1994; Taket, 1994; Vahl, 1994; Midgley *et al*, 1996). Community OR of this type may sweep in a wide range of stakeholders, but the focus will always be on either statutory agencies or partnerships between the statutory sector and community organisations.

A second role might be for community OR practitioners to support the construction of pluralism within local communities, and freedom of movement between community groups. Pluralism might be further developed by the provision of OR support to life-style movements (i.e., people who come together because of a common interest or characteristic). Some OR techniques (e.g., those concerned with financial and publicity planning) might be useful in helping local life-style organisations manage their affairs. To encourage free movement between life-style groups, agencies that publicise and provide support services to them (e.g., libraries, local adult education authorities, people designing contact pages on the internet) might be glad of some help. As far as we are aware, no practitioners have yet engaged in such activities and called them community OR, but several projects have been conducted with umbrella organisations in the voluntary sector that point to some of the kinds of OR techniques that might prove useful—e.g., designing methods of self-evaluation and participative organisational restructuring (Gregory and Jackson, 1992a,b; Andrew, 1994).

A third role for community OR in support of welfare liberalism could be in helping with campaigns for the rights of citizens, through quantitative analyses of key issues and/or the facilitation of groups that are planning campaigns. Interestingly, all of the work on supporting campaigns that we are aware of in the literature is orientated to securing participative democratic planning, or the improvement or maintenance of services (Mar Molinero, 1993; Friend, 1994; Moulin, 1994; Wainright, 1994)—while we know from our personal contacts that some community OR practitioners do indeed support citizen rights campaigns, we are not aware that any of these projects have been written up for public consumption.

A final possible role, to some extent linked to campaigning, is for community OR practitioners to aid community groups and voluntary organisations in monitoring the wider community so as to identify vulnerable or disadvantaged groups that might need statutory help before they can help themselves. This could involve surveys, the modeling of population dynamics and/or cost-benefit analyses. Bringing those who are socially excluded to the attention of the authorities would provide a valuable service to the liberal Welfare State, and indeed community organisations might usefully contribute to the design of new support services through facilitated participative planning.

A common feature of most of the possibilities listed above is community OR support for the involvement of community organisations in the provision of information (about quality, disadvantage, etc.) to appropriate statutory agencies so that effective remedial action can be taken.

4.2.2 Radical Liberalism and Community OR

While the focus of welfare liberalism is largely on the State provision of services, in radical liberalism the State's only role is to guarantee individual freedoms. For this reason, radical liberal community OR practitioners would not want to support quality initiatives or user involvement in publicly-run services. An alternative that would help advance the move from welfare to radical liberalism would be to provide voluntary organisations and community groups with key expertise (e.g., group facilitation, financial planning, statistical analysis) in order to help them develop bids for funding from statutory agencies so that they can run services themselves. This would sometimes put them in competition with other similar groups (as well as business organisations), and may create conflicts of interest when voluntary organisations traditionally have campaigning roles directed at statutory agencies. Nevertheless, such activities would certainly help in the drive to minimise the State machinery. Having said this, however, transferring welfare services from the State to voluntary organisations is only a half-way house: at some point there would be a need for full privatisation to satisfy the most radical liberals.

Another activity that would be valuable in support of radical liberalism is the provision of help for campaigns for the individual rights of citizens (see the last section). This is an overlapping area with welfare liberalism. A further contribution could be made in aiding life-style groups, but (unlike in welfare liberalism) not in support of State co-ordination mechanisms like publicly-funded libraries and educational institutions. The only other role for community OR practitioners in support of radical liberalism, if one accepts the argument that the arts (especially literature) generate human solidarity and head off rampant selfishness, might be in working with community arts groups to improve the organisation of their affairs—as long as this doesn't involve helping them with applications for public funds.

We are not aware of any community OR practitioners working in the areas described above. However, the potential is obviously there, either for someone to choose to do so, or for community OR practice to drift this way if the prevailing political climate moves further in a radical liberal direction.

4.3 Marxism

Let us now move on to discuss the second tradition, Marxism. We should first of all acknowledge that this, like the other traditions, contains a variety of positions, some of which are referred to in the literature as 'neo-Marxist' in that their authors share some core assumptions with Marx (1887), but part company with him on one or more important issues.

It is central to the analysis of Marx (1887) that inequalities of wealth are no accident: the continuation of capitalism depends on the accumulation of wealth in a few hands so that this can be invested in the development of new products, and provide incentives for entrepreneurship. Because capitalism is now a global force, it contributes to massive global inequalities where five out of the six billion inhabitants of the world are said to be living in poverty (Ramonet, 1998). For those who believe that material equality is an ideal worth striving for, this makes capitalism an anathema.

Mandel (1975), developing the work of Marx, produces an incisive analysis of the boom-bust cycles that capitalist economies inevitably go through: companies market innovative products; they make profits; others see that there is money to be made, so make similar products; as competition increases (while the size of the market remains the same) each company makes less and less profit; eventually, some of the companies go bust; wealth begins to concentrate in the hands of a few companies again; reinvestment in the development of innovative products becomes possible; the cycle resumes. If the various industries were at different points of the cycle at different times, there might not be a major problem—but the tendency is for whole economies to go through the cycle together, as companies are so tightly interconnected. Every time there is a bust, there are massive social consequences in terms of unemployment and poverty for large numbers of workers, and for Mandel (1975) and many others this is quite simply unacceptable.

For Marxists, liberalism is just as much a target as capitalism because the former is seen as a false ideology that prevents workers from realising their common interest in changing the political-economic system. As long as personal wealth remains a goal for so many (far more than could ever achieve it), and as long as individuals only look out for themselves in pursuit of this goal, the liberal/capitalist orthodoxy will remain in place.

Let us now look at the solution proposed by Marx and his followers. First, Marx and Engels (1888) believe that the end of capitalism is as inevitable as the end of feudalism (which capitalism replaced). Some busts are more severe than others, and a particularly severe bust (or series of busts) will leave so many people in poverty that they will realise that their interests do not lie in rescuing the current system, but in creating a revolution and instituting a new system of common ownership—socialism. Socialism is said to be a system where laws are necessary to ensure common ownership and prevent a slip back into capitalism. The idea, however, is that socialism is only a transitional phase: once all the people come to realise the benefits of common ownership, the need for laws to enforce it will recede. The end result will be a communist society in which people implicitly understand how to live in non-exploitative relationships with one another—there will be no need for laws regulating economics to enforce this.

An essential aspect of Marxist analyses is to view capitalist social relations in terms of class. There are those who own the means of production, and those who have no option if they are going to survive but to work for these owners. There is therefore a capitalist class and a 'working class'. The revolution is in the hands of the latter who constitute the vast majority of the population. According to Marx (1887), it is the responsibility of those who can already see the need for socialism to help bring it about by fostering a sense of class consciousness amongst the working class. Although revolution is viewed as inevitable because of economic reasons, it can be hastened and made easier by such activism. Here we see a stark contrast with liberalism: for Marxists, the interests of the working class are the prime consideration, while for liberals it is individuals who have priority.

There are several key points in the above analysis with which neo-Marxists of one kind or another disagree. Gramsci makes the important point that the 'working class' is not a unitary phenomenon (Nemeth, 1980). He argues that there are a variety of different interests expressed by people in the general population. To create change, it is necessary for activists to create a 'hegemony' of interests—to build bridges between interest groups so that they begin to see that they also have a common interest in the achievement of socialism. Perhaps the most 'radical' dissenter is Hobsbawm (1985) who, along with others calling themselves neo-Marxists, argue that Marx never anticipated the birth of the Welfare State and the rise in living standards that capitalism has actually allowed to trickle down to the majority of the population (at least in the West). As a result, it is futile to hope for a revolution: even in times of bust, most people still have enough invested in the system to preserve it. Hobsbawm (1985) takes a similar line to Gramsci in that he focuses on the need for a hegemony of interests—interests in feminism, anti-racism, workers' rights, participative democracy, local improvement, gay liberation, the rights of older people, etc. Note that only a small minority of these interests entail the status of 'worker'. The purpose of mobilising a hegemony is not to ferment a revolution, but to move straight to the constitution of a participative socialist government through democratic means that will address both the common and individual interests of all the participating groups.

So far, in this account, the term 'community' has not been mentioned. In the original work of Marx (1887) and Marx and Engels (1888), 'community' could be interpreted in terms of class interests: a community could be said to exist in so far as the working class have a consciousness of their collective identity. Implicitly, then, community relates directly to production: only those without recourse to the means of production (and without facilities to maintain themselves through trade) can constitute a working class community. There is also an ideal of community in communism, the goal to which Marxists aspire. It is a perfect community that can perpetuate itself through the common acceptance of non-exploitative practices without the need for this to be enforced in law. For later writers, especially Hobsbawm (1985), the sole focus on production is dropped. Here, community takes on a different meaning: the consciousness of a common interest amongst diverse social groups that does not entail the removal or reduction of diversity, but is focused on mobilising action for the benefit of all—as a collective, and as groups with different interests within the collective.

Moving on now to discuss how community OR practitioners might act in support of Marxism, we should differentiate between 'classical Marxism' (e.g., Marx, 1887) and 'neo-Marxism' (e.g., Hobsbawm, 1985).

4.3.1 Classical Marxism and Community OR

The focus of a classical Marxist community OR practitioner would no doubt be the creation of a class consciousness amongst workers. One way of doing this would be to lend their support to workers' organisations, such as trade unions. Potentially, the whole gamut of OR techniques could be useful to support trade unions in their planning, campaigning and negotiating activities. However, in doing this, it would be important to ensure that the role of community OR is not subverted so that it becomes solely focused on satisfying the temporary needs of workers within the narrow boundaries of a single enterprise. To protect against this possibility, it would be important for community OR practitioners to think explicitly in oppositional terms: if they are providing quantitative analyses, then their client should be the union alone (not the employer); and if they are facilitating a meeting that involves the employer as well as the union, then it may be advisable to use a method that is based on dialectical debate between clearly opposed positions (e.g., Mason and Mitroff's, 1981, strategic assumption surfacing and testing) that will allow the workers to build a critique of the employer's position (thus incorporating consciousness raising as part of the process). If the workers fail to achieve what they are seeking, then the opportunity can be taken to support them in asking what could be done next time to make success more likely. Scenario planning (e.g., Shoemaker, 1991) might be useful in this regard, and a focus on boundary judgements (Churchman, 1970; Ulrich, 1983; Midgley, 1992; Midgley *et al*, 1998) could also help workers identify who they might connect with in other organisations to make a stronger stand in future. This way, the idea of a common interest across workers' groups may take on personal meaning. As far as we are aware, nobody has yet worked with a trade union and called this community OR.

If workers' organisations are the key 'community groups' from a classical Marxist perspective, then we might ask what other workers' organisations could usefully receive support from community OR? Answering this question could depend on what interpretation is put on the term 'common ownership of the means of production', which is the goal of many socialists. If this is interpreted as nationalisation, then working to support nationalised industries might be a key task for a classical Marxist community OR. However, if moving to 'common ownership' is interpreted as the formation of workers' co-operatives, then these might be worthy of support instead. A great deal of operational research has been conducted with the UK nationalised industries (especially in the 1950s), and many of those who undertook this work were reportedly socialist in their political leanings (Rosenhead, 1987). It should be noted, however, that this kind of practice has never been described as *community* OR. In contrast, several projects conducted with workers' co-operatives *have* been explicitly identified as community OR (Mason, 1994; Ritchie and Townley, 1994).

A final role for a classical Marxist community OR might be to seek out new forms of common ownership that are just getting started, and give them support so that they flourish and provide models for others. An example is local exchange trading systems (LETS schemes) which use a nominal currency to allow the exchange of goods and services largely outside the confines of capitalist enterprise. This could be defined as common ownership in the sense that the community of participants controls the currency—there is no ownership in the formal sense, and a LETS scheme is not a business with any value. Again, we know of no community OR practitioners working in this area.

4.3.2 Neo-Marxism and Community OR

Some of the above activities may be of limited interest to the neo-Marxist community OR practitioner, but the sole focus on the realm of production will be regarded as problematic because it does not take into account the need for hegemony across diverse sectors of society. There are essentially three forms of practice that could support a neo-Marxist drive for socialism: the first is working with socialist local government organisations to support hegemonic development; the second is to work directly with community groups and voluntary organisations to support efforts of campaigning and local community co-ordination; and the third is to support consciousness-raising initiatives in local communities. Each of these will be discussed in turn.

There are several local government organisations in the UK run by left-wing Labour Party groups. Many of these have adopted the hegemonic pursuit of socialism, and (since the late 1970s) have sought the involvement of community groups in designing and running services. One mechanism to do this, popularised by the Greater London Council in the late 1970s and early 1980s (Campbell and Jacques, 1986), is to advertise a pot of money and ask for applications from community groups and voluntary organisations with ideas for service development. Two ideas inform this practice: first, community groups are closer than local government officers to the people, and therefore have a better idea of what is needed; and second, the visible investment of resources in the locality, with projects largely under the control of community groups, engenders feelings of empowerment, and therefore enhances political commitment to the Party which provided the money. The result, according to the theory, is both better services and the development of hegemony. Community OR practitioners could support such initiatives by helping local government bodies plan decentralisation and resource allocation to local communities. They might also provide evaluation services, with an emphasis on formative methods (involving the provision of on-going feedback) so that community groups perceive evaluations as supportive rather than judgemental. Summative evaluations, where the practitioner presents an 'expert' report at the end of a period of research, have been criticised for their demoralising effects in organisations (Patton, 1978, 1980; Guba and Lincoln, 1989). Finally, OR services could be provided directly to the community groups in receipt of funding to minimise project mismanagement and failure. Before ending this discussion of local government support, we should note that some countries have no socialist local governments of this kind, and even in the UK there has been so much central government action to restrict local government spending in the last fourteen years that opportunities to provide new money to community groups have been strictly limited. Depending on the locality, this may therefore be a difficult option for community OR practitioners to follow, and we are not aware of any such initiatives being supported under the banner of community OR to date.

In the absence of enlightened local government, community OR practitioners may prefer to work directly with community groups and voluntary organisations. Remembering the importance of hegemony to neo-Marxist theory, it will be vital, not simply to problem-solve within the boundaries of the local group, but to stress the need for networking and co-ordination within the community. Thus, community OR may help in forging alliances between community groups with common interests. It will also be particularly important to support the campaigning efforts of community groups, as this is one means of their politicisation. There are several examples in the literature of community OR projects which have helped to build alliances and common agendas between community groups (often involving statutory agencies too) (Barr and Vangen, 1994; Gregory and Midgley, 2000). There are also examples of the contributions that community OR practitioners have made to the planning of campaigns (Mar Molinero, 1993; Friend, 1994; Moulin, 1994; Wainright, 1994), although it should be noted that, if a campaign fails, the result may be feelings of disempowerment unless the community OR practitioner can support the group in developing a wider agenda, linking in with other groups, during the process of developing the campaign (Midgley, 1993).

One caveat to the idea of working across the boundaries of community groups is that the community OR practitioner would have to be wary of the involvement of statutory authorities. Remembering that the Welfare State is part of the infrastructure of the liberal/capitalist orthodoxy, it would be preferable to work towards hegemony amongst community groups and voluntary organisations to the exclusion of statutory agencies. This rule might be broken, however, if the practitioner and other participants are confident that the involvement of the statutory sector will not simply result in the community groups being drafted into its service on its own terms. A good example of community groups having their agenda shaped by liberal government is provided by Ochoa-Arias (1996), who documents the role of neighbourhood associations in Venezuela.

Finally, the neo-Marxist community OR practitioner may wish to support consciousness-raising groups in the community. This can be done within local action groups with members who recognise the need to reflect on the political implications of their activities, and want their planning to be informed by these reflections. An example of this kind of consciousness raising is provided by Ochoa-Arias (1996) who used the methodology of interpretive systemology (Fuenmayor, 1991a–c; Fuenmayor *et al*, 1991; López-Garay, 1991) in his work with Venezuelan neighbourhood associations to support community activists in reconsidering their political role. Another example is Ritchie and Thunhurst's (1990) use of 'conscientization' to inform their community OR practice (conscientization is Freire's, 1972, 1973, method of generating political awareness through explorations of the language used by people to describe their activities).

4.4 Communitarianism

Having discussed liberalism and Marxism, we can now proceed to the third tradition of interest in this paper, communitarianism. Of the three traditions, this is probably the most diverse. Indeed, many writers who have been labelled 'communitarian' do not actually use the term themselves. Following Booth Fowler (1991), we accept the term 'communitarianism' with the understanding that there are multiple communitarian positions with substantial overlaps between them, and there are also significant and complex debates on the normative nature of community within each of these. Booth Fowler found it impossible to capture all this variety in a good-sized book, and we are not even going to attempt to do so in a single chapter, except very superficially. Given this caveat, let us begin by discussing the communitarian critique of liberalism (which is often portrayed as 'the enemy'). This analysis will allow us to surface the main issues upon which most communitarians agree. We will then go on to discuss four distinctive (albeit overlapping) positions within communitarianism: participative democracy, historical communitarianism (our term for works which examine pre-industrial models of community), religious belief and the politics of ecology. These are four of the six positions proposed by Booth Fowler (1991), although we have renamed some to enable a narrower focus. We have not covered the other two because one is peculiar to the US context (Republicanism), and we want the discussion to have transnational interest, while the other is a proposal for a new way of thinking about communitarianism based on existentialism that has not yet been fully elaborated (Booth Fowler, 1991).

First, many criticisms have been raised against the liberal idea of the autonomous individual with the capacity for rational thought (for two particularly well known critiques, see Sandel, 1982, and MacIntyre, 1985). The basis of their argument is that individuals are socially constructed beings who, as users of shared symbolic languages, simply could not exist independently from communities. Liberals are also accused of overemphasising rationality, largely to the exclusion of affective commitments to tradition (MacIntyre, 1985). Thus, in the eyes of communitarians, the claim by Rawls (1971) and others that individuals are in some sense 'presocial' is simply wrong. While communitarians have tended to overemphasise the extent to which liberals adhere to the idea of the autonomous individual (most, such as Caney, 1992, stress that autonomy is only partial), there is nevertheless a substantial difference between liberals and communitarians on whether or not individuals are able to choose their community affiliations. Communitarians stress that communities are held together by implicit, socially shared meanings that take a considerable time to form: they cannot just be chosen. Communities are essentially *traditions* which form the basis for understanding and are not easily opened up to internal critique—although critique is possible when traditions become really problematic and outside influences are sought (MacIntyre, 1985).

The different understandings of community proposed by liberals and communitarians inform opposing normative views on the nature of the good society. For communitarians, the good of the community as a whole comes before the good of individuals within it—indeed, the latter is defined in terms of the former. According to Sandel (1982) and others, this is both a normative commitment and an empirical fact that results from the socially constructed nature of individual identities.

Perhaps one of the most powerful critiques of liberalism, shared (as far as we are aware) by all communitarians, is that liberal society is dangerously fragmented, and this fragmentation stems directly from the neglect of community cohesion in favour of individualism. The growth of an 'underclass' with no loyalties to the wider society; self-serving consumerism, regardless of the needs of others; the increasing rate of family dissolution, with serious emotional consequences for many thousands of children; a massive rise in criminal activity, together with a wide-spread fear of crime—these are all cited by communitarians as evidence of fragmentation (MacPherson, 1973; Lasch, 1977; Robertson, 1980; Bellah *et al*, 1985; Oldenquist, 1986). The accusation is that liberalism has allowed morality to become relative: there is no longer a commonly accepted notion of the social good (Bell, 1978; Bloom, 1978; Sandel, 1982; Neuhaus, 1984; Harrington, 1985; Cochran, 1989).

The liberal focus on rights also comes in for criticism. Communitarians believe that the emphasis should be on the duties we have to each other and the wider community (Sandel, 1982; MacIntyre, 1985): if people had a (positive) understanding of their duty to others, it is argued that there would be far less need for a (negative) defense of rights—although, as Etzioni (1996) points out, by no means all communitarians agree with MacIntyre (1985) that the discourse on rights should be abandoned.

While liberalism comes in for criticism in and of itself, its association with capitalism is also a bone of contention. Capitalism (and hence liberalism) is blamed for rampant consumerism and the dominance of corporations in establishing values (Kirk, 1962; Goodwyn, 1978; Lustig, 1982; Harrington, 1985). The vast majority of communitarians see the antidote to this as being community participation in the generation of shared values, upon which co-operative decision-making can be based (e.g., Pateman, 1970; Sale, 1980; Mansbridge, 1983; Barber, 1984; Bowles and Gintis, 1986)—although there are some who think that this view overemphasises human rationality (e.g., Cochran, 1989).

It is in the reaction to liberalism (described above) that we see the common ground that makes it meaningful to talk of communitarianism as a tradition: a belief in the normative primacy of the community over the individual (although some de-emphasise the individual more than others); a focus on social virtues or duties rather than individual rights (although not always to the exclusion of rights); and, for most, an emphasis on the power of participative decision-making to regenerate community cohesion. Having looked at the common ground, let us now examine some of the diversity that is present in the communitarian movement (for further examples, see Booth Fowler, 1991).

One line of thought that has been labelled 'communitarian' (Booth Fowler, 1991) is pursued by the participative democracy movement (Pateman, 1970; Mansbridge, 1983; Barber, 1984; Bowles and Gintis, 1986). The basic idea here is that commitment to community can be fostered through genuine participation, so that everybody feels that they have a stake in the system. Participation has to be local and relatively small-scale if it is to be both meaningful and manageable (Mansbridge, 1983). The emphasis of this kind of communitarianism is on movement towards value consensus and accommodation in decision-making. The participative democracy movement has spent much of its energies on promoting workplace participation, exemplified by the detailed studies conducted by Pateman (1970), but it is generally believed that participation is needed in most social activities, not just work.

In contrast with the immediately practical focus of the participatory democrats, MacIntyre (1985) has been interested in what we might term 'historical communitarianism'. His search for community stretches back to the study of ancient Greek texts. Historical analysis is important for MacIntyre because he believes that the liberal emphasis on the individual and his or her desires has become so pervasive that it is now almost impossible for people to think in terms of social virtues. Indeed, he claims that all we have left are the fragments of older moral orders in which individuals automatically saw the needs of the community and their own needs as one. Given the almost insurmountable difficulty of thinking about present issues in anything but liberal terms, he advocates historical investigation to begin to piece together the fragments and develop a new language of community.

Interestingly, despite MacIntyre's reputation as one of the most radical communitarians, in terms of his rejection of individualism (Cochran, 1989), his vision of an ideal community is not one which is entirely closed to outside influences and critical reflection by community members. Indeed, while he believes that communities should evolve transcendent notions of the common good, he also believes that multiple communities with different notions can and should co-exist. Learning about other communities is both possible and desirable in order to inform the development of one's own community understanding. In this sense, it might be appropriate to call the understandings of social good that MacIntyre wants to see developed *quasi*-transcendental (as opposed to fully transcendental), as they are not accepted as absolutes. Because of this desire to refuse absolutes and preserve the notion of reflection, MacIntyre (along with Bell, 1978, who also resists absolutism) has attracted negative criticism from some other communitarians (Booth Fowler, 1991).

This discussion of absolutes naturally leads on to a consideration of the work of several writers who base their communitarianism on religious foundations. These are of two types: those writing about small-scale religious communities (e.g., Wallis, 1976, 1981; Palmer, 1981), and those who believe that, if society as a whole is to take a communitarian turn, this has to be based on religious values (e.g., Hauerwas, 1981; Diggins, 1984; Bellah *et al*, 1985; Tinder, 1989). The latter are not utopian thinkers, but believe in God and argue that communities simply cannot survive without fully transcendental value systems. As soon as such systems are undermined, we face a descent into relativism and inevitably individualism.

One final perspective that is worthy of note has been developed by writers involved in the politics of ecology. For many such writers, the need to re-establish cohesive communities is an immediate and vital concern. Their analysis is quite different from the usual critique of the liberal fragmentation of society: they argue that the liberal/capitalist orthodoxy has not only created social problems, but (through unsustainable economic growth) has brought us to the brink of ecological catastrophe (Meadows *et al*, 1972, 1992; Mesarovic and Pestel, 1974). Indeed, the fate of our entire species is at stake. As Booth Fowler (1991) points out, there are two different shades of green activist: some argue that we need to scale down our technology and establish small-scale, participative communities which aim to live within the capacities of the ecosystems they are based in (Sale, 1980; Goldsmith, 1988), while others argue that salvation lies in the establishment of a world community where we become more aware of the effects of localised political decisions on others, and act in the interests of the whole human population and the other forms of life we are co-dependent on (Mesarovic and Pestel, 1974; Brown, 1978; Stavrianos, 1981). At least one author (Ophuls, 1977) argues for both planetary government *and* downsizing communities. This is not contradictory if one accepts the principle of subsidiarity (Knight, 1996; Backhaus, 1997): the idea that the lowest possible level of government should have responsibility for managing issues (e.g., combating global warming needs to be an international concern, but planning a new housing estate might be dealt with best at a local community level). However, one thing that is common to most green activists is an assumption that the certainty of our predicament provides the foundation for building new transcendental values upon which sustainable communities can rely.

While there are broad similarities between communitarians, we see that there are also great differences. Indeed, an argument could be created (see the conclusion to this paper) for a separation of some of the above positions into a new category. Therefore, in considering the types of community OR that might support communitarianism, we will sub-divide the discussion into participative democracy, historical communitarianism, the religious perspective and green politics.

4.4.1 Participative Democratic Communitarianism and Community OR

The obvious point to start with when discussing how community OR could support the participative democracy movement is that only *participative* community OR will help in this regard: intervention in the mode of expert (keeping calculations hidden in a 'black box' and presenting a summative report) can have disempowering results (White and Taket, 1993), and may work against participative democracy. We say that this point is obvious, but the line between 'participative' and 'non-participative' practice is not always clear: it is usually possible to point to a stakeholder group who has been excluded, and there are of course different modes of participation (Arnstein, 1969; Oakley, 1991; Mumford, 1996).

One way of thinking about modes of participation is to use Arnstein's (1969) 'ladder', which distinguishes between non-participation, tokenism and citizen power. Setting aside non-participation, tokenism includes informing people, consulting with them and attempting to placate them. In contrast, citizen power may include partnerships, delegation of authority, or full citizen control. We suggest that only these forms of citizen power are sufficient to support a move to participative democracy. The implication for community OR is that projects which do not, at the very minimum, involve partnerships between stakeholders should not be undertaken (consultation, for example, is not enough). It is interesting that, when we examine the community OR literature, we find that participation is a principle that is almost universally accepted. While it could be argued that some attempts at participation slip back into consultation, the will is certainly there to create a participative practice (see Ritchie *et al*, 1994, for some examples).

It is not only the mode of participation that is important, however, it is also vital to consider the boundaries of inclusion and exclusion. Is it sufficient, as with most cases of worker control in industry, to include those within the organisation but leave out environmental stakeholders (those outside the organisation who are affected by its activities)? And what about representing non-human elements of the environment? Questions like these are not easy to answer in the abstract, as there may be specific circumstances on the ground that demand justifiable compromises on the extent of participation (Ulrich, 1983; Ho, 1997). For community OR practitioners to deal with this issue well enough to claim that they are helping to advance participative democracy, we suggest that they must at least reflect on boundary judgements in the context of practice (where possible in communication with stakeholders), and justify inclusions and exclusions. For a review of the theory and practice of boundary critique, including a discussion of methods, see Midgley *et al* (1998).

So far, we have discussed the modes and boundaries of participation. In the context of welfare provision, however, it is also useful to distinguish between 'consumerist' and 'critical' participation (Thompson, 1995). When stakeholders outside a welfare organisation are able to participate, we have to ask who sets the agenda? With citizen control, this is obvious. But in partnerships and the delegation of authority, the terms of reference can be set by the organisation to restrict the activities of stakeholders in ways that they would not ideally wish to accept. When this happens, Thompson (1995) suggests it indicates that a consumerist mentality is in operation—and, as we saw earlier, consumerism is closely associated with welfare liberalism. It is acceptable for organisations to set prior agendas within the tradition of welfare liberalism because welfare is ultimately the responsibility of the State, not ordinary citizens. The alternative to consumerism, which she calls 'critical' (we might say communitarian), is either to start with stakeholder agendas from outside the welfare organisation, or to build partnerships that allow the meaningful discussion of stakeholder perspectives even when these contradict the organisational agenda. If participative democracy is to be promoted, and slippage back into liberalism avoided, then community OR practitioners would be advised to ensure that any projects they undertake with welfare organisations have a critical foundation. A number of community OR projects have been conducted using this understanding of critical participation, many of which have employed methodologies from the critical systems literature (Cohen and Midgley, 1994; Gregory *et al*, 1994; Midgley *et al*, 1996, 1998; Gregory and Midgley, 2000).

A final question before we move on is, what constitutes a community organisation from a participative democratic point of view? The answer seems to be any organisation that is participative in nature. Business and welfare organisations count, as the emphasis is not on the purpose of the organisation or the ownership (which is a common concern of Marxism and radical liberalism), but on the nature of control. This therefore gives community OR an extremely wide brief indeed.

4.4.2 Historical Communitarianism and Community OR

Participative democratic and historical communitarianism are at opposite ends of the spectrum of communitarian politics: the former is concerned with beginning to practice communitarianism in the here and now, while the latter claims that this is impossible without slipping back into liberalism. The historical communitarian argument is that we only have fragments of a community-orientated morality left over from older ways of thinking, and so what is needed is historical investigation and the construction of a new community morality relevant to the present day.

The implications of this argument for community OR practitioners wishing to support the construction of a new community morality is that community OR should set aside most of the methods and methodologies currently in use and concentrate on the investigation of understandings of community as represented in the literature from pre-industrial, and especially pre-Kantian, epochs (according to Fuenmayor, 1994, the work of Kant, 1787, marks the beginning of the present liberal epoch). There is an approach called interpretive systemology (Fuenmayor, 1991a–c; Fuenmayor *et al*, 1991; López-Garay, 1991; Ochoa-Arias, 1996) that might be relevant to this task. Interpretive systemology asks the investigator to take a problematic phenomenon and identify alternative understandings of it, revealing different interpretations of the grounds for its existence. Some of these understandings may be generated from historical study. This has been used twice in the context of community OR: once to surface different understandings of Venezuelan community organisations (Ochoa-Arias, 1994, 1996), and once to investigate the phenomenon of community OR itself (Weedon, 1992). Only once a greater understanding is generated of what it means to participate in a moral community would it be legitimate, from a historical communitarian perspective, to consider how community OR can be reconstructed as a directly interventionary practice.

4.4.3 *Religious Communitarianism and Community OR*

In looking at the implications of religious communitarianism for community OR, we need to differentiate between those promoting small-scale religious communities, and those who believe that, if society as a whole is to take a communitarian turn, this has to be based on transcendental religious values.

It seems fairly obvious that, to promote small-scale religious communities, the focus of community OR would need to be on direct support for these. Such communities can, in a sense, be viewed as organisations—which are no doubt susceptible to familiar problems amenable to solution using OR methods of one kind or another. The question is, would the movement of OR methods into the religious community be enough, or would there also be a need to develop new problem structuring techniques that directly reflect the religious convictions of those who will participate in their use? The only way to answer this question would be to see what happens in practice and, as far as we are aware, nobody has yet worked with a religious community in the name of community OR.

The second form of religious communitarianism, which is seeking to build a religious society, would no doubt have a different use for community OR. The focus of the religious community OR practitioner wishing to promote this form of communitarianism could be in support of the planning and campaigning work of evangelical movements that are spreading their gospel. Methods like soft systems methodology (Checkland, 1981; Checkland and Scholes, 1990) might be useful in developing strategies upon which action can be based.

There is one remaining question to be answered, however, with regard to religious communitarianism and community OR. Is the use of OR techniques to 'do good' in society (when the practitioner is religious) sufficient to qualify as the promotion of religious communitarianism? While many religious community OR practitioners do indeed use their skills in this way (we know this from personal contacts), in order to promote religious values people on the receiving end of the OR support must understand that the practitioner follows a particular religion; recognise that his/her practice is informed by it; and respect what the practitioner is doing. While the community OR practitioner cannot easily control whether s/he is respected, s/he can make her religion explicit. If s/he keeps it implicit, then she will have the satisfaction of knowing that she is 'doing good' (and will provide an example of 'good' for others to follow), but it would be difficult to claim that she is promoting *religious* communitarianism: the values she would be promoting could be divorced from their religious base and thereby lose their transcendental character.

4.4.4 Green Communitarianism and Community OR

The final form of communitarianism to consider is that promoted by the two kinds of green activist: those focusing on the desire to downsize communities and promote local government in the interests of sustainability (we might call them localists); and those focusing on the desire for a world community, including those who advocate subsidiarity, again in the interests of sustainability (globalists).

In examining the implications of localism for community OR, we should first ask whether any form of operational research that promotes local community cohesion and does not obviously perpetuate or promote unsustainable practices can have the label 'green' applied to it. Our answer is that these are only the minimum criteria for the legitimacy of OR within a localist green perspective. In addition, the community OR practitioner should be committed to a 'steady state' economics (Daly and Cobb, 1989), or refuse the discourse of economics altogether on the grounds that 'ecology into economics won't go' (McBurney, 1990); seek to reduce energy consumption; and resist technological fixes (which are regarded with suspicion because of the side-effects of previous uses of technology to solve our problems).

In the view of localist green thinkers, our predicament is extremely urgent. This means doing something in communities right now, not focusing all our energies on yet more debates about the pros and cons of ecologically sustainable lifestyles. Stokes (1981), for instance, advocates a wide variety of local initiatives, including home and community gardening and the local generation of energy. The cornerstone of Stokes's philosophy is community self-help, and this is the key to understanding what the community OR practitioner should do if she wishes to support localist green communitarianism. She should offer support to any local project that generates cohesion and meets the green criteria listed above (or others derived from reflection on the literature about green localism). A variety of community OR approaches could be relevant here, from planning methods—especially those, like interactive planning (Ackoff, 1981), that evolve a vision of a desired future—to support for financial management. We are not aware of any community OR practitioners currently working on green projects of this kind, but there is clearly scope for development here.

When it comes to the globalist green vision, however, the focus of community OR would have to be quite different. Here, 'community' refers to shared understandings of virtuous practice that span the whole world. One important aspect of globalist green communitarianism is the idea that, to deal with many issues that transcend national and continental boundaries, there is a need for some form of world government (Brown, 1978; Hauerwas, 1981; Stavrianos, 1981; Goldsmith, 1988). Community OR practitioners could help contribute to the formation of this in several ways. First, by getting involved in international relations, seeing how planning methods developed for use in organisations transfer to international negotiation (and also learning from those in other disciplines already working at this level of community intervention). Second, community OR practitioners could offer their services to existing world organisations, such as the United Nations (UN), with the specific intention of fostering enhanced mutual understanding and support through project work. Getting access to organisations like the UN is not always easy, and it might be necessary to look for other bodies to support international bridge-building work in the first instance, with the intention of publicising the results of projects and thereby establishing a reputation in the global arena. Finally, it would be appropriate for community OR practitioners to work with local, national and international campaigning organisations (e.g., Greenpeace and Friends of the Earth) who are seeking to influence international policy on environmental and other global issues. These organisations might find OR support useful, especially in planning campaigns, developing information systems and maintaining good quality internal communications (some of these campaigning organisations can be quite large, with offices spread throughout the world).

However, world government is not the only focus of globalist green communitarians. Another is subsidiarity (Knight, 1996; Backhaus, 1997): building a hierarchy of government from global to local, with issues being dealt with at the lowest possible level. The principle of subsidiarity is already accepted by the European Community, although in practice there is a great deal of work still to be done to make it a reality. An important thing to realise about subsidiarity, however, is that it is not always automatically obvious what level of government is the best to deal with a particular issue. Because issues are seen quite differently by different people, one group or nation may regard an issue as a regional or national concern, while another may see it as having global significance. Therefore, as subsidiarity comes to be operationalised in Europe (and almost certainly elsewhere), there are going to be many debates over a variety of complex issues at the international, national, regional and local levels. This should be regarded as an opportunity by globalist community OR practitioners, as they can offer their services, both in the facilitation of planning, and in the modelling of complex issues to help clarify strategic possibilities. Systems approaches of various kinds should be of particular value here because of their focus on boundaries (e.g., Churchman, 1970; Ulrich, 1983; Midgley, 1992; Midgley *et al*, 1998); hierarchies in modelling (e.g., Wilby, 1994); designing multi-layered organisations (Beer, 1985); and producing holistic plans (e.g., Ackoff, 1981; Checkland, 1981; Checkland and Scholes, 1990; Flood and Jackson, 1991; Flood, 1995; Midgley, 1997). However it will be important when engaging in this kind of activity to ensure that, whatever levels of government are being considered, the global context of the debate is always understood: planning must be more than merely the instrumental resolution of problems—it must foster a sense of global community and an ethic of mutual care.

One other possible arena for community OR intervention might also have relevance here. While the focus so far has been on governance, there is another world-wide institution that has the potential for fostering a global consciousness: the internet. Community OR practitioners with knowledge of the workings of the internet could support green organisations in developing materials and discussion groups. There might also be a role for setting up and facilitating international technology support networks so that internet facilities can be introduced into the least developed countries where access is still virtually non-existent. Note the difference between the localist and globalist strategies here: initiatives like the introduction of internet services into developing countries would most likely be seen as an anathema by localists because of the cultural homogeneity it could foster, and the use of electricity which it demands (although electricity does not have to be provided through energy-inefficient national grid systems, like in the developed countries, but can be generated locally in much more efficient, albeit less reliable, ways).

4.5 Conclusion

In this chapter, we have argued that every community OR practice has a normative vision of community embedded in it. If practitioners do not reflect on the different visions that it is possible to support, then there is a danger that they will default to the understanding of community that is implicit in the liberal/capitalist tradition currently dominant in the West (and increasingly most of the rest of the world). There are alternatives, however. To clarify some of the choices available to community OR practitioners, we reviewed three major political traditions (liberalism, Marxism and communitarianism), as well as sub-divisions of these, and asked the question: what kind of community OR practice would support each one? In all, we identified eight different (sometimes overlapping) forms of community OR practice, each of which would be capable of promoting a different normative vision of community. There are therefore substantial political choices open to community OR practitioners.

It is our view, however, that the options presented here do not exhaust the possibilities for community politics that could be constructed. There is, for example, a significant tension within the communitarian tradition between authors who advocate a transcendental system of values that are placed beyond critique (Lasch, 1977; Palmer, 1981; Diggins, 1984; Tinder, 1989), and those who want to preserve the notion of critical reflection within the community context (Bell, 1978; MacIntyre, 1985). The latter appear to be in the minority, but we follow Frazer and Lacey (1993) in arguing that there is a substantial reservoir of ideas in the 'critical' camp (as well as elsewhere) that can be used as the basis to develop a new and distinct position. While some authors, such as Booth Fowler (1991), claim that any notion of critical reflection is irredeemably liberal in character, we are not convinced that this is the case. If there is a cultural shift away from the liberal/capitalist orthodoxy in the next millennium, we suggest that it is likely to be towards a form of communitarianism that includes within it some of the ideas from our present epoch (such as critical reflection), but transformed by the new context in which they are embedded. In our view, it is also likely that it will have a distinctly green complexion. While it is possible to simply exercise choice between the political alternatives that currently exist, we are excited by the possibility of contributing to the construction of a new, normative vision of community in which critical reflection and community cohesion will be viewed as complementary concepts—and which will provide the grounds for making community OR a significant force as we enter the new millennium.

Acknowledgement

We would like to thank the British Council for the funding they provided for an exchange programme between our two Universities, without which this chapter could not have been written.

References

Ackoff, R.L. (1970). A black ghetto's research on a University. *Operations Research*, **18**, 761–771.

Ackoff, R.L. (1981). *Creating the Corporate Future*. New York, Wiley.

Andrew, C. (1994). Total systems intervention and the management of complexity: A practical experience. MA Thesis, University of Hull.

Arnstein, S.R. (1969). A ladder of citizen participation. *Journal of the American Institute of Planners*, **35**, 216–224.

Backhaus, J.G. (1997). Subsidiarity and ecologically based taxation: A European constitutional perspective. *Public Choice*, **90**, 281–310.

Barber, B. (1984). *Strong Democracy: Participatory Politics for a New Age*. University of California Press, Berkeley.

Barr, C. and Vangen, S. (1994). Investigating effective collaboration. In *Community Works: 26 Case Studies showing Community Operational Research in Action*. Ritchie, C., Taket, A., and Bryant, J. (eds.). Pavic Press, Sheffield.

Beer, S. (1985). *Diagnosing the System for Organisations*. Wiley, Chichester.

Bell, D. (1978). *The Cultural Contradictions of Capitalism*. Basic Books, New York.

Bellah, R.N., Madsen, R., Sullivan, W.M., Swidler, A., and Tipton, S.M. (1985). *Habits of the Heart: Individualism and Commitment in American Life*. University of California Press, Berkeley.

Bloom, A. (1987). *The Closing of the American Mind*. Simon and Schuster, New York.

Booth Fowler, R. (1991). *The Dance with Community: The Contemporary Debate in American Political Thought*. University Press of Kansas, Lawrence, Kansas.

Bowles, S. and Gintis, H. (1986). *Democracy and Capitalism: Property, Community and the Contradictions of Modern Social Thought*. Routledge and Kegan Paul, London.

Brown, L. (1978). *The Twenty-Ninth Day: Accommodating Human Needs and Numbers to the Earth's Resources*. Norton, New York.

Campbell, B. and Jacques, M. (1986). Goodbye to the GLC. *Marxism Today*, **30**(4), 6–10.

Caney, S. (1992). Liberalism and communitarianism: A misconceived debate. *Political Studies*, **40**, 273–289.

Checkland, P. (1981). *Systems Thinking, Systems Practice*. Wiley, Chichester.

Checkland, P. and Scholes, J. (1990). *Soft Systems Methodology in Action*. Wiley, Chichester.

Churchman, C.W. (1970). Operations research as a profession. *Management Science*, **17**, B37–B53.

Cochran, C.E. (1989). The thin theory of community: The communitarians and their critics. *Political Studies*, **37**, 422–435.

Cohen, C. and Midgley, G. (1994). *The North Humberside Diversion from Custody Project for Mentally Disordered Offenders: Research Report*. Centre for Systems Studies, Hull.

Daly, H.E. and Cobb, J.B. (1989). *For the Common Good: Redirecting the Economy Towards Community, the Environment and a Sustainable Future*. Green Print, London.

Diggins, J.P. (1984). *The Lost Soul of American Politics: Virtue, Self-Interest, and the Foundations of Liberalism*. University of Chicago Press, Chicago.
Dworkin, R. (1989). Liberal community. *California Law Review*, **77**, 479–509.
Etzioni, A. (1996). Positive aspects of community and the dangers of fragmentation. *Development and Change*, **27**, 301–314.
Flood, R.L. (1995). *Solving Problem Solving*. Wiley, Chichester.
Flood, R.L. and Jackson, M.C. (1991). *Creative Problem Solving: Total Systems Intervention*. Wiley, Chichester.
Frazer, E. and Lacey, N. (1993). *The Politics of Community: A Feminist Critique of the Liberal-Communitarian Debate*. Harvester Wheatsheaf, New York.
Freire, P. (1972). *Pedagogy of the Oppressed*. Harmondsworth, Penguin.
Freire, P. (1973). *Education: The Practice of Freedom*, Bergman-Ramos, M. (ed.) (trans. by). Writers and Readers Publishing Cooperative, London.
Friedman, M. (1963). *Capitalism and Freedom*. University of Chicago Press, Chicago.
Friend, J. (1994). Community involvement in health strategy in Tower Hamlets. In: *Community Works: 26 Case Studies Showing Community Operational Research in Action*, Ritchie, C. Taket, A. and Bryant J. (eds.). Pavic Press, Sheffield.
Fuenmayor, R. (1994). Systems science: Addressing global issues—the death rattle of a dying era? *Systemist*, **16**, 110–157.
Fuenmayor, R.L. (1991a). The roots of reductionism: A counter-ontoepistemology for a systems approach. *Systems Practice*, **4**, 419–448.
Fuenmayor, R.L. (1991b). The self-referential structure of an everyday living situation: A phenomenological ontology for interpretive systemology. *Systems Practice*, **4**, 449–472.
Fuenmayor, R.L. (1991c). Truth and openness: An epistemology for interpretive systemology. *Systems Practice*, **4**, 473–490.
Fuenmayor, R.L., Bonucci, M., and López-Garay, H. (1991). An interpretive-systemic study of the University of Los Andes. *Systems Practice*, **4**, 507–525.
Goldsmith, E. (1988). *The Great U-Turn: De-Industrializing Society*. Green Books, Hartland.
Goodwyn, L. (1978). *The Populist Movement: A Short History of the Agrarian Revolt in America*. Oxford University Press, New York.
Gregory, A.J. and Jackson, M.C. (1992a). Evaluating organizations: A systems and contingency approach. *Systems Practice*, **5**, 37–60.
Gregory, A.J. and Jackson, M.C. (1992b). Evaluation methodologies: A system for use. *Journal of the Operational Research Society*, **43**, 19–28.
Gregory, W.J. and Midgley, G. (2000). Planning for disaster: Developing a multi-agency counselling service. *Journal of the Operational Research Society*, **51**, 278–290.
Gregory, W.J. and Walsh, M.P. (1993). Quality, ideology and consumer choice—health care standards and stakeholder participation. In: *Managerial Issues in the Reformed NHS*. Malek, M., Vacani, P., Rasquinha, J., and Davey, P. (eds.). Wiley, Chichester.
Gregory, W.J., Romm, N.R.A., and Walsh, M.P. (1994). *The Trent Quality Initiative: A Multi-Agency Evaluation of Quality Standards in the National Health Service*. Centre for Systems Studies, Hull.
Guba, E. and Lincoln, E. (1989). *Fourth Generation Evaluation*. Sage, London.
Harrington, M. (1985). *The Politics at God's Funeral: The Spiritual Crisis of Western Civilization*. Penguin, New York.
Hauerwas, S. (1981). *A Community of Character: Toward a Constructive Christian Ethic*. University of Notre Dame Press, Notre Dame, Indiana.
Ho, C.H. (1997). A critical process for the evaluation of methodology. PhD Thesis, University of Hull.
Hobsbawm, E. (1985). The retreat into extremism. *Marxism Today*, **29**(4), 7–12.

Jackson, M.C. (1988). Some methodologies for community operational research. *Journal of the Operational Research Society*, **39**, 715–724.
Kant, I. (1787). *The Critique of Pure Reason*, 2nd edn, 1933 (trans.) Smith, N.K. Macmillian, Basingstoke.
Keys, P. (1987). Management and management support in community service agencies. *Dragon*, **2**, 19–45.
Kirk, R. (1962). *A Program for Conservatives*, revised edn. Regnery, Chicago.
Knight, W.A. (1996). Towards a subsidiarity model for peacemaking and preventive diplomacy—making Chapter VIII of the UN Charter operational. *Third World Quarterly*, **17**, 31–52.
Lasch, C. (1977). *Haven in a Heartless World: The Family Besieged*. Basic Books, New York.
Locke, J. (1689). *Two Treatises of Government*, 1967 edn, Laslett, P. (ed.). Cambridge University Press, Cambridge.
López-Garay, H. (1991). An interpretive-systemic study of the Regional Planning Corporation of Los Andes in Venezuela. *Systems Practice*, **4**, 491–505.
Lustig, R.J. (1982). *Corporate Liberalism: The Origins of Modern American Political Theory, 1890–1920*. University of California Press, Berkeley.
MacIntyre, A. (1985). *After Virtue: A Study in Moral Theory*. Duckworth, London.
MacPherson, C.B. (1973). *Democratic Theory: Essays in Retrieval*. Oxford University Press, Oxford.
Mandel, E. (1975). *Late Capitalism*. New Left Books, London.
Mansbridge, J.J. (1983). *Beyond Adversary Democracy*. University of Chicago Press, Chicago.
Mar Molinero, C. (1993). Aldermoor School: The operational researcher on the side of the community. *Journal of the Operational Research Society*, **44**, 237–245.
Marshall, T.E. (1950). *Citizenship and Social Class and Other Essays*. Cambridge University Press, Cambridge.
Marx, K. (1887). In: *Capital: A Critique of Political Economy*, Vol. I, 4th edn, Engels, F. (ed.). (trans.) Moore, S. and Aveling, E., Lawrence and Wishart, London.
Marx, K. and Engels, F. (1888). *The Communist Manifesto*, revised edn. Lawrence and Wishart, London.
Mason, N. (1994). Workers' co-operatives and common ownership in the UK. In: *Community Works: 26 Case Studies Showing Community Operational Research in Action*. Ritchie, C., Taket, A., and Bryant, J. (eds.). Pavic Press, Sheffield.
Mason, R.O and Mitroff, I.I. (1981). *Challenging Strategic Planning Assumptions*. Wiley, New York.
McBurney, J. (1990). *Ecology into Economics Won't Go: Or, Life is not a Concept*. Green Books, Hartland.
Meadows, D.H., Meadows, D.L., Randers, J., and Behrens, W.W. (1972). *The Limits to Growth*. Universe Books, New York.
Meadows, D.H., Meadows, D.L., and Randers, J. (1992). *Beyond the Limits: Global Collapse or a Sustainable Future?* Earthscan, London.
Mesarovic, M. and Pestel, E. (1974). *Mankind at the Turning Point: The Second Report to the Club of Rome*. Dutton, New York.
Midgley, G. (1992). The sacred and profane in critical systems thinking. *Systems Practice*, **5**, 5–16.
Midgley, G. (1993). From the ashes of despair. *Journal of the Operational Research Society*, **44**, 1165–1167.
Midgley, G. (1997). Mixing methods: A critical systems perspective. In: *Multimethodology: Towards Theory and Practice for Combining Management Science Methodologies*. Mingers, J. and Gill, A. (eds.). Wiley, Chichester.

Midgley, G. and Milne, A. (1995). Creating employment opportunities for people with mental health problems: A feasibility study for new initiatives. *Journal of the Operational Research Society*, **46**, 35–42.

Midgley, G., Kadiri, Y., and Vahl, M. (1996). Managing stories about quality. *International Journal of Technology Management*, **11**, 140–150.

Midgley, G., Munlo, I., and Brown, M. (1998). The theory and practice of boundary critique: Developing housing services for older people. *Journal of the Operational Research Society*, **49**, 467–478.

Mill, J.S. (1859). *On Liberty*, 1974 edn. Penguin, Harmondsworth.

Moulin, M. (1994). Getting planners to take notice: Influencing the design of a new maternity unit. In: *Community Works: 26 Case Studies Showing Community Operational Research in Action*. Ritchie, C., Taket, A., and Bryant, J. (eds.). Pavic Press, Sheffield.

Mumford, E. (1996). *Systems Design: Ethical Tools for Ethical Change*. Macmillan, Basingstoke.

Nemeth, T. (1980). *Gramsci's Philosophy: A Critical Study*. Harvester Press, Brighton.

Neuhaus, R.J. (1984). *The Naked Public Square: Religion and Democracy in America*. Eerdmans, Grand Rapids.

Noad, N. and King, L. (1977). Area co-ordination: Some examples compared. *Linkage*, **2**, 10–13.

Nozick, R. (1974). *Anarchy, State and Utopia*. Basic Books, New York.

Oakley, P. (1991). *Projects with People: The Practice of Participation in Rural Development*. International Labour Organisation, Geneva.

Ochoa-Arias, A.E. (1994). The possibilities of community OR in a Third World country. *International Transactions in Operational Research*, **1**, 345–352.

Ochoa-Arias, A.E. (1996). An interpretive-systemic study of community organizations in Venezuela. PhD Thesis, University of Hull.

Oldenquist, A. (1986). *The Non-Suicidal Society*. Indiana University Press, Bloomington.

Ophuls, W. (1977). *Ecology and the Politics of Scarcity: Prologue to a Political Theory of the Steady State*. W.H. Freeman, San Francisco.

Palmer, P. (1981). *The Company of Strangers: Christians and the Renewal of America's Public Life*. Crossroad, New York.

Parry, R. and Mingers, J. (1991). Community operational research: Its context and its future. *Omega*, **19**, 577–586.

Pateman, C. (1970). *Participation and Democratic Theory*. Cambridge University Press, Cambridge.

Patton, M.Q. (1978). *Utilization-Focused Evaluation*. Sage, Beverly Hills.

Patton, M.Q. (1980). *Qualitative Evaluation Methods*. Sage, London.

Pindar, S. (1994). Planning a network response to racial harassment. In: *Community Works: 26 Case Studies Showing Community Operational Research in Action*. Ritchie, C., Taket, A., and Bryant, J. (eds.). Pavic Press, Sheffield.

Ramonet, I. (1998). Lost childhoods. *Le Monde Diplomatique*, January.

Rawls, J. (1971). *A Theory of Justice*. Harvard University Press, Cambridge, MA.

Ritchie, C. (1994). Community OR—five years of organised activities and beyond. *International Transactions in Operational Research*, **1**, 41–49.

Ritchie, C. and Thunhurst, C. (1990). How useful are operational research methods and techniques to community organisations. In *Towards the 21st Century—Proceedings of the 1990 Conference of the Association of Voluntary Action Scholars*.

Ritchie, C. and Townley, C. (1994). Sheffield Turkish Bath Company. In *Community Works: 26 Case Studies Showing Community Operational Research in Action*. Ritchie, C., Taket, A., and Bryant, J. (eds.). Pavic Press, Sheffield.

Ritchie, C., Taket, A., and Bryant, J. (eds.) (1994). *Community Works: 26 Case Studies showing Community Operational Research in Action*. Pavic Press, Sheffield.

Robertson, J.O. (1980). *American Myth, American Reality*. Hill and Wang, New York.

Rorty, R. (1989). *Contingency, Irony and Solidarity*. Cambridge University Press, Cambridge.

Rosenhead, J. (1986). Custom and practice. *Journal of the Operational Research Society*, **37**, 335–343.

Rosenhead, J. (1987). From management science to workers' science. In: *New Directions in Management Science*. Jackson, M.C. and Keys, P. (eds.). Gower, Aldershot.

Rosenhead, J. and Thunhurst, C. (1982). A materialist analysis of operational research. *Journal of the Operational Research Society*, **33**, 111–122.

Sale, K. (1980). *Human Scale*. Coward, McCann and Geoghegan, New York.

Sandel, M. (1982). *Liberalism and the Limits of Justice*. Cambridge University Press, Cambridge.

Schoemaker, P.J.H. (1991). When and how to use scenario planning—a heuristic approach with illustration. *Journal of Forecasting*, **10**, 549–564.

Simpson, P. (1994). Liberalism, State and community. *Critical Review*, **8**, 159–173.

Stavrianos, L.S. (1981). *Global Rift: The Third World Comes of Age*. William Morrow, New York.

Stokes, B. (1981). *Helping Ourselves: Local Solutions to Global Problems*. W.W. Norton, New York.

Taket, A. (1994). Evaluation in the health service. In: *Community Works: 26 Case Studies Showing Community Operational Research in Action*. Ritchie, C., Taket, A., and Bryant, J. (eds.). Pavic Press, Sheffield.

Thompson, J. (1995). User involvement in mental health services: The limits of consumerism, the risks of marginalisation and the need for a critical approach. Research Memorandum No. 8. Centre for Systems Studies, University of Hull, Hull.

Tinder, G. (1989). Can we be good without God? *Atlantic Monthly*, December, 68–85.

Tomasi, J. (1994). Community and the minimal State. *Critical Review*, **8**, 285–296.

Trist, E. and Burgess, S. (1978). Multiple deprivation: A human and economic approach. *Linkage*, **3**, 8–9.

Ulrich, W. (1983). *Critical Heuristics of Social Planning: A New Approach to Practical Philosophy*. Haupt, Berne.

Vahl, M. (1994). *Improving Mental Health Service in Calderdale: An Evaluation of 5 Schemes Funded through the Mental Illness Specific Grant*. Centre for Systems Studies, Hull.

Wainright, T. (1994). From protest to project management. In: *Community Works: 26 Case Studies Showing Community Operational Research in Action*. Ritchie, C., Taket, A., and Bryant, J. (eds.). Pavic Press, Sheffield.

Wallis, J. (1976). *Agenda for a Bibical People*. Harper, New York.

Wallis, J. (1981). *The Call to Conversion*. Harper and Row, New York.

Weedon, C. (1992). *An Interpretive Systemic Study of Community Operational Research*. MA Dissertation, University of Hull.

White, L. and Taket, A. (1993). The death of the expert. *Journal of the Operational Research Society*, **45**, 733–748.

Wilby, J. (1994). A critique of hierarchy theory. *Systems Practice*, **7**, 653–670.

Wong, N. and Mingers, J. (1994). The nature of community OR. *Journal of the Operational Research Society*, **45**, 245–254.

II

Local Action for Community Development

5

A Black Ghetto's Research on a University

Russell L. Ackoff

The worst thing that can happen to operations research is that our conception of what it ought to be becomes equivalent to our conception of what it is. In a changing world, even equilibrium must be dynamic.

OR has been dynamic. It has made considerable progress in the last two decades, most of it technical. Continued technical progress will be necessary but not sufficient if OR is to thrive, not merely survive. To thrive, OR will have to increase its ability to deal with critical social problems and expand its relevance to strategic decision making in private, as well as public, domains. The frontiers of the Seventies [*this chapter was originally published as a journal paper in 1970—eds.*], it seems to me, ought to be new contexts for OR's application, or at least more systematic and comprehensive exploration of areas whose boundaries may have been crossed, but which have not yet been adequately settled.

Among other things, it is in such explorations that we are most likely to find critical shortages and inadequacies of our techniques. Thus such exploration should be a major stimulus to continued technical development. But more important is the fact that such exploration is likely to produce an enlargement of our concept, methodology, and philosophy of OR that will increase its relevance and potency.

This is a story about such an exploration. From a technical point of view it is uninteresting; but from a conceptual and philosophical point of

Originally published 1970.

Russell L. Ackoff • 1021 Lancaster Avenue, Suite 201, Bryn Mawr, PA 19010, USA.
Community Operational Research: OR and Systems Thinking for Community Development, edited by G. Midgley and A.E. Ochoa-Arias, Kluwer Academic/Plenum Publishers, New York, 2004

view I have found it to be one of the most rewarding efforts with which I have been involved.

First, the story; then I will reflect on its significance.

5.1 *The Story*

The problems associated with the black ghettos in urban America are familiar to all. Perhaps less familiar are the numerous unsuccessful attempts that have been made by most urban universities in the United States to contribute towards their solution. It was exposure to some of these failures that led Professors Robert B. Mitchell (City Planning), William Gombero (Industrial Relations), and me in 1967 to engage in a series of discussions that in turn led to the development of what we believed to be an untried approach to university aid to the ghetto.

This approach was based on a few simple assumptions. First, we assumed that inhabitants in the black ghettos should be given an opportunity to solve their own problems in their own way; that they will not, and should not accept 'white solutions' because whites have demonstrated no particular competence in solving the blacks' problems. Furthermore, we believed that blacks could learn more from their own failures than they could from white successes. Thus, we concluded that the best the white community can do to help the black community is to enable it to solve its problems in the way it, the black community, wants to. Our task, then, was to try to make the resources of the Management Science Centre and more generally, those of our University, available to the black community to use as *it, not we,* saw fit.

Our approach required receipt of a request for aid from a black ghetto. Fortunately, in February of 1968, just after coming to this conclusion, a modest request for drafting assistance came to us from Forrest Adams, a black planner employed by the newly formed Mantua Community Planners (MCP), which was a coalition of the five most active organisations in the neighbourhood just to the north of our University.

Mantua covers about eighty city blocks and has a population of approximately 22,000, 98 per cent of whom are black. By almost any standard, it is a critical poverty area. Almost 25 per cent of its housing units are overcrowded, and more than 50 per cent of them are in substandard condition. Its male unemployment rate falls between 15 and 20 per cent, more than three times higher than the rate in Philadelphia as a whole. Thirty-seven per cent of its families earn less than $3,000 per year. The educational level is very low: more than a third of Mantua's residents who are over 25 years old have had less than eight years of education. Nearly 40 per cent of its minors receive some type of public assistance, more than six times the city's rate. Sixteen per cent of its population from 7 to 17 years old were arrested in 1964, nine times the rate in the city as a whole. Its adult crime rate was more than twice that of the remainder of the city.

The disadvantaged and underdeveloped state of Mantua is obvious.

The three professors arranged for a meeting with Forrest Adams and Herman Wrice, President of a group called the Young Great Society (YGS). Mr. Wrice was the most prominent of the emerging leaders in the community. At that meeting we offered to employ any three people from the community selected by Mr. Adams and Mr. Wrice to work on the development of their community in any way that those so employed saw fit. The three were selected within a day. They were Andy Jenkins, President of the Mantua Community Planners and Vice-President and co-founder with Herman Wrice of the Young Great Society, Richard Hart, and Mrs. Doris Hamilton, Treasurer of MCP. A fourth employee on the staff of YGS was added a short while later. Those employed were given office space (which they never used), secretarial aid (which they used occasionally), and a graduate student to serve as an assistant (whom they used extensively). It was made clear to the Mantua team members that the faculty involved would volunteer nothing but were available to help them as they saw fit. They were told that they would be completely self-controlling, even with respect to hours, location, and content of their work. They had no need to come to the University except to pick up their pay checks.

Funds for support of this activity were obtained from the Anheuser-Busch Charitable Trust, which enthusiastically supported the concept and the effort based on it. But the initial grant was small. In order to stretch it to cover nine months of work, the three faculty members involved provided their services at no cost. This turned out to be a blessing in disguise. Subsequently, other faculty members, none of whose time could be bought for participation in 'paying projects', were willing to give some of it to this effort.

The original three on the Mantua team were hired on a Friday. They asked us to meet with them on the following Monday. At that time they presented for comment a programme for their activities that they had developed over the weekend, and they proposed regular weekly meetings with us in order to review their progress and discuss their problems. Before I review that progress, let me cite a few important events that occurred subsequently.

Within a few months it became apparent that in order to meet and coordinate all the requests for aid that were coming to us from the team, and all the people from the University who became involved in providing the aid, a full-time senior member of our staff was required. Marvin Rees took on this job. More of the successes and fewer of the failures of this project are due to him than to any other member of the University.

At the end of about the sixth month of the effort, the Mantua team and the University group collaboratively prepared a proposal for continued support of their joint effort. The proposal was submitted both to the Anheuser-Busch Charitable Trust and to the Ford Foundation. It was the Mantua team's decision that any additional funds that might be obtained continue to come to it through the University. Much to our delight, each of the foundations came through with $50,000 per year for each of the next two years.

Since the project began in February of 1968, the Young Great Society and the Mantua Community Planners have grown considerably in size and strength. They and their leaders, Herman Wrice and Andy Jenkins, are by far the most significant development forces in Mantua. Herman Wrice has attained prominence not only in Philadelphia but at the state and national levels as well.

Let me review the part of what Mantuans have done for themselves in which we have been of some help.

YGS has set up nine manufacturing firms that grossed a little more than $1.5 million in 1969 and employed about 125 people from the community. The most successful of these firms makes electronic circuit boards and is supplying such companies as IBM, General Electric, and Leeds and Northrup. These nine firms and several others are being brought together physically and spiritually in an industrial complex managed by the Mantua Industrial Development Corporation, a subsidiary of YGS. The facilities just acquired for this complex will also serve as a hot-house for new business enterprises and will provide a wide variety of types of service and assistance to the businesses that it nurtures.

With University aid, MCP and YGS have been instrumental in obtaining loans from banks for the establishment and operation of a number of small businesses in Mantua and in providing these enterprises with needed technical and managerial assistance. Much of the latter has been provided by Wharton graduate students.

Both MCP and YGS provide employment services and have placed several hundred Mantuans in the last two years. Furthermore, these two organisations themselves employ about 330 people in activities that they manage. Together they solicited and brought about $600,000 for development programmes into the community last year.

MCP established the Mantua Community Federal Credit Union which Mantuans can join for 25 cents and have savings accounts that permit deposits with no lower limit, and from which they can obtain loans at relatively low interest rates.

YGS and MCP operate an Architectural and Planning Centre and a Joint Workshop that are staffed by University faculty and students as well as by members of the community. The Centre and Workshop have produced neighbourhood development plans to which about $6 million has already been pledged. The Centre has also planned the renovation and rehabilitation of more than thirty houses in Mantua in 1969 and about 120 in 1970. The reconstruction work has been done by local contractors using indigenous labour, most of which was trained on the job. The Centre has also done the architectural planning required for the many facilities used in the many activities of the two sponsoring organisations.

We recently obtained from a major oil company a contract for the Architectural and Planning Centre to design a community service station. The uniqueness and imaginativeness of the resulting design led to a subsequent contract for preparation of detailed working drawings and specifications for the station. Steps are now being taken to build a prototype in Mantua, one that will be operated by the community. This station will include an auto-, motorcycle-, and bicycle-parts store, a do-it-yourself repair and maintenance facility, a training centre, recreational facilities, and offices for rent.

Plans for the community are frequently reviewed at open meetings. Attendance at these meetings has grown constantly. Block groups have been organised to help develop enlightened public opinion on issues of importance to the neighbourhood. This development has been enhanced by a weekly hour-long broadcast by MCP over one of the city's radio stations. YGS has successfully launched two weekly newspapers that are distributed and widely read beyond Mantua. Both yield a profit derived from advertising income. In addition, YGS recently initiated a weekly newsletter from which I shall reproduce a selection below.

Their educational program is extensive. A school for 150 children in their fifth to eighth grades was opened in September of 1968. This mini-school, as it is called, is completely controlled by a neighbourhood board that selects its own teachers and designed its own curricula. It is supported by both the city's Board of Education and a grant from the Rockefeller Foundation. The school is characterised by innovative pedagogy and relevance of content for disadvantaged children. Unfortunately, the school's building was recently destroyed by fire and hence it is currently operating in temporary quarters. Three new scattered mini-schools are being planned. Eventually there will be a network of them covering the community.

This coming September the Benjamin Banneker Urban Centre will open in Mantua. It will be a school with no entrance, only exit requirements, directing its major effort to rehabilitation and prevention of dropouts. It will begin with the seventh grade and add a year each year until it provides an undergraduate college degree.

A group of faculty volunteers from the Wharton School, together with MCP, YGS, and black groups from neighbourhoods other than Mantua conduct an evening programme in business education at the University. This program is open to the disadvantaged of the entire metropolitan area. It is now in its second year and has more than a hundred students enrolled in it. Several of those who attended the first semester were given scholarships in the regular day-time degree program of our college. These evening students are also provided with counselling and employment services.

Scholarships to private schools in the suburbs have been obtained for several children from Mantua. Tutorial programmes for children in local public schools and programmes for orienting and assisting new teachers in these schools are provided.

The most recent and perhaps most exciting educational project has been the Urban Leadership Training Programme. The following description of this programme was written by Ronald E. Thompson, one of its members, for the Mantua Community Newsletter of April 1970.

> On February 9 started the greatest event that ever happened in urban history. The event I'm talking about is the starting of the Urban Leadership Training Program. The program started with 21 gang leaders from the Mantua community. The Young Great Society and Mantua Community Planners are affiliated with the program. YGS funds the program. The 21 leaders were picked from different corners which consist of 36th Street, 39th Street, Lancaster Avenue, 41st Street, and 42nd Street.
>
> A few months ago, before the program started, these corners were at war with each other. Many times before the start of ULT, social workers tried to gather the corners together, but the problem that would result would be more conflict between the young men. As always, somebody would end up getting hurt.
>
> So far the program is doing very good. The great thing about the program is that the young men have unity among each other. You know yourself that it is good because without unity you do not have anything.
>
> The University of Pennsylvania provides space for the young men. Members of the University faculty and community workers of various fields volunteer to teach the young men different courses. Some of the courses the young men are taking consist of criminology, sociology, black studies, community health services, community planning, housing rehabilitation, radio and TV, and communications. The purpose of these courses is to prepare for future black leadership in all fields.
>
> After this program is ended, the program will be repeated with 21 more black brothers.
>
> The ULT brothers are working on several projects. One of these is the Leader Movie project. The Leader Movie, which is located at 41st Street and Lancaster Avenue, has been closed for some time. Police Commissioner Frank Rizzo wants the movie to be made into a PAL (Police Athletic League) building but we know it would just result in a conflict between the little kids. I hope that with the help of the community we could fight Rizzo mentally. The ULT

> brothers would like to own the Leader and turn it into a place where the younger kids in the community could have something to do besides training to fight one another.
>
> Urban Leadership Training is turning out to be a community organization to fight the problems of the ghetto area of Mantua.
>
> The young men do other kinds of field work. Recently the ULT brothers did a housing survey of the Mantua area for the Architecture and Planning Center of YGS. The young men also take trips in various parts of the US. One trip was taken in the Harlem area of New York where the young men took surveys and analyzed the ghetto neighborhood problems there and compared them with the problems in Mantua. We found out that the way they are living in Harlem is the same identical way we're living.
>
> The program is the first I have known to ever understand gang problems.
>
> I used to be affiliated with one of the gangs before I went into the service. I have been home from the service since January. I was in Vietnam. I compare the fighting in Vietnam with the fighting in the streets and find out it is almost the same. The only thing is that in the streets you are fighting your own brother.
>
> The brothers are not fighting now. That is why the program might be the greatest event in urban history.

Physical and mental health services have been initiated within the community. These include two medical centres, a mobile clinic that goes to the people, and a drug-addiction treatment centre. The Department of Community Medicine of our University and members of the community are now actively engaged in developing more extensive health services that they have collaboratively designed for Mantua.

MCP arranged for the City's Department of Recreation to convert two city-owned parking lots into recreational areas. MCP sponsors a basketball league that uses these facilities. They are also used for outdoor shows, dances, food sales, raffles, and other community activities. MCP has held a number of outings for neighbourhood children; the most recent was attended by more than 800 children. Use of the University's indoor swimming pool by neighbourhood children was arranged for the summer. MCP sponsors dramatic, music, and dance classes in the community. Last year the students of the dance put on an evening programme in the University's auditorium that was exciting in conception and professional in execution. It also made a profit.

I could go on and mention the recently opened Day Care Centre, the Nursery School, the free legal, social, and welfare services, and many others, but I think I have gone far enough to provide a picture of what has been and is happening. Those from the community that we support in turn support a very large number of others, all of whom are working on community development. Our own services continue to be available to any individual or group in the community that asks for them. At times we have had as many as thirty members of the University's faculty and student body involved in such support.

Initially we had to hide the fact that we were supporting the members of the Mantua team lest they be rejected by their community as pawns of the University or as 'Uncle Toms'. But by their behaviour and accomplishments they have established their independence of our influence and can now openly acknowledge our support, and even use it to their advantage. Our relationship is now taken for granted in the neighbourhood. The friendships we have formed with Mantuans extend beyond the boundaries of the project and have helped integrate us into their community.

We have learned much more about the nature of the ghetto by being involved in it under the direction of its members than we could have learned by any traditional type of research. More important is the fact that our knowledge and understanding of the ghetto is now acknowledged by its leaders. I know of no other way by which such a state could have been brought about. One consequence of it is that, whereas we were initially asked to help only on problems involving external relations, for some time now we have also been deeply involved in problems internal to the ghetto. A second consequence is that our special knowledge of, and relation to, the ghetto is acknowledged by our University's administration. Hence we are used as advisors on University-community relations and have helped design and operate new functions and units within the University that are concerned with these relations.

5.2 The Implications

It is easy to become absorbed by the humanistic aspects of this effort, but I believe the scientific and managerial implications of it are also important. Let me consider a few of these.

First, note that we, the researchers, designed a relation with the 'subjects' in which the subjects conducted research on us. It was the community members' task to find out how to use the University and our Centre effectively, and, by so doing, to solve the researchers' problem of determining how to be useful. Not only would we have been rejected by the community if we had tried to do research on it, but even if we had not been rejected, we would not have had such access to it as we have had.

Little of what we have done is OR, or even research in the conventional sense. But we have never rejected a request for aid because it did not involve research or because it required action that was beneath our dignity. In addition to such activities as chauffering, acquiring empty oil drums and pond water for a fish tank, we have conducted research, for example, to determine the cost of racial segregation to the whites in the

Philadelphia area and we have done economic-development and land-use planning for the neighbourhood.

At first glance, this relation of researcher to researched may seem to be relevant only to the specific type of situation involved in this particular project. But is it? Although universities have had some success in serving governmental and industrial organisations, they have not been nearly as successful as many pretend. I suspect the reasons for their lack of success are related to my remarks about the ghetto.

We in OR do not know government and industry as well we think we do, and administrators and managers know this. Therefore, they tend to use us in restricted and highly structured ways, on specific problems on which we have propagandised them into believing that we have some competence. But much of its use of our services is only token. For example, I recently reviewed the work of a management science group in one of this nation's largest corporations. Although its fifteen professionals have been carrying out studies for more than a decade, not one had been implemented. Nevertheless this group insisted on and had complete control over the projects it undertook.

Perhaps we should not tell administrators and managers how to use us, but rather involve them in systematic efforts to find out what we can best do and how they can best use us. This may well change our conception of our skills, but in return we may be given the opportunity to become a more integrated part of organisational decision making, planning, and development.

Secondly, the types of planning problems in which we have been traditionally engaged involve organisations that already have enough resources at their command to attain further growth and development. Hence, allocation of resources, rather than generation of new resources, preoccupies our planning efforts. In ghettos (and in most underdeveloped countries) this is not the case. Few resources are available. Hence, planning in the ghetto is very sensitive to potential sources of resources and to the uses for which they can be obtained. This requires development of plans that are more integrated with those of the larger system of which it is a part than is usually the case in most governmental and industrial planning. I have never seen planners as sensitive to, and aware of, the plans made by the larger units of which they are a part, as I have seen in the ghetto. Ghetto leaders insist on knowing what is going on 'up there' and they often understand the implications of higher-level plans better than do those who prepare them. How can we reproduce this state within government and industry?

The ghetto cannot use the annual budget as an instrument of or substitute for planning because it never knows what resources will be available to it over even short planning periods. Hence its planning must be continuous and adaptive. Its long-range plans cannot be built up out of arbitrary fiscal blocks. Furthermore, because plans cannot be imposed on a black ghetto from above, they must be supported from below. Thus its planning must also be participative.

Planning in government and industry is seldom continuous, adaptive, or participative. But shouldn't it be? The pervasive preoccupation of managers and administrators with making a good showing in the current fiscal year leads to an imbalance of concern with short- and long-range performance, in favour of the short. Where planning is continuous, concerned with both the long and short of the future, and adaptive, this imbalance does not occur. Promotions do not motivate ghetto leaders because there is not a hierarchical structure in which to advance; they are motivated by a desire to increase the scope and importance of their jobs, and this is largely under the control of the individual involved. He is not bound by charters and job descriptions. He truly 'manages by objectives'. Might not a little more of this in government and industry be desirable?

We have tended to build our institutions in such a way as to minimise errors of commission, but in doing so we have increased the frequency of errors of omission. The emerging structure of the ghetto has reversed this state. Ghetto leaders are more concerned with not doing what should be done than with doing something that does not need to be done. This orientation, it seems to me, is more likely to yield progressive development than is the converse orientation of most of our public and private institutions.

The ghetto is not a highly structured system, let alone a hierarchical one. Hence its leaders can only survive by effectively serving both their constituents and the larger community that contains them. Without support of the larger community, ghetto leaders cannot get the resources and programmes that they require to serve their constituents. Without serving their constituents they cannot retain any followers. Hence only effective leaders can survive. This is not true in either government or industry, where Peter's Law—everyone rises to his level of incompetence—seems to hold almost universally.

The ghetto leader has neither rank nor authority vested in him from above. Therefore, he cannot use these to get others to do what they do not want to do. He must know what his followers want, what they can be persuaded to do, and how he can persuade them to do it. If these skills were available to those who currently direct others by virtue of rank and authority vested in them, imagine how much more effective they would be.

I know of no better way to develop such skills than by the democratic process of giving collective control over an authority to those who are individually controlled by that authority. Such control can even be built into bureaucracies by use of participative-management schemes. Such schemes have been suggested by many but tried by few.

We are witnessing an almost world-wide pressure for participative democracy, in public affairs and in universities. Can business, industry, and governmental agencies be far behind? I do not think so. Therefore, current leaders in these institutions would do well to study and gain understanding of the successful ghetto leader. They too may have to convert from practicing the art of management to practicing the art of leadership.

Finally, let me deal with the reluctance of the ghetto, like governments and industries, to change and to exploit the few resources that are available to them. Ghettos do not push their leaders; they are pulled by them. Ghettos have to be led into internal revolutions. In Mantua we have observed the development of an effective way of doing so. The motto of the Mantua Community Planners is *'plan or be planned for'*. This recognises an important fact that is both obvious and ignored: it is not true that if nothing is done, nothing will happen. A great deal will happen and most of it will be undesirable. Ghetto leaders know that most current trends in the larger community that includes their neighbourhood are detrimental to their neighbourhood's interests. These trends are the subject of their constant study and analysis. They educate their constituents to an awareness of these trends and their consequences. Thus, the effective ghetto leader has a knowledge of and concern with his environment that few governmental or industrial leaders have. The latter tend to take their environment for granted and feel that by perceiving changes in it and adapting to them, their organisations can at least survive and may even grow. Hence they take little responsibility for what happens in their environment. Not so in the ghetto. Its leaders cannot afford to let the environment take its own course. However limited his effect on the larger environment may be, the ghetto leader uses all his power to push it in a direction that is compatible with his aspirations for his neighbourhood. He believes in active intervention in the larger community and its future because he knows that his neighbourhood cannot thrive unless the larger community thrives. He not only knows this, but he acts on it. Can we say the same about leaders of other types of institution?

Operations researchers also have a great deal to learn from effective ghetto leaders. They should develop the same attitude towards OR as these leaders have towards their communities. They should not be willing to let OR develop only in response to changing external conditions. They should want to help create a world in which the capabilities of OR are considerably extended but in which the need for OR is diminished. Like the ghetto leader whose objective it is ultimately to dissolve the ghetto by having it absorbed into the main current of the culture of which it is a part, so OR's objective should be the dissolution of OR as an autonomous and segregated activity by having it absorbed into every aspect of the organisational life of which it is a part.

Acknowledgement

The original paper was based on an address to the Philadelphia Section of the Operations Research Society given on June 5, 1970, and it was written when the author was working at the Wharton Business School, University of Pennsylvania, Philadelphia, Pennsylvania, USA.

6

Housing in the Dearne Valley: Doing Community OR with the Thurnscoe Tenants Housing Co-operative

Charles Ritchie

6.1 Introduction

This chapter outlines some elements of the involvement that the Community Operational Research Unit had with the Thurnscoe Tenants Housing Co-operative over the period 1989 to 1995. The Co-operative was the first client of the Unit and a highly important one in view of the range of work undertaken and the mutual benefits that came out of the relationship.

It begins with a short account of the salient events in the formation of the Co-operative, explaining the role of community operational research (OR). A fairly full account of the various stages of the involvement is given which attempts to cover issues of process and content as well as methods and techniques. The changing role of the Unit, from one of offering 'external' consultancy to that of a closer continuing working relationship, is explored. Finally, some conclusions and evaluation of the overall involvement are offered.

Final version accepted for publication July 2002.

Charles Ritchie • Department for Education and Skills, Moorfoot, Sheffield, S1 4PQ, UK.

Community Operational Research: OR and Systems Thinking for Community Development, edited by G. Midgley and A.E. Ochoa-Arias, Kluwer Academic/Plenum Publishers, New York, 2004

6.2 The Thurnscoe Tenants Housing Co-operative

Thurnscoe is a mining village in South Yorkshire. It is located in the Dearne Valley, an unemployment hot-spot, which sits at the intersection of the boroughs of Barnsley, Rotherham and Doncaster. Like many others in the area, the local pit, which was more than 100 years old, had been closed for over two years at the time of the Unit's first involvement. The residents of Thurnscoe were predominantly ex-miners, their widows and their families.

Progressively since the 1960s, British Coal had been divesting itself of its housing stock. In January 1987, the Tenants of Thurnscoe discovered, by accident, that their homes were to be auctioned at a hotel in London. The tenants hurriedly intervened to have the sale stopped.

Stopping the sale was a major success, but it only provided temporary respite. British Coal was still committed to its policy of surrendering its role as landlord. Following considerable activity and protracted negotiations, the tenants formed themselves into a Co-operative and bought the 361 houses with a mortgage of £1.75 m. At a stroke, they became the largest co-ownership housing co-operative in the country.

A condition of their mortgage was that housing management should be carried out by an established housing association, the Yorkshire Metropolitan Housing Association (YMHA) for the first three years. This condition they accepted, albeit a little reluctantly. However they always intended to take over all management tasks at the earliest opportunity.

The Co-operative finally took possession of their homes in August 1988. The Community Operational Research Unit was formally opened in September 1988. By this time though, there was already a considerable history of joint working.

6.3 First Contacts with the Tenants Co-operative

The Thurnscoe tenants were fortunate from the outset, in that they found themselves in contact with a handful of willing supporters, anxious to see them succeed and, if possible, to assist them in their fight to retain control of their own housing. The Leeds-based regional officer for Shelter provided valuable assistance at the time of the Co-operative's creation. The Co-operative Development Worker with the Barnsley Metropolitan Borough Council provided continual general support in a way that skilfully combined gentle cajoling and encouragement with building up the skills and the self-confidence of Co-operative members.

To all who came in contact with them at the time, the potential and the limitations of the membership were immediately evident. Here was a group passionately committed to the maintenance of the integrity and cohesion of the houses and of the area within which they lived. Initially fearful of control passing to a remote anonymous landlord, their motivation and conviction were overwhelming. At the same time, though, here was a group whose collective managerial experience of running a large financial undertaking was zero. Indeed, even their experience of working within small Committee organisations was limited. Some basic rudiments of trade union workings were familiar to them; but the management of a whole housing estate was a very different order of problem with a very diverse collection of issues and problems.

Located within the Dearne Valley, the Co-operative was eligible for assistance from the Northern College of Residential Adult Education under its specially funded Dearne Valley Project. Prior to contact with the Community Operational Research Unit, this assistance had already been enlisted to develop some basic skills in Committee procedures and workings; to provide women members of the Committee with some assertiveness training; and to engage in some elementary strategic planning.

Thus, at the time of the Unit's first involvement, the members of the Committee had already enjoyed a number of successful engagements with the Northern College. And it was as a natural extension of this existing series of activities that the Unit was first introduced to them. Following a number of preliminary meetings, at which the Unit consultant (Colin Thunhurst) acquired familiarity with the Co-operative's current interests, concerns and problems, and at which the Committee acquired familiarity with the consultant, a workshop was held at the College on 29 February 1988.

Within the workshop, the history and the objectives of the Community Operational Research Unit were explained to the Co-operative members, with emphasis placed on the action research philosophy and its organic relationship to the work of the College, rather than on the special contribution of OR. In essence, the Unit was trading on the goodwill that had already been established between the Co-operative and the College.

Out of the workshop came two specific proposals for further work between the Unit and the Co-operative. In the short-term, a further workshop was to be held at which the issue of record keeping was to be considered. In the medium-term, a brief was compiled for a three-month interim project, to be conducted by an external Unit consultant, pending the appointment of the Unit's first full-time research officer.

After a number of delays, meetings and communications between the Unit and the Co-operative, the second workshop took place on 30 June 1988. The outcome of the workshop was that members of the Co-operative decided that the computerised recording of information was not the best route for them to follow at that particular time. It was a decision that they had now taken out of knowledge rather than ignorance, and a decision that they felt they could confidently review as circumstances changed.

6.4 *External Consultancy—Problem-Structuring Workshops*

The role of consultant, to work with the Co-operative prior to the appointment of the Unit's first full-time research officer, was taken by John Friend, a freelance OR consultant who had spent over 20 years in the Institute for OR within the Tavistock Institute of Human Relations. Here he had played a leading part in developing the 'strategic choice approach' to participative planning (Friend, 1989).

From early July onwards, he attended some of the weekly Committee meetings as an observer. This allowed him to get to know its members and for them to get to know him. This also allowed him to get a feel for their decision-making culture, as well as for some of the issues that they currently faced.

The spirit was one of an imminent success within a long crusade; but there were also some underlying tensions. There was an immediate burning issue over whether the Co-operative should take over the houses that were currently vacant. Some of these were in very poor condition and therefore a potential financial liability for the members. There were tensions too in the Committee's relationship with their principal professional advisers, the YMHA.

The idea of further problem-structuring workshops gained ready acceptance among the Committee members. It was agreed that the first of these should be held on the theme of maintenance and improvement of the houses. On 6 September 1988 a group of nine members went to Northern College for a workshop on maintenance and improvement issues. John Friend acted as principal facilitator with Peter Long from the Sheffield City Polytechnic acting as co-facilitator. The Co-operative Development Worker from Barnsley also participated as a supporting resource. It was quickly agreed to focus on middle-range maintenance issues, as opposed to running repairs or major improvements. Flip charts and post-it notes were used for an initial brainstorm of issues and uncertainties. By the end of the morning, it had been agreed to focus on decisions concerned with the training of maintenance workers; the choice between putting work out to tender and building up a direct work force.

It was agreed to compare the pros and cons of two contrasting strategies. One of these involved a 'crash programme' of training and early take-over of responsibility by the Co-operative, while the other was labelled a 'slow but steady' approach. It was clear that the members of the Co-operative were ideologically in favour of the former strategy. So the facilitators found themselves testing whether the 'slow but steady' approach could have any countervailing advantages and drawing out the main sources of uncertainty getting in the way of a clear decision in favour of a crash programme.

Out of this discussion came a review of the practical actions that could be considered to reduce the most important of the uncertainties; and the skeleton of an immediate actions programme (or a 'commitment package', in strategic choice language) was agreed. It was clear that there were many investigations to be made, and that the co-operation of the YMHA would be vital if the crash programme were to stand any chance of success. The motivations of the YMHA in this respect were themselves a major source of uncertainty for the Committee. Therefore, it was agreed that a second workshop would be held in a few weeks, and that this time the YMHA would be invited to participate.

Continuing contacts between the facilitators, the Co-operative and the managing agents led to a design for a second workshop in mid-November which differed in several respects from the first session. This time, it was agreed to work for a day and a half, starting with an update of the position reached at the end of the first workshop. The YMHA's newly appointed local manager was invited to be present, both on the first afternoon and at the end of the second day.

It was proposed to work using a combination of strategic choice (Friend, 1989) and the strategic assumption surfacing and testing approach (Mason and Mitroff, 1981). This seemed appropriate because of the importance in this workshop of bringing to the surface the working assumptions of the Committee, the YMHA and other significant stakeholder groups, for which these techniques have been noted to be particularly suited (Jackson, 1988). The facilitating roles were shared between Jim Bryant and Peter Long of the Sheffield City Polytechnic, and John Friend.

A new strategy of starting up a local Community Business for maintenance was by now finding favour among members of the Co-operative, even though there were as yet few successful examples of this form of organisation outside Scotland. It was therefore decided to compare this strategy with the continuation of the *status quo*, involving a higher level of dependency on the YMHA. Flipcharts were again used to build up a systematic record of the assumptions relating to the various stakeholder groups that appeared to support each strategy. As before, a list of actions to be followed through was drawn up. Further development of this area is described below.

Key actions at this stage concerned the drawing up of a business plan and research into funding opportunities, and the testing of whether it was possible to generate closer involvement in the work of the co-operative on the part of tenants at large. At this stage, a significant change was discernible in the relationship between the Committee and the other tenants. The Committee were no longer merely the champions of the Thurnscoe community in winning control over their homes: they had become the managers and in effect the landlords, who had to exercise responsibility over such unpopular decisions as rent increases and policy in relation to the recovery of arrears.

6.5 Working More Closely with the Co-operative

Shortly after this series of workshops, the first full-time member of staff, Charles Ritchie (the author of this chapter), joined the Unit as Senior Research Officer. The relationship between the Unit and the Co-operative developed from one based specifically around individual projects and workshops to a much closer one involving more day-to-day contact and involvement on *ad hoc* problems and issues in addition to the more clearly identifiable project work. The Senior Research Officer was able to attend fortnightly Committee and other meetings and was therefore in a position to contribute to current debate and, more importantly for the Unit, was able to spot and initiate specific OR projects and sessions.

During Committee meetings members of the Unit did not attempt to play a facilitating role but rather attended in an 'advisor' capacity; able to comment on issues, bring additional information or suggest that it might be valuable to hold a special session dedicated to a particular topic. Although the Committee members had had training in the conduct of meetings and their various roles, meetings of the Committee were rarely as structured as those in most traditional organisations; frequently there was no written agenda. Therefore, it was common practice that discussion of important and recurring topics and development of strategy took place at specially convened meetings that were facilitated by staff from the Unit or other development and support workers.

As well as working on issues of major importance (such as sales policy, financial planning and community participation), these many dedicated sessions covered recurring issues such as the organisation of events and public meetings; the production of a newsletter; minor spending issues; and how to respond to requests from tenants, the managing agents and statutory bodies. These sessions, often using flip charts and drawing on elements of OR methods such as strategic choice and decision analysis, encouraged a structured approach to decision making and led to a more systematic recording of decisions and arguments.

During this phase of our involvement with the Co-operative, members of the Unit worked concurrently on a number of issues. Rather than attempting to trace the chronological development of this work programme, the following sections concentrate on particular topics or types of involvement.

6.6 *Policy and Decision-making*

6.6.1 Thurnscoe Community Repairs Business

One of the tasks which came out of the earlier workshop involved finding out the potential interest in forming a community business to undertake repairs work on the estate. It was decided to hold an evening public meeting in a local pub. Invitation was mainly by word of mouth and approximately 30 men turned up. A brief report on the idea of a repairs business was made by a member of the Committee. This was followed by members of the Enterprise Department of Barnsley Council who spoke about the possible help which was available from the Council. After questions, people were invited to return to a meeting two weeks later if they still felt that they were interested and prepared to get involved in the work of setting up such a business.

About 25 men came to the second meeting, where I talked about the work involved in carrying out a feasibility study and ultimately drawing up a business plan. Again, people were invited to return to a first 'working'. Only six men came to this meeting, of whom three were employed and the other three were unemployed or on training schemes. Some of the basic concerns of the group were recorded on a flip chart. These included the amount and sources of work, job security, responsibility and management. There was also a great deal of concern about who would actually be involved in the business: Thurnscoe is a close community and individuals hold strong opinions about others. We also looked at ways of clarifying some of these issues.

In all, six meetings were held with the group. A great deal of debate concentrated around the amount of work that would be available. The Unit undertook to carry out some basic analysis of repairs. This showed that the Co-operative was likely to provide 50–75% of the amount that would be required to sustain the business. There would also be work from the development of a local church. Therefore, the business would possibly need to generate additional income. This compounded issues of responsibility and job security. The area has a long history of dependence on a single large employer, British Coal, who had offered reasonable job security but little in the way of training or encouragement in management, administration and entrepreneurial skills.

Throughout, the group remained unenthusiastic about the use of the flipchart and were resistant to any structured approaches to considering decisions. Although there was no distrust of Unit staff—indeed a good relationship was developed on a personal level—the group were clearly unhappy about some of the other bodies involved and felt that the concept of a community business was being foisted on them. They were also unused to the idea of applying for grants and felt that the church work was, in some way, charity.

At the sixth meeting it was mutually agreed that the existing group did not want to proceed with setting up a business. With greater and more concentrated support from the Unit or other development workers, it is quite possible that the business would have been established. However, the level of support that was needed was far greater than that which the Unit could supply at that stage.

6.6.2 Facilitating Policy-Forming Discussions

As noted above, the Committee was now perceived by many as the new landlords. They were accosted in the street and informed of detailed repairs needed, and people also made representation to them about issues such as wanting to buy their homes. Therefore, despite any misgivings that they had and the efforts made to involve members, the Committee did have to consider policy options from a 'detached' standpoint and make recommendations to members, who ultimately decided policy by voting at public meetings. This was not an easy task for them. They were aware that they had their own prejudices and concerns; and also, they knew that many members would ultimately hold them responsible for any decisions—regardless of the democratic process. Consequently, it was not difficult to persuade Committee members of the need for analysis and debate of important policy issues.

6.6.3 Developing a Sales Policy

There were many sessions to develop policy on specific issues, such as the employment of a worker and development of allocations and transfer policies. Of specific note, and particularly illustrative, were the sessions on developing policy on sales.

The sale of houses, in particular to sitting tenants, remained a highly contentious issue throughout (and beyond) the Unit's involvement.

There were differing and strongly held views amongst members of the Committee and the wider membership. It encapsulated the whole clash between economic and social goals as well as the divide between individual and community welfare.

Before the Co-operative was established, British Coal had offered all tenants the chance to buy their homes at a substantial discount. Many in the area had accepted. The remainder—those who went on to form the majority of the original Co-operative—had refused for both financial and social reasons. However, because of its financial situation, the Co-operative had had to sell houses that it could not afford to repair. Also, individual members' financial circumstances had changed. This led to a number of members wanting to buy their own homes from the Co-operative. Further, they felt that they should be able to buy at the discounted price originally offered by British Coal. Opinions were not clearly divided between those who wanted and could afford to buy, and those who could not or chose not to. The complex community interrelations meant that many had divided loyalties and sympathy with others. It presented a situation where clear informed debate and analysis was vital. Several sessions were held to debate the sales issue. The early sessions were focused on the financial implications of any sales policy. In particular, information was compiled and presented (on flipcharts) about the economics of selling an individual house. It was necessary for the Committee to understand that it made 'financial sense' to sell some houses and not others, largely depending on their state of repair and market value.

Subsequent meetings considered the 'social' implications of selling houses. The original aims of the Co-operative had included providing cheap rented housing, keeping the community together and using the size and financial power of the Co-operative to act as a spur to employment in the area. Selling houses detracted from all of these. However, the 'economic realities' of the 1980s and their financial position weighed heavily on them. At times there was a tendency to forget non-financial aspects. It was decided that the Co-operative needed to be more explicit about its aims and in particular to be more concerned about different stakeholders when considering issues. A list of stakeholders was drawn up which included groups such as current tenants, potential future tenants, those who could afford to buy their homes, and those who could not. This list was then used to consider and rank possible sales options.

The issue of sales price was hotly debated. Many believed that the Co-operative should not make a profit out of original members by selling to them at market price; others felt that everyone had had a chance to buy at discount and that the Co-operative was now a business with equal responsibility to all members. Others felt that there should be no sales at any price. Houses had not been individually valued when the Co-operative took over, so that all had a notional equal price despite very obvious differences in current market value. On the other hand, the market value of an individual property was often a reflection of the work and investment of the current sitting tenant. In order to help to resolve the problem, a chart was drawn up showing the value of hypothetical properties at different stages of the Co-operative's existence, together with assumptions about repairs spending by the Co-operative and the sitting tenant. The impact of selling at different prices was then analysed in terms of 'profit' to different stakeholders. This had the effect of 'distancing' the issue and allowing a calmer and more rational debate. It also raised the possibility of members receiving some credit for their own investment.

Eventually, the Committee decided to recommend a final opportunity for sitting tenants to buy, but at market value. After that, only properties that the Co-operative could not afford to repair or re-let would be sold. In effect this meant that it was unlikely that any sitting tenant would buy since those who had expressed an interest had expected a discount. The issue continued to be debated over the years. However, the decision achieved consensus and a satisfaction that such a contentious issue had been debated clearly and completely and that it was possible to look back and see the justification for the decision.

Almost all of these meetings were facilitated by staff at the Unit. Extensive use was made of flipcharts to record stakeholders, criteria and uncertainties, as well as displaying key facts and information. A systems diagram showing the relationship between different financial elements and also incorporating some non-financial effects was developed, and this proved to be helpful as a continuous reference. Its development was also a very useful focus for ensuring that everyone followed and understood the debate as it progressed. In addition, at any time, it showed how far we had progressed in considering the various interrelated elements. Having established the interrelations, it was then possible to refer to the diagram when attempting to assess the relative importance of different effects.

6.6.4 Strategic Review—A Second Round of Workshops

In August 1989 it was agreed to hold a further workshop to review the strategic plans of the Co-operative. It was decided to use the strategic choice approach, as in the earlier sessions with John Friend. Through the Unit's closer working involvement with the Co-operative during the intervening period, it was clear that issues other than the arrangements for maintenance had been coming to the fore, and it was therefore agreed to start the workshop with no prior assumptions as to what its focus of concern might be.

From an initial agenda building exercise, there emerged a cluster of diverse but interrelated issues. These were concerned with policy in relation to arrears, sales to tenants, possibilities for extending the role of the Co-operative into management of other facilities, and ways of increasing the involvement of tenants. As in the earlier workshops, it was necessary to cut corners in the use of OR methodology in order to emerge with a package of short-term actions.

6.7 Financial Planning

Interspersed with the many one-off sessions, there were a number of clearly identifiable areas of project work. The first area concerned financial planning within the Co-operative.

6.7.1 Financing the Co-operative

When the Co-operative had purchased their homes from British Coal they had had indications that grant money would be available through the Housing Corporation and other sources to carry out the major modernisation works required. None of these sources materialised and all work to be carried out on the housing stock had to be financed by the Co-operative itself. The only sources of major income were rents and the proceeds from any sales. The scale of modernisation and repair work required on the houses was such that the level of rental income was completely inadequate to fund any major programme. Estimates for modernisation ranged from £15,000 to £40,000 per house, which had each cost around £4,500 to buy. Thus, the rents required to fund such a programme were well beyond the resources of the current tenants. Profits from sales had been in the range of £4,000 to £10,000 per house. There also existed the possibility of major refinancing using deferred interest or index linked loans.

All of this raised complex interrelated issues of rent levels, scale and rate of modernisation, sales policy, risk and many other factors. These issues were complicated further since they could not be regarded as purely financial matters, given the Co-operative's social aspirations. The Unit was closely involved in all aspects of financial planning, considering both short-term issues such as setting the rent level, and long-term issues such as consideration of a major refinancing of the Co-operative.

6.7.2 Understanding Financial Information

Underpinning all of this was some basic work to help people understand financial accounts and figures and to help them incorporate these into the decision making process. Some of this involved just reading through accounts with members of the Committee, answering questions and explaining the meaning of particular items such as 'depreciation'.

However, there was also important work done in helping people to get beyond the mere understanding of financial accounts to developing a real 'feel' for the numbers. The most important element of this work involved presenting financial information in different ways. Most obviously, much use was made of simple charts and diagrams to show breakdowns of where money came from and how it was spent. The second type of presentation was to 'translate' overall Co-operative finances to show the implications for individual households on both a weekly and an annual basis. Many people found it difficult to manipulate and relate to figures in tens and hundreds of thousands of pounds, yet found it relatively straightforward to discuss the same information in terms of a few pence or pounds rent per week. The essential bridge between overall and individual finances was the use of 'key figures'. These were developed specifically to aid the process of discussing the house sales policy, described above.

Potentially useful facts were written on flipcharts and displayed on the wall of the Committee room. Facts such as: 'one house sale is equivalent to ten pence per week on everyone's rent' or 'a one pound a week rent rise is worth £17,000 per year' or 'a one pound a week rent rise would let us borrow an extra £150,000'. These facts could then be referenced during the discussions. The charts remained displayed on the walls for several weeks and were referenced frequently during many subsequent meetings.

6.7.3 Using Financial Models

The Unit's work—in developing both the general understanding and the actual decision-making process—involved the use of financial planning models on computer spreadsheets. The Co-operative had been successful in obtaining a microcomputer, printer and software through sponsorship. It was decided that a computer-based financial model would be both useful and usable by members of the Committee.

A simple model showing one year's finances was developed. In the first instance it was used by members of the Unit with Committee members to produce the 'key figures' referred to above. It was also used to test out different rent levels to allow a number of options to be considered and put forward to members. At this stage some basic risk analysis was carried out, in particular to consider the impact of changes in interest rates. This model was transferred to the Co-operative's own machine and a number of members of the Committee were taught how to use the model so that they could 'play' with it to get a better feel for the financial situation. A few of the Committee did try the model and were pleased with the results. The majority preferred that runs of the model be carried out by the Unit and the results presented on flipcharts or as 'key figures'.

Two further models were developed. The first was in response to investigations into the possibilities of major refinancing using deferred interest and fixed rate loans. The issue was highly complex, in financial terms alone, and expert advice was sought from brokers in the money market. Therefore, the computer model was used exclusively by members of the Unit to produce possible scenarios for discussion by the Committee. As before, results from runs were transferred to flipcharts which formed the focus of discussion by the Committee. In this way the Committee was able to identify and investigate the main risk. However, they also requested that the model be used to look at the impact of differential rents on improved and unimproved properties.

The second model, which was developed and used in partnership with the YMHA, was more closely focused on deriving a comprehensive modernisation package for the whole estate, involving house sales, rent increases and limited further borrowing. The model was developed mainly by Joanna Bennett and Helen Dunn, students from the Birmingham M.Sc. OR course, as their summer project. Their report of the development of the model clearly demonstrated their own learning and also highlighted some of the issues of working with community clients (Bennett and Dunn, 1995).

At the end of this sustained work on financial issues, the Committee were relatively happy to deal with information produced by computer models and felt able to question the results and request additional runs and materials. Individual members of the Committee became closely involved in the financial side of Co-operative business and were actively involved in advising on the development of the model. They still shied away from running the models themselves, but were certainly active participants in deciding on what runs should be done and interpreting the results.

6.8 *Promoting Community Involvement*

A sustained parallel area of engagement was to stimulate and structure the involvement of the wider membership in decision-making and other

Co-operative activities. The Co-operative was formed out of adversity. Therefore, the Committee had to make major efforts to communicate the ethos of being part of a co-operative and to try to encourage participation in all Co-operative activities. Some of the Unit's involvements in this area were necessarily of the 'making jam' variety: helping in the planning and organising of events, such as open days for members, and of other public meetings, as well as helping with the production of a newsletter.

This work also allowed members of the Unit to work alongside workers from the Yorkshire Co-operative Housing Services (YCHS), a secondary housing co-operative based in Barnsley. Like the Thurnscoe tenants, members of the secondary co-operative had had no exposure to OR methods but had considerable community development experience. This afforded the opportunity to combine elements of both areas of expertise.

6.8.1 Structured Street Meetings

Within the general activities described above there were a number of more structured OR involvements. The first one was the organisation and running of a series of street meetings. The need for these sorts of meetings was highlighted during the sessions on sales policy when it became clear that no valid discussion could be held without having a clearer picture of the members' views. It was felt that information could not be gathered in a 'traditional' public meeting since members needed some information before being able to have a sensible discussion.

During planning sessions with the Committee and the worker from the YCHS it was decided that the meetings should be informal and entertaining, and should incorporate a quiz (about co-ops) and bingo, as well as the discussion session. The discussions were structured using an approach based on 'nominal group technique' (Delbecq and van de Ven, 1971), whereby tenants would work first on their own, then in small groups, and finally in plenary, focusing on extreme statements about the Co-operative. This 'reverse cascade' approach is highly participative and allows each person to feel that they have contributed to the overall debate, and indeed they are able see their own private views and comments featured in the overall report of the meeting.

A series of meetings was arranged inviting about 30 members to each. Almost three-quarters of the meetings were cancelled because of poor attendance. However those which did run were regarded as being very successful, generating enthusiasm and a good level of debate. Although they had been intended to concentrate on the issue of house sales, it was felt, in view of the problems with attendance, that they should look first at basic issues of what the Co-operative was about, its problems and what it could achieve. They clearly demonstrated one of the major problems facing the Co-operative (and indeed many community initiatives), which is the low level of expectations concerning active involvement, particularly in activities concerned with seemingly abstract tasks of planning and decision-making. Conditions of passivity had become deeply ingrained, and required an immediate sense of urgency for them to be breached. However, a number of good ideas did emerge and were taken up; and those who attended felt that they had learnt a great deal about the Co-operative.

6.8.2 *Action-Methods Based Meetings*

Another of the strands of promoting community involvement was based around using a range of participatory approaches, commonly termed 'action methods'—in particular, sociodrama (Sternberg and Garcia, 1989). Although the facilitation of many sessions and workshops were based around the use of fairly conventional approaches and tools, such as discussion and flip charts, we also used some less common approaches. For instance, during many discussions we would use objects—anything that came to hand—to represent elements of the discussions: the main protagonists, ideas, structures. This had the advantage of providing another focus so that those who felt least happy speaking could join in the 'fun' activity of moving some of the objects around or speak 'on behalf of the tea mug'. This became a very natural part of many discussions.

We were able to build on this use of objects to run a couple of sessions using more advanced action methods. One of these involved an open meeting that was aimed at exploring possible futures for the Co-operative. Ron Weiner (an independent sociodrama consultant) and I facilitated it. The day was held in a local community centre and the morning was spent 'building Thurnscoe'. Local people started by chalking out the streets and then representing the main focal points with relevant objects—the local pubs were represented with beer cans of the appropriate brews. As different people added different elements, Ron would speak to them, asking them to describe to the others what they were doing and eliciting relevant stories to add to the whole picture. These were often further developed by others in the group. This seemed to feel relatively unthreatening and a wealth of material was generated. Afterwards people commented that they had learned a great deal—partly memories being evoked, but also development of an appreciation of how others felt about the area.

In the afternoon, participants were encouraged to take part in short activities to explore particular aspects revealed in the morning. Some of these focused on possible physical developments and others explored relationships between different groups. The activities went beyond conventional role-play and used elements of sociodrama techniques, such as 'doubling' and 'the empty chair'. Throughout, people entered into the spirit and seemed largely happy with the novel approaches used. Certainly, the majority who attended did actively participate and most said what fun it had been and asked when could we do it again.

The issues raised during the day were mapped and formed the basis for future work of the Committee. Several of the episodes were regularly quoted at future meetings to illustrate feelings on key issues. Use of action methods became a key element of many involvements by the Unit over the years. The combination of creativity and involvement created by action methods and the strengths of OR structuring techniques proved powerful on many occasions.

6.8.3 A Co-operative Survey

A major distinct piece of work carried out with the Co-operative was a survey of the estate to gather demographic, housing management and other information from tenants. The impetus for the survey came from the need to be able to supply factual information about the Co-operative membership when applying for funding from outside sources. An environmental survey had been carried out before the Co-operative came into existence, which showed powerful evidence of need and deprivation in the area. However, it had been based on a small sample, provided an incomplete picture and was felt to be rather dated. It was also clear that the Co-operative had insufficient data about the make-up and condition of the housing stock. It was decided that a survey of the entire estate was required.

The survey was carried out between February and March 1990. Members of the Co-operative Committee formed a team of interviewers. Pauline Booker, who was working for three months at the Unit on secondment from the Polytechnic of Central London, co-ordinated the team.

The team spent many hours working on a questionnaire which had been devised with students from Sheffield City Polytechnic. The team worked to ensure that all the interviewers interpreted each question in the same way and discussed the general techniques of interviewing. There was some apprehension within the team about going into other peoples' houses and asking rather a lot of questions, some of which were about sources of income. The original questionnaire was radically overhauled.

Members of the team interviewed each other as the modifications were made. The questionnaire was also pre-tested on two households.

This led to yet more revisions and some concern at the length of time that each interview took—even without any tea drinking!

A further radical rethink led to a proposal for splitting the questionnaire in two. A complete census was to be conducted for part of the questionnaire, mainly for those questions where the location of the household needed to be known; and a random sample of 40 households was to be used for those questions where it was only necessary to estimate the proportion of tenants in particular categories. This resulted in a census questionnaire of 20 questions and a supplementary schedule of 19 questions.

The team divided into three pairs and each took responsibility for a set of streets. The Committee members knew most of the estate fairly well and could warn each other where there might be difficulties. It was decided that four households should be avoided, mainly because they were in dispute with the Committee for various reasons.

Interviews started in earnest on 6 February 1990. Lunchtime team meetings were held and proved useful in various ways—to compare notes on how the questionnaire was working out and to make minor adjustments. Also, there were discussions of the needs of particular groups of tenants such as the elderly. It was felt that some needed assistance with window cleaning, decorating, gardening, etc. Others needed specific advice on benefits and the name of a local Welfare Rights Officer became indispensable. The lunchtime meetings also provided the opportunity to boost each other's morale—especially on days when it was cold and wet.

Interviews continued for the next few weeks, interrupted amongst other things by illness and family commitments; it was mid-March before the end was in sight. Contact was made with 299 out of 342 households. Members of the team were able to supply information on many of the non-responding households, including details of family size and composition.

The initial apprehensions about interviewing soon evaporated. It was surprising how easily people parted with information, even the rather detailed question on benefits. The question of the sales policy came up from time to time and people said that it was being considered by the Committee and would be brought to the tenants in due course. Very strong feelings were expressed at times, since some tenants said that the rents had increased but that there was not much to see for it. These proved difficult occasions for the Committee members to handle, but everyone coped.

The questionnaires were taken to the Unit at Northern College for checking, coding and keying. After the survey a number of tenants attended a meeting to discuss the ways in which they might get involved in Co-operative activities. It was clear that their efforts would have to be co-ordinated and that some existing Committee members would have to provide this function. It was difficult to see this momentum being maintained without the help of a full-time worker. The Committee had to be both policy makers and implementers—all on a voluntary basis. This is difficult to sustain for long periods of time. Employment of a worker was an issue that was considered at a further, separate workshop.

The results of the survey were later reported back to the membership at public meetings and through the Co-operative's newsletter, and were used to consider issues such as future housing requirements and in submissions to the Housing Corporation and charitable trusts. The survey had produced a substantial quantity of very valuable, robust information about the Co-op, but it had also been a key element in developing the skills and confidence of members of the Committee. The learning generated by the direct experience of doing the research had a number of further benefits for the Co-op and the Committee, and also provided the platform for another important area of work for the Unit: Community-led research. This involved working with a wide range of communities, geographical or otherwise, helping them to define and undertake a range of research activities. It also led to a major programme of teaching and training.

6.9 Some Conclusions

In view of the timescale and importance of the involvement of the Unit with the Thurnscoe Tenants Housing Co-operative, it is appropriate to offer a few reflections on the work in terms of the scope and impact of community OR.

6.9.1 Eclecticism of Problems, Methods and Practices

First, the nature of the funding for the Unit allowed us to tackle strategic as well as tactical problems, using whatever tools and techniques seemed to be appropriate. The action research approach helped to ensure that the Unit was involved in areas that the Co-operative regarded as being of prime importance and open to an OR-type analysis. While the Unit had to be aware that it was necessary to use OR and related techniques in assisting the Co-operative, the motivation was not to find problems to which an OR technique could be applied, but to work with the Co-operative in identifying their problems and then to help them tackle these.

This meant that the Unit had to be eclectic in its methods and working practices. There is perhaps little in this that is uniquely distinctive to community, as opposed to traditional, operational research; all OR units must, to a greater or lesser degree, exhibit flexibility. However, the drawing of rigid boundaries around the subject matter of OR is less easily understood by groups unfamiliar with the demarcations of professional and academic disciplines. Whilst not dismissing the risks of 'undefining' operational research (Sims, 1981), these demarcations seem less important than announcing to a group that you cannot help because 'that's not OR'. Further, it is of little value to a community group to tell them that what they really need are accountancy skills when there is no possibility of these becoming available.

The Unit was particularly successful in combining highly participative techniques, such as action methods and sociodrama, with powerful OR structuring approaches. Both sets of techniques can appear daunting and we rarely applied any technique in its entirety. In particular we lifted and combined key elements of many OR techniques (e.g., rich pictures and 'CATWOE' from Checkland's, 1981, soft systems methodology; and classifications of uncertainty, comparison areas and commitment packages from Friend's, 1989, strategic choice). Similarly, we borrowed different bits of a range of action methods to help generate ideas and discussions. Not only did the combinations seem to work but they also provided for exciting and energising sessions.

6.9.2 Collaborative Working

There were implications for the practice of community operational research in terms of forming working relationships with other professionals. The Unit co-operated successfully with community development, community education and community housing workers. Architects and planners offer their services to community groups through the Association of Community Technical Aid Centres (ACTAC) and other bodies. The involvement of the Unit showed that it was possible to incorporate an OR input into these wider activities.

6.9.3 Hard and Soft OR

Within the boundaries of OR, the Unit had to demonstrate a range of modelling talents. Although the early balance of the work tilted towards the newer, so-called 'soft' problem-structuring methods, the Unit needed to employ more than 'a drop of the hard stuff'. Whilst we would not claim to have applied anything approaching the complete repertoire of OR techniques, we were able to use and test the applicability of a wide range, in a variety of contexts. Experience suggested that even the tools of classical OR have a role to play in different problem contexts and that perhaps the process of involvement and application—the craft skill of operational research—is of greater importance than the techniques used.

6.9.4 Contrasts with Traditional OR

Perhaps what was most striking was the similarity between community OR and OR as practised in the most successful OR groups in traditional organisations. Many similar criteria applied—responding to client needs, achieving results in an acceptable timescale, involvement in a range of problems at different levels in the organisation, indicating new areas for consideration, involvement in implementation and maintaining communication appropriate to the structure of the organisation. It is in these last two areas where there were perhaps the greatest differences. The democratic decision-making structure of the Co-operative, the level of informal contacts and the significant cultural differences between members of the Co-operative and the OR consultants meant that far greater attention had to be paid to the process of working and the politics of involvement. The relative importance of 'soft' techniques and the importance of group dynamics probably called on a wider range of skills than is necessary within a traditional organisation, while the need for high-level mathematical skills or knowledge of 'classical' techniques was reduced.

6.9.5 Importance of Personal Relationships

Also significant were the personal relationships built with members of the Co-operative. The involvement of Unit staff in activities beyond project work or workshops, as well as social contacts, inevitably led to the development of a closer relationship between client and worker than would be the case with most traditional clients. Clearly there is the potential for 'role conflict' as the OR worker moves from colleague or friend to facilitator or advisor. There is also potential for disagreement over politics and values.

6.9.6 Potential Clash of Roles

Potential conflicts were negotiated throughout the involvement: they were not ignored. However, there was no clash of politics and values so great that either side wanted to terminate the relationship. The mutual respect and trust that was built up throughout the involvement ensured that the transition between, and combination of, roles was unproblematic. Indeed, our role of facilitating at meetings was welcomed, not least because everyone wanted to have their say, and not be constrained by 'chairing' a meeting. Frequently, when facilitating, we were asked 'what do you think?' It was difficult, but necessary, to try to restrict an answer to emphasising the need to consider arguments, even though members may well have been aware of our own sympathies—and indeed may have been asking the question to get us to support their own position.

It was important to remember at all times that the problems and decisions were the Co-operative's and not the Unit's. Our greatest concern was to help them make the decisions that they felt were best for them. Our role was to help the process of making those decisions by providing information and analysis and using OR-type methods to clarify and structure the decision issues. It was not our role to persuade the Co-operative to accept our recommendations. If the clash of values on which decisions were based had been too great, it would have been necessary for the Unit to consider its role and, possibly, to withdraw.

6.10 A Final Comment

Throughout our involvement with the Thurnscoe Tenants, they struggled with two complementary problem areas. The first was the process of establishing decision-making machinery and methods that would permit the full articulation of the Co-operative members' desires and aspirations. As well as the obvious primary goal of personal and family housing security, the tenants cared intensely about the social atmosphere and infrastructure of their community. On the other hand, the Thurnscoe Housing Co-operative was a multi-million pound financial enterprise.

In both these areas I believe that OR served the Co-operative well. However, the relationship between the Unit and the Co-operative extended beyond the timescale of the involvements described above. Eventually it became necessary to reduce the amount of effort that the Unit could devote to the Co-operative, but we remained in contact for much longer. Members of the Co-operative would contact the Unit to ask advice on specific issues and problems and contact would also be maintained through members of the Co-op attending courses at the Northern College. It was clear to all who had contact with the Co-operative that their management and decision making skills had developed considerably over the years. I would certainly not claim that this was entirely due to the involvement of the Unit; a huge amount of development and learning took place through the sheer necessity of day-to-day involvement in running the Co-op. However, it was equally clear that the Unit had made a valuable long-term contribution to the Co-op and, perhaps more importantly, to the individuals involved in it. It was a pleasure working with them all.

Acknowledgements

At the time the intervention discussed in this chapter was completed, the author was Director of the Community Operational Research Unit (CORU) at Northern College.

This chapter draws heavily on two papers jointly written by the author with Colin Thunhurst, John Friend and Pauline Booker (Thunhurst and Ritchie, 1992; Thunhurst *et al*, 1992). Those papers also acknowledged the contributions of a number of OR practitioners who assisted in the work described, particularly Peter Long and Jim Bryant. In addition, my thanks go to a number of other practitioners whose names are noted in the chapter, and others (particularly at the Northern College) who are not named.

Above all, my sincerest thanks go to the tenants and the Committee members of the Thurnscoe Tenants Housing Co-operative. An account such as this cannot do due credit to the personalities and enthusiasm of Co-operative members. I would not want to single out individuals; every member of the Committee has worked closely with the Unit, but particular thanks go to Elsie, Janet, Anne, Nancy, Maureen, Joe and Tommy—some of whom are no longer with us. The Co-operative is.

References

Bennett, J. and Dunn, H. (1995). Renovate or deteriorate? An analysis of the problems facing the Thurnscoe Tenants Housing Co-operative. In: *In at the Deep End*, pp. 91–112. Bowen, K. (ed.). Pavic Press, Sheffield.

Checkland, P. (1981). *Systems Thinking, Systems Practice*. Wiley, Chichester.

Delbecq, A. and Van de Ven, A. (1971). A group process model for problem identification and programme planning. *Journal of Applied Behavioural Science*, **7**, 446–492.

Friend, J. (1989). The strategic choice approach. In: *Rational Analysis for a Problematic World*. pp. 121–157. Rosenhead, J. (ed.). Wiley, Chichester.

Jackson, M. (1988). Some methodologies for community operational research. *Journal of Operational Research Society*, **39**, 715–724.

Mason, R. and Mitroff, I. (1981). *Challenging Strategic Planning Assumptions*. Wiley, New York.

Sims, D. (1981). The practice of OR: Operationalising Ackoff. *Management Decision*, **19**, 39–47.

Sternberg, P. and Garcia, A. (1989). *Sociodrama: Who's in Your Shoes?* Praeger, New York.

Thunhurst, C. and Ritchie, C. (1992). Housing in the Dearne Valley: Doing Community OR with the Thurnscoe Tenants Housing Co-operative. Part 2—An Evaluation. *Journal of Operational Research Society*, **43**(7), 677–690.

Thunhurst, C., Ritchie, C., Friend, J., and Booker, P. (1992). Housing in the Dearne Valley: Doing Community OR with the Thurnscoe Tenants Housing Co-operative. Part 1—The involvement of the Community OR Unit. *Journal of Operational Research Society*, **43**(2), 81–94.

7

Looking Forward with a Council for Voluntary Service

Andy Marlow and Jim Bryant

In the UK the work of many voluntary and community organisations is brought together on a locality basis under the umbrella of a Council for Voluntary Services (CVS). So a CVS seeks to respond to the collective needs of the many voluntary groups in a geographical area (often roughly coterminous with a local authority) and to act as a single voice for them, especially in discussions with statutory bodies. This chapter gives an account of some work to shape a forward plan that was conducted with a specific CVS—we shall call it 'Caldburn CVS' to preserve its anonymity—located in North-West England. The development of the plan revealed some dysfunctional characteristics of the CVS operation, and these were addressed in a supplementary study that is also reported here.

7.1 Background

Traditionally each CVS has fulfilled a number of roles:

- To co-ordinate the activities of local voluntary organisations;
- To offer a resource and information point as well as support services for voluntary organisations;

Final version accepted for publication July 2002.

Andy Marlow • Northumbria University, Carlisle Campus, Paternoster Row, Carlisle, CA3 8TB, UK.
Jim Bryant • Sheffield Hallam University, Stoddart Building, Sheffield, S1 1WB, UK.

Community Operational Research: OR and Systems Thinking for Community Development, edited by G. Midgley and A.E. Ochoa-Arias, Kluwer Academic/Plenum Publishers, New York, 2004

- To provide a link between the voluntary and the statutory sectors;
- To represent local voluntary organisations in decision-making; and
- To assist in the identification of social needs and in the creation of strategies to address them.

These roles have been extended by the social and economic policies of successive governments that have increased the involvement of the voluntary sector in service delivery. In turn this has placed additional demands upon each CVS in terms of its processes of governance and accountability, since funding agencies have a duty to assess and evaluate the use of monies made available. It was largely in response to these pressures that Caldburn CVS determined to draw up a second strategic plan to guide its work during the mid-1990s.

Its Council, which consists of representatives from all member organisations, formally directs the work of the CVS, but practically policy is shaped and made by a 17-strong Executive Committee that includes 9 elected members of the CVS Council, 4 co-opted Local Authority Councillors and 4 paid CVS staff. The latter includes a full-time General Secretary, a part-time Liaison Officer and two part-time Administrative Assistants. The Executive Committee meets six times each year and is supported by two standing committees (Management/Finance and Public Relations) plus various ad hoc groups.

At the time of the work reported here, the CVS was coming towards the end of the period covered by its first 5-year strategic plan and it was felt that the rapid changes taking place in the sector meant that the CVS would benefit from the creation of a fresh forward plan, provided it was flexible and responsive to the new environment. Furthermore, in contrast to the previous plan which was developed by an employee of the Social Services through a short series of formal meetings with an ad hoc group from the CVS, it was felt that the new plan should involve quite wide consultation with relevant stakeholders and have a far broader attack. It was at this point that one of us (Marlow) fortuitously made first contact with the CVS and so became involved in the project described here.

To take the work forward, a Future Planning sub-committee was established to review and consider the development of the CVS, and to organise the creation and delivery of the new plan for consideration by the Executive Committee 6 months hence, with the intention that the plan would then be launched publicly some 3 months later.

7.2 *Scoping*

So that the consultation process would take account of the widest possible range of relevant perspectives, an informal listing of key stakeholders was created with the assistance of the General Secretary. Briefly this consisted of the following groupings:

- Voluntary groups affiliated to the CVS;
- Bodies funding the CVS (Social Services, Health Authority, Local Authorities);
- Customers of the CVS (i.e. its members and the recipients of voluntary help);
- Public voluntary sector organisations (e.g. the National Association for CVS [NACVS] and the National Council for Voluntary Organisations [NCVO]); and
- Related Government Departments (Health & Social Services and Environment).

Fact-finding meetings were arranged and held with representatives of these stakeholder groups. In each case the discussion was fairly informal but a diversity of issues were raised.

The three statutory bodies—Social Services, Health and Local Authority—raised quite similar matters. High on the Social Services agenda was the need for tighter monitoring and evaluation of CVS expenditures (for budgetary transparency, rather than because of any specific concerns). The nature of the contracting relationship also required attention. For the Health Authority, accountability for expenditure was also an issue, but so too was the need for the CVS to be able to operate within changing health service delivery frameworks. The Local Authority shared these agendas.

The interviewees from the voluntary groups themselves took a rather different view of the key issues. They were apprehensive of the prospect of having to maintain stronger clerical and administrative systems—to 'act as puppets' for the statutory bodies—which they felt betrayed the essential ethos and style of voluntary work. This was especially daunting for the smaller groups that lacked the skilled volunteers or spare resources to undertake such administrative tasks.

For the national voluntary sector organisations the blurring of the boundary between statutory provision and the traditional voluntary provision of services was a major current shift of emphasis. This was driving a need for each CVS to determine a clear focus for its activities amongst prioritised needs. Points about the increased scarcity of funding and around tendering and contracting were also mentioned.

Taken together, all these points (and others) were summarised in the 'rich picture' (a technique borrowed from Checkland, 1981) presented in Figure 7.1. This shows some of the main pressures at work in the arena as well as the relationships between the different groups and organisations.

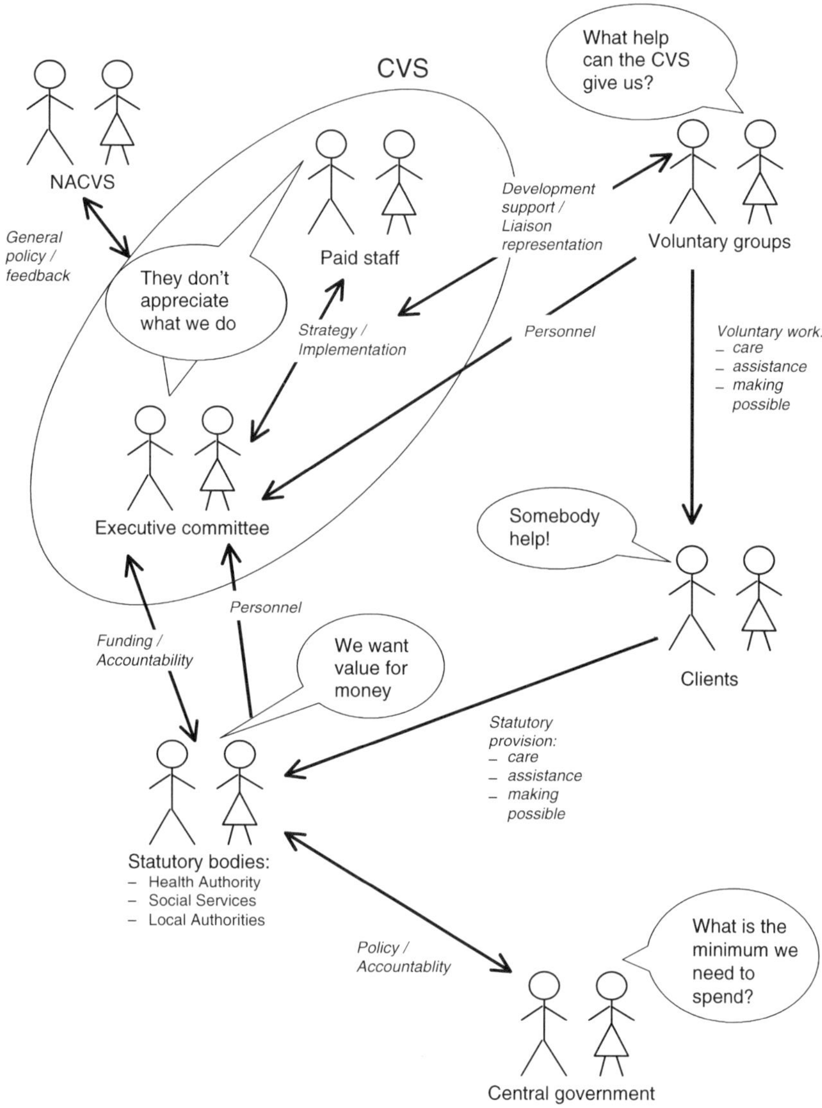

FIGURE 7.1 Rich picture

7.3 Directions for the Future

The creative generation of ideas for the future of the CVS was seen as an essential starting point for the production of the strategic plan. Since it was clear that the widest possible view should be taken, it was decided to preface the special Executive Committee meeting that had been arranged to commence the planning process by pre-circulating a short questionnaire to members to gather views about the current activities of the CVS. This was intended to save members time—always at a premium—during the meeting itself, and to provide an organising framework for the brainstorming that would be conducted there.

The questionnaire was organised into sections under the four key roles of the CVS (development, support, liaison and representation) as well as one concerned with the internal management of the CVS. Respondents were asked to indicate on a 7-point scale the importance they attached to each of a dozen or so activities within these sections.

The 50% response rate achieved for the questionnaire gave enough useful guidance, but at the same time was a sharp indication that lack of time was constraining the commitment and contribution of committee members. Analysis confirmed the value of the 4-role classification used in the questionnaire, so the brainstorming session facilitated at the special Executive Committee meeting was also shaped around these themes. This meeting, attended by all 17 of the members, was introduced with a short explanation of the principles of the brainstorming process: suspend judgement; freewheel; cross-fertilise. This was essential, as the process is often abused and the term loosely attached to any form of group creativity, whereas to succeed it must be a disciplined and explicitly structured activity with clear rules to achieve the maximum payoff.

Over 50 distinct ideas were produced in the brainstorming session and these were taken together with the questionnaire responses to provide a good sense of members' hopes and apprehensions. In particular they fed strongly into a SWOT (Strengths, Weaknesses, Opportunities, Threats) analysis carried out shortly afterwards by the General Secretary and the facilitator. The first two aspects (SW) focus upon internal factors and so this analysis also drew upon historical annual and financial reports; a review of organisational structures and processes; an audit of the physical resources of the CVS (premises, equipment, etc.); and a consideration of present relationships with external bodies. The second two aspects (OT) of the SWOT analysis look at an organisation's environment, and so here involved consideration of changes in funding procedures; legal requirements; boundaries between related organisations; and the CVS catchment area (demographics, infrastructure, etc.).

This review highlighted the high level of uncertainty that the CVS faced, and suggested that it would be very hard to create a meaningful or explicit plan with a 5-year time horizon. Consequently agreement was sought and obtained from the Future Planning sub-committee to create a 2-year plan, with the idea that this be revisited and amended annually thenceforth on a rolling basis. From this position, a 3-part plan, looking at the past, present and future in a linked manner, was produced in sections. These were called 'where we've been and where we are now', 'where we are now and where we want to be' and 'where we want to be and how we are going to get there'. While the content of the first two of these was available from the reviews carried out so far, the third section was produced only in draft form, and then developed through an iterative process over a period of several months with members of the sub-committee as well as through wider consultative meetings with the membership.

However the eventual plan will not be discussed further here, for the production process that has been described revealed some quite fundamental problems in the management of the CVS, and it is to these that we now turn.

7.4 *The Reconciliation Intervention*

During discussion for the preparatory work on the strategic plan, it soon became clear to Marlow that relationships between individuals within the CVS were neither as constructive nor supportive as possible. People were pulling in different directions and dissipating the limited resources available. To address these matters, a complementary investigation was proposed to, and agreed by, the General Secretary, whereby some of the underlying issues might be surfaced and the differences and emergent conflicts resolved. Success in this task would also assist in the eventual implementation of the strategic plan.

The problem of poorly functioning management and structures in the CVS manifested in a number of ways. These included uncertainty about areas of responsibility; divergent accountabilities (e.g. to the CVS and to the member organisations); style and means of activities (for example, because of the very diverse organisational cultures represented); levels of communication; and relationships to external bodies. However there was no clear agreement as to what was wrong or what needed to be done: it was rather a 'mess' (to use Ackoff's, 1981, term) of highly interrelated concerns, activities and processes, which sharpened into different 'problems' depending upon the viewpoint taken. Indeed it was so lacking in explicitness that while others could share Marlow's sense of unease, neither they nor he could pin it down much more definitely.

Flood and Jackson (1991) have proposed a three-phase methodology—Total Systems Intervention (TSI)—for drawing upon the strengths of complementary systems methodologies to address such complex situations as that facing the CVS. This framework was ideal for application in the situation we encountered in Caldburn, though we used it as a guide rather than prescriptively.

The first phase of TSI—'creativity'—involves using systems metaphors to access fresh ways of thinking about what is or might be going on in a situation. Specifically the aim is to identify a dominant metaphor that provides the most appropriate and generative perspective for conceptualising the situation. In the present case it was clear that there was strong potential in viewing the CVS as an organism, a brain, a culture and a coalition. Indeed the issues seemed to intertwine around the relationships (*coalition*) between plural interests (*culture*) in the context of an organisation that has grown organically (*organism*) and that must learn to function (*brain*) within a rapidly-changing ecological system! Now it is clear that while any of the metaphors might prove useful, the role of the project—to reconcile issues around roles and responsibilities within the CVS—privileged the culture metaphor, though drawing on the other three. It is interesting to note in passing that, in a somewhat similar example of work with a CVS, Flood and Jackson (1991) made the same selection. However, we diverged from them in the 'choice' phase of TSI, for though we too considered that we faced a 'complex-pluralist' problem context, we selected a different methodology with which to address it.

Typifying the problem context in this second phase of TSI was not without difficulty, since though it was clear that we were dealing with a 'pluralist' case here (multiple participants, with compatible interests, but mildly divergent values and beliefs, who were nevertheless prepared to work together in consensus decision processes) there were arguments for regarding the CVS system as 'simple' (relatively unchanging structure; good sense of purpose; quite coherent). However, the challenging and uncertain future facing the CVS was giving rise to internal stresses and the informality and ambiguity of many of the working arrangements (which were the subject of this intervention) both suggested that it might better be better considered as a 'complex' system. This diagnosis suggested that we required a methodology that would mesh with the metaphors we had chosen earlier, and would work within a complex-pluralist context.

Now, TSI embraces only a limited repertoire of systems methodologies from which such a choice can be made, and we were not convinced that those it indicated would be practicable or productive here. Instead our own experience suggested that the strategic options development and analysis approach (SODA) (Eden, 1989) would be wholly compatible with the need to translate the dominant vision of the CVS, its form and orientation, into a specific agenda for organisational development. SODA would not only make explicit the beliefs and values of stakeholders in and around the CVS, but it would also provide a process for the negotiation of alternative views about the management and structures of the organisation.

In SODA, cognitive maps, representing in explicit form the personal belief systems of individual actors in a situation, are created and then merged into a composite picture—termed a 'strategic map'—that subsequently provides a catalyst to shape a portfolio of actions to which the group can commit its energy. Maps show concepts linked by arrows that signify causal theories. At Caldburn, interviews were conducted with nine people close to the core of the CVS operation, including both paid staff and committee members, and each was recorded using mapping notation. Figure 7.2 shows a broad schematic view of the overall strategic map produced from these components. Key themes were internal communication; the conduct of meetings; conflict between paid staff and the executive committee; issues around teamwork and committee members' capabilities; and tensions between the CVS and its constituent voluntary agencies. Importantly the map also shows (at the 'head' of the chains of argument) some of the practical solutions which people felt might alleviate these problems.

In the event it proved impractical to convene a full meeting of the Executive Committee to develop the strategic map, but its implications were discussed quite fully with the General Secretary, who took them on board with understanding. Indeed informal discussions with the General Secretary of a neighbouring CVS had revealed that many of the issues discussed were common ones, and that some of the solutions suggested by the interviewees were valid and had been implemented with some success elsewhere. The critical common strand demonstrated in the neighbouring CVS was a firm steer from the General Secretary, for instance in improving relationships between his committee and the paid staff, and in enhancing the level of commitment and contribution of the committee members. This gave the Caldburn Secretary added confidence that she should similarly attempt to push change through in her organisation. For this reason, airtime to discuss the strategic map was allocated within one of the regular meetings of the Executive Committee.

Rather than attempt to review the range of issues uncovered by the strategic map, it was decided that the short time available would be best used in addressing a single focal area, leaving other matters to be dealt with elsewhere. The chosen area was one closest to home: the functioning of the Executive Committee itself. Matters covered included:

- Digressions from the agenda;
- Introducing and debating complex issues;

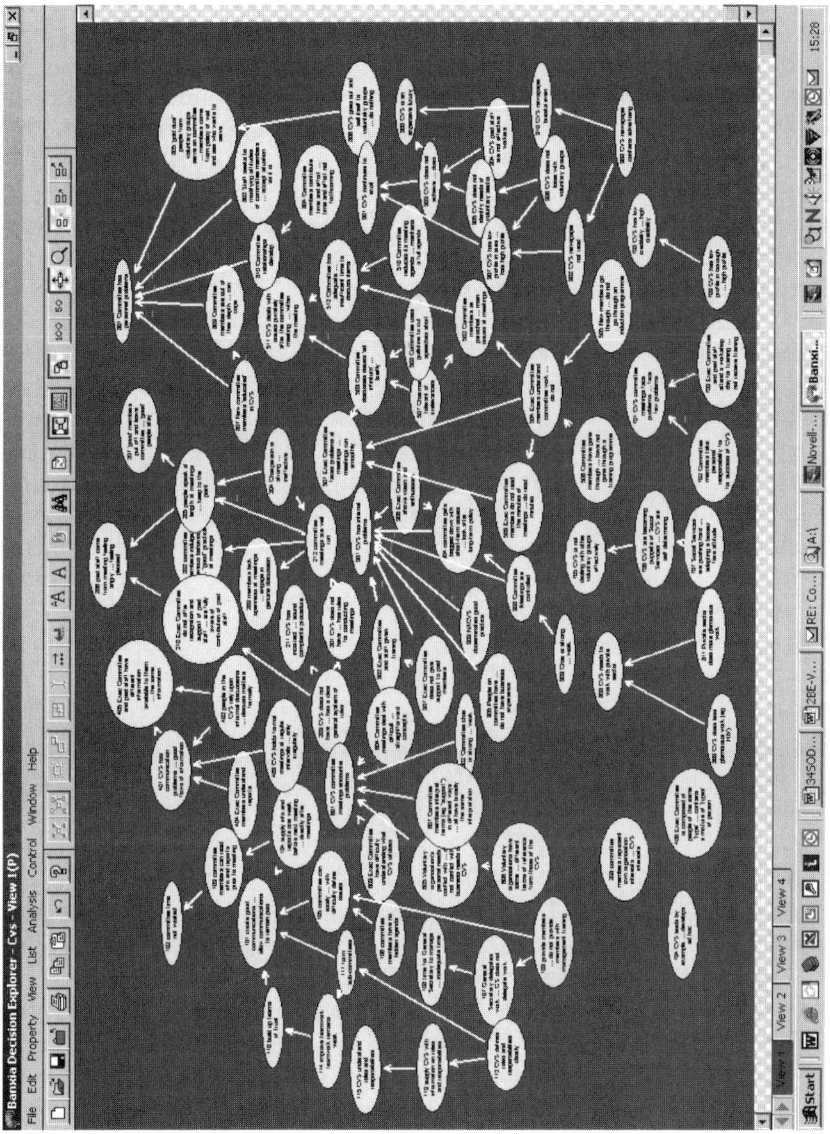

FIGURE 7.2 A broad schematic view of the overall strategic map

- Maintaining a CVS focus rather than discussing the needs of specific voluntary bodies; and
- Preventing slippage of discussion between topics.

Common-sense action plans were determined to deal with these and other matters. These related, for example, to pre-circulating briefing papers; tighter chairing of meetings; and the use of a timed guillotine in discussions. While all of these were 'obvious' and quite simple matters to implement, the need to do so highlighted the fact that the issues concerned had never previously been directly confronted. Here, as so often, unspoken assumptions or accepted ways of working had previously been allowed to dominate, to the detriment of the management of the CVS.

The process, though limited, was encouraging and liberating, and the General Secretary felt able to press ahead on a number of other fronts stemming from the strategic map. These included the introduction of management training, which had not previously been taken seriously, but which was now given a high priority at her insistence. She also determined to increase her efforts to 'trawl' the voluntary groups for 'gold-dust' (new members to join the CVS Executive), rather than continuing vainly to hope that such individuals would 'emerge' spontaneously, which had not happened in the past! While none of these ideas in themselves seem revolutionary, or even mildly radical, they represented a fresh approach to management for the CVS.

7.5 Review

This chapter has described an involvement that lasted for just under a year from the time of initial contact until the launch of the Strategic Plan and the final round of meetings. All concerned in the project felt that in different ways it delivered benefits for them. In this section we shall conclude by reflecting a little upon these benefits, but first we shall consider some of the features of this project that we suggest typify community operational research practice.

Serendipity is a common feature of community OR work. In the present project, it was fortuitous that contact was made with the Caldburn CVS just as they were seeking to develop a strategic plan, and it was also fortunate that they saw what we had to offer here as being relevant to this task. The fact that our formal strategic management techniques were only relevant and eventually applied to a limited degree was unimportant, but our credentials (knowledge of the techniques) provided an entry.

As the above account makes clear, another key requirement for operating in this field is flexibility and open-mindedness. It would have been perfectly possible simply to have contributed to the planning process, but the 'reconciliation intervention' that was prompted by our perspective as outsiders commenced a process that in the long run was arguably at least as important for the CVS as the strategic planning. This willingness to engage with emergent issues and to provide appropriate and timely support for the client organisation is a valuable attribute for the OR worker. So too is a preparedness to redesign on-the-hoof intended methodologies. Community OR is not an area for the purist who inflexibly seeks to implement each step of some carefully crafted, multi-stage process. The chances are that time will be too short; arrangements will fall through; new issues will surface; participants will be unexpectedly absent; or other conditions will render inappropriate or non-viable the 'methodological master plan'.

Community clients are invariably hard-pressed for the time to engage in extended explorations of issues, yet paradoxically are among the most able to see the importance of taking a view that includes and respects many perspectives. Similarly, as we encountered here, they are often very willing to try out what to them are novel processes, provided that these processes align sufficiently with the climate of involvement and democratic participation that usually pervades such organisations. This is a further contrast with much 'mainstream' work, where strong management hierarchies or parochial attitudes may prove insurmountable barriers to the use of methods or processes that depend upon frankness and honest debate.

However, for all the differences from more conventional OR work, there are strong similarities. In the present project, for instance, the principal benefits accrued from the breaking of assumptions about how things were being done: from the insights provided by a fresh vision, or from giving consideration to previously unspoken views that were 'voiced' through their inclusion in the strategic map. And from the consultant's point of view, when using an overarching approach like TSI, the practitioner is still well advised to 'do what s/he does best' rather than simply using the meta-methodology as a rule-based schema for selecting an approach. This is in line with Flood and Jackson's (1991) view that their meta-methodology should be regarded as a means for encouraging critical reflection on the strengths and weaknesses of different methodologies; it should not be used to *determine* methodology choice.

The case also illustrates a number of other features that are common in community OR. These include the careful balancing of responsibilities to multiple stakeholder groups; the difficulty of planning in a setting where the broad parameters are established by wider, more powerful bodies; the precariousness of operating with personnel whose very involvement may be withdrawn in a fit of pique; and the challenge of operating to the professional standards of partnering organisations, while relying upon unpaid volunteers. These and other difficulties that frustrate and enliven all community OR projects have also been explored elsewhere (e.g., Ritchie *et al*, 1994).

Finally, some characteristics of community OR that arise specifically because of working with the voluntary sector may be noted. These became obvious when we looked at the reasons for the various tensions between the parties involved in our project: between the voluntary workers and the paid staff; between members from the diverse constituent bodies; between the joint 'project' of the CVS and the preferences of the voluntary organisations; and between the CVS agenda and the demands of external bodies (e.g. funders and government). These tensions derive much of their energy from a combination of the caring passion with which voluntary participants hold, and are likely to express, their beliefs, and the resource-poor setting in which they almost inevitably operate. However the underlying differences of perspective often stem from radically divergent worldviews (what Checkland, 1981, calls '*weltanschauung*') that give rise to more fundamental problems of communication, collaboration and achieving consensus. Imbalances in power, in perceived legitimacy or in the ability to articulate ideas all exacerbate the problems of creating effective partnerships. To work with any assurance of success in this problematic arena, ways of acknowledging and capitalising upon diversity while managing conflict and promoting co-operation are essential. Taket and White (2000) have given some valuable pointers in this direction.

Having pointed to some significant issues for community OR that are typified by our work, we now turn to a short evaluation of the Caldburn project. There were benefits both to the client organisation and to others involved in the intervention. For the CVS, the obvious outcome was a well-structured, defensible and coherent strategic plan. Less visible was the start of a change in internal management processes and a fresh climate enabling the CVS to capitalise more effectively upon the strengths of its members. Like all CVSs, Caldburn CVS had always had to cope with its awkward mediating role between the 'grassroots' voluntary bodies and statutory institutions: the intervention did much to clarify and develop this position and to enable it to operate with more assurance in the future. For the consultant, the outcomes included not just the experience of applying methods and techniques in an unfamiliar setting, but the recognition of wider lessons about the management of the consultant–client relationship. And the payoffs still go on if you, the reader, reflect on what we have said and engage with it in your own practice.

References

Ackoff, R.L. (1981). *Creating the Corporate Future*. Wiley, New York.

Checkland, P. (1981). *Systems Thinking, Systems Practice*. Wiley, Chichester.

Eden, C. (1989). Using cognitive mapping for strategic options development and analysis (SODA). In *Rational Analysis for a Problematic World: Problem Structuring Methods for Complexity, Uncertainty and Conflict*, Rosenhead, J. (ed.). pp. 21–42. Wiley, Chichester.

Flood, R.L. and Jackson, M.C. (1991). *Creative Problem Solving: Total Systems Intervention*. Wiley, Chichester.

Ritchie, C., Taket, A., and Bryant, J. (eds.) (1994). *Community Works: 26 Case Studies Showing Community Operational Research in Action*. Pavic Publications, Sheffield.

Taket, A. and White, L. (2000). *Partnership and Participation: Decision-making in the Multiagency Setting*, Wiley, Chichester.

8

Facilitation as Fair Intervention

Wendy J. Gregory and Norma Romm

8.1 Introduction

This chapter elucidates a view of facilitation in which explicit attention is paid to the way in which the facilitator of group processes may adopt a role as (temporary) participant in the process, at times contributing content-suggestions to the discussions. Our argument is relevant to the practice of community operational research ('Community OR' for short) in that it considers what it might mean to develop a community towards enriched dialogical competence. Midgley and Ochoa-Arias (1999) indicate that when practitioners undertake work labelled as Community OR, there are often implicit visions of community that they bring to bear as they proceed. In this chapter, we pay explicit attention to the way in which we operated in terms of a specific conception of 'community' when we acted as facilitators in the context under consideration. Our approach draws to a large extent on Habermas's (1984, 1987) concern with enhancing 'communicative rationality' in social relationships. We spell out a pragmatised version of Habermas's argument as it relates to validity checking, but we

Accepted for publication May 2001.

Wendy J. Gregory • Institute of Environmental Science and Research (ESR) Ltd., Christchurch Science Centre, 27 Creyke Road, PO Box 29-181, Christchurch, New Zealand. Centre for Systems Studies, Business School, University of Hull, Hull, HU6 7RX, UK.
Norma Romm • Centre for Systems Studies, Business School, University of Hull, Hull, HU6 7RX, UK.

Community Operational Research: OR and Systems Thinking for Community Development, edited by G. Midgley and A.E. Ochoa-Arias, Kluwer Academic/Plenum Publishers, New York, 2004

also extend/modify the argument to take account of the continued fragility of the dialogical process.

Jackson (2000) indicates that, for Habermas, 'our progress toward a rational society is measured to the extent to which communicative competence is realised in society' (p.34). But he remarks that Habermas's 'social theory is at quite a high level of generality' (p.34). In this chapter, in an effort to shift the focus of Habermas's argument somewhat, we do *not* posit that the role of the facilitator is to orient people towards *consensual understanding* (in Habermas's sense of the regulative ideal of consensus guiding inter-subjective encounters). Instead we aver that the role of the facilitator can be to facilitate a greater awareness on the part of participants of the ways in which their position may be challenged by alternatives. The facilitator thus hopes to open the space for adjustments of position on the part of participants, and these adjustments in turn allow for negotiated views of 'the situation' and options for action therein to be considered.

The chapter outlines the theoretical basis of our approach and also draws on an example of a research study in which we employed our proffered mode of facilitation. We did so as our way of exploring the issue of standard setting in the provision of health services in a northern region of the UK. The National Health Service (NHS) in the region funded our work, and it involved a number of organisations. These included a purchaser of health services, two provider Trusts, local health authority representatives, and users from a local branch of the Royal National Institute for Blind people (RNIB) and a Visually Impaired Persons group (VIP). A paper written prior to us embarking on the facilitation of the workshops (Gregory and Walsh, 1993) described the project rationale and research objectives, which basically entailed providing a dialogic arena in which problems of communication associated with the purchaser–provider–user division might be addressed. Workshops were consequently organised which involved both peer-group and multi-agency discussions. These were facilitated individually by Wendy Gregory, Norma Romm and Mike Walsh in turn.

Our starting point for the deliberations in this chapter is a number of comments made during some of the workshops, which suggested that, as facilitators, we were not employing a value-neutral stance. Commentators said that we were 'taking sides'; that we *should* be 'neutral facilitators'. These comments led us to reflect on what lay behind them: a view of what facilitation is about. We realised that in order to proceed with our facilitations, we had to indicate to participants that we were not proposing to follow a supposedly 'neutral' stance of merely aiding the clients (participants) to 'assess their own conditions and generate and evaluate alternatives', as Chensult (1989, p.50) puts it. Instead we realised that we concur more with Pasmore's (1993) suggestion that the facilitator, as partner in the design effort, may make suggestions regarding the design of 'knowledge systems' (pp.79–80).

In considering the question of whether the facilitator's involvement in this fashion may lead to unwelcome 'imposition' and subsequent feelings of 'non-ownership' on the part of the participants, we argue that, to prevent undue imposition, facilitators need to be aware of their involvement. They should also take account of this as part of their critical self-reflection on their intervention. Our view of this requirement is detailed in the chapter.

The aim of the chapter is to propose the possibility of ethical intervention in which the facilitator admits his or her own involvement in offering ideas and views on the issues raised (and indeed on issues that *s/he* may raise). Our idea is that this can be regarded as legitimate, provided that s/he tempers this involvement with a recognition of the need to submit any inputs made to critical reflection and self-reflection. S/he can operate fairly insofar as s/he is aware of the need, not only to discursively question/challenge the statements made by participants in groups, but also of the need to be open in turn to other people's challenges as part of the same dialogical process. But before offering our ideas on how facilitation may proceed in this fashion, we offer a brief overview of how we see the relevance of our discussion on facilitation to Community OR practice (and indeed OR practice more generally).

8.2 The Relevance of Facilitation for OR Practice

In his paper entitled 'The Future of Operational Research is Past', Ackoff (1979a) insisted that the identification of OR solely with the development of mathematical models to find 'optimal' solutions to definable problems, portrayed a 'past-oriented' outlook. He argued that OR would have to undergo a radical transformation to (at least) include a concern with the ways in which managers, and those affected by managerial decisions, could become involved in designing and implementing what he called 'purposeful systems' (Ackoff, 1979a, p.96). He emphasised that this implied the need for people to develop their skills to review their way of addressing so-called problematic situations.

Ackoff's suggested proposals for the renewal of OR were not only directed at OR practitioners working in not-for-profit enterprises. He averred that his suggestions could be suitably applied in a myriad of situations where the emphasis is on aiding people to structure ways of making decisions inter-subjectively, whether in more or less profit oriented contexts (see also Parry and Mingers, 1991, pp.582–583). In our approach to facilitation, we concur with some of the points originally stressed by

Ackoff with regard to the importance of facilitating participation as a means of generating 'ways forward' in the design of plans. However, we highlight the facilitator's responsibility in contributing to 'fair' discussions of plans; and we focus on critical self-reflection as including awareness of the ways in which facilitators' interventions may serve to direct the process. Further thoughts on the responsibilities of interveners are offered in Gregory (1996, 2000), Romm (1996a,b, 2001a,b), and Gregory and Romm (1996, 2001). Also, Midgley (2000) provides an exploration of the meaning of intervention from a critical systemic point of view.

As regards Ackoff's (1979b) plea for generating a design process that is less expert-dependent than the more traditional focus of OR on expert-modelled 'optimal' solutions, his concern echoed many of the initiatives that were being undertaken in other fields at that time. In planning design, for instance, it is noteworthy that Rittel (1972) suggested that:

> Because this [spread of] expertise is frequently controversial, and because of what can be called the 'symmetry of ignorance'—i.e. there is nobody among all these carriers of knowledge who can guarantee that his [sic] knowledge is superior to any other person's knowledge with regard to the problem at hand—the process [of addressing planning] should be organised as an argument (Rittel, 1972, p.320).

Rittel (1972) contends that the feature characterising argumentative-oriented design methods is that:

> Statements are systematically challenged in order to expose them to the viewpoints of the different sides and the structure of the process becomes one of alternating steps on the micro-level; that means the generation of solution specification toward end statements and subjecting them to discussion of their pros and cons. This process in turn raises questions of a factual nature and questions of an … ought-to-be nature (p.321).

Rittel points out that, in the discussion of plans, questions concerning possible (empirical or experienced) consequences of plans, as well questions of their moral desirability, should be part of the debate. Rittel's proposals here offer an avenue to open the discussion on statements of 'experience' as well as on so-called 'ought' statements. Both have to be subjected to processes of argumentation. But having indicated the importance of methods aimed at facilitating argument, Rittel does not offer much detail on the role of the facilitator in the process. Apart from suggesting that part of the process of facilitating argument involves the 'stimulation of doubt' (Rittel, 1972, p.325), and that the facilitator should concentrate on controversy as measured by weight of importance and variance of judgement on issues, he offers no further suggestions.

Rittel's (1972) conceptualisation of the possibility of generating design methods to facilitate argument was taken forward by others. For example,

Mitroff and Emshoff (1979) suggested the importance of rendering the assumptions of participants transparent in order to allow them to be challenged. This can be seen as a development of Rittel's suggestions. However, it can be claimed that their tool for enriching argument still does not concentrate specifically on the *facilitator's possible involvement* in the argumentative process in order to 'raise' its quality.

In response to this lacuna, certain researchers began to concentrate on exploring the role of facilitators as involved persons in animating discussion. White and Taket (1993) make the point that we should recognise the 'interpreter'/facilitator as being involved in an interpersonal process 'that aims to "shape perception," where this is done consciously rather than inadvertently' (p.23). They argue that, while acknowledging the innovatory capacity of clients, OR practitioners may help to 'sketch out options and take part in the debate ... the views of the interpreter are valuable only if the debate is personally meaningful to all involved' (White and Taket, 1994, p.735). They suggest that in this process it is important that facilitators are 'critically reflective' about their 'role as facilitators or animateurs [animators of change]' (p.746).

White and Taket's (1993, 1994) suggestion can be seen to concur with Bleicher's (1982) argument that the involvements of researchers in social life become raised to a more self-reflexive level. Romm (1991) states the argument thus:

> Their [researchers] entry into the 'field' can never leave either themselves or their 'subjects' unchanged. Given the unavoidable involvement of the researcher [it is suggested that] this involvement becomes raised to a more 'self-reflexive' level, in which the researcher ... accepts responsibility for his or her intervention (Romm, 1991, p.170).

But in the light of the above, we are still faced with the question of how facilitators may enhance the quality of their interventions while ensuring that they do not become *unduly impositional*. White and Taket (1994) seem to offer an 'answer' by arguing that the interventions are valuable only if the debate is personally meaningful to all those involved. Are there other ways in which we may temper the problem of undue partiality in the way that facilitators 'add input' into the discussions?

Our chapter spells out another answer to the question of the fairness of intervention by arguing for a specific notion (and practice) of expressed openness to discourse. We articulate our conception of fair intervention by drawing on (and interpreting and extending) Habermas's four 'validity' checks for discursive engagement (1984, 1987), in which participants, including the facilitator as participant, are called upon to render their own and others' views open to critical confrontation. Many may associate our argument with what has been labelled a 'critical modernist' position, in that it sees the possibility of 'progress' in developing 'workable' negotiations of visions-and-actions through communicative competence (in the way in which we define it). Nevertheless, wherever our argument is located, we wish to make an input into ongoing discussions about what is (can be) involved in facilitation. As we have indicated earlier, the option of playing a role as facilitator has become a relevant feature of operational research practice—especially when it is deemed suitable to employ 'softer' approaches to address so-called 'messy' issues (cf. Jackson, 2000, p.273). We believe that our notion of facilitation offers a conception of how facilitators can offer guidance while not unfairly imposing a favoured direction on the group discussion.

8.3 *Involvement in Both the Process and Content of Discussion*

One way to begin considering our conception of facilitation as fair intervention is to discuss it in relation to the notion of process consultation developed by Schein (1987) and advanced by others (e.g. Burke, 1987; French and Bell, 1990). Schein points to process consultation as being especially relevant to situations in which the consultant—as helpful intervener—perceives that there is something to be done, but does not wish to prescribe measures or solutions to clients and instead wishes to activate their own goal-setting process. The emphasis of process consultation is that the facilitator strives to improve group dynamics or to provide guided learning to help participants gain skills of an interpersonal nature whilst transforming the pattern of communications. Process consultation is less concerned with a specific *what* (is done), it is more concerned with managing *how* things are done (Schein, 1987, p.39).

The role we are defining for a facilitator is different to that proposed by Schein (1987) who suggests 'that consultants and managers have a common problem: how to influence situations toward desired goals in the human systems in which they intervene' (p.1). Here, Schein seems to be taking for granted that organisations (and more importantly, people within organisations) have shared goals that need not be challenged by the facilitator. He specifically argues that process consultation should not address *what* is to be done, but rather *how* it is to be achieved (Schein, 1987, p.39). This introduces something of a paradox, since in focusing on the processes whereby agreed goals are implemented, process consultation actually provides an arena for the discussion of goals; i.e., it (paradoxically) recognises that goals may not be shared. Without explicitly requiring that goals be subjected to a process of review, process consultation enables discussions to occur that may include, as part of their content, the (re)consideration of goals.

The main difficulty that we see in utilising Schein's model of process consultation as it stands, is that it assumes that communicative practice can be improved without the need to challenge the bases on which certain statements are made. Yet facilitators may wish to help participants to learn how communications can be challenged in order to unearth underlying assumptions and values. Also, what if we, as facilitators, have reason to believe that a particular group of stakeholders have been 'marginalised' (Midgley, 1992, 2000)? How might we help to ensure that communication becomes grounded in legitimate expressions of viewpoints, rather than on who has the power to influence?

Our answers to these questions are guided by our understanding/interpretation and extension of Habermas (1984, 1987), who has suggested that communicative competence may be achieved through the use of four validity checks on any statements people make. In the workshops upon which we report, we set out to apply a pragmatised version of Habermas's validity checks, which required that individuals learn to use them in debate. This raised further questions, mainly concerning how we were to help individuals to gain 'communicative competence'. We see the process of gaining communicative competence as being an educative one—which involves 'education with' in the sense that participants are able to explore aspects of their context while being guided in the process of learning in order to improve competence. Gregory (1993, p.205) provides an indication of the 'doing with' approach to educational encounters; and McKay and Romm (1992), as well as Romm (1997, 2001a), offer an indication of how this may be achieved in an accountable fashion. This leads us onto a consideration of how interveners wishing to account for themselves as contributing to enriching processes of dialogue may defend their interventions as 'fair'.

We believe that an important principle of what we take to be fair intervention in group discussion is the need for facilitators to become critically aware—that is, to operate what has sometimes been called critical self-reflection (cf. Gouldner, 1973, 1975; Bernstein, 1991; Berry, 1993; White and Taket, 1993, 1994). This suggests to us that the facilitator should keep in consciousness that s/he will, as a matter of course, bring attitudes, presuppositions, values and beliefs to bear in the process of facilitation. S/he needs to bear in mind that the process of facilitation is not undertaken in some value-less vacuum, and that every intervention of a facilitator draws upon some ground. Now, what does a facilitator need to do in order to become critically aware, and to be able to act more ethically in group situations? Our suggestion is that, by subjecting his or her thoughts (unspoken observations) to the Habermasian validity checks, a facilitator may become more aware of assumptions influencing his or her responses. However, a purely monological inquiry would be insufficient and in danger of failing to reveal much of significance. Therefore, the facilitator must be *prepared to articulate these observations in order to open them to any challenges arising from within the group*, again using the four validity checks.

In addition, we contend that a critically self-reflective facilitator is in a better position (since s/he is aware of the need for discursive checking of claims) to intervene in situations to contest certain statements which have passed unchallenged by the group. *This means that the group learning may be enhanced by the interactions of the facilitator who no longer simply focuses on how the process is working, but is also able to comment on the content of the process.* A fuller discussion of the challenging of prevailing ideologies as they express themselves in group discussions can be found in Gregory (1992) and McKay and Romm (1992).

In effect, we want to suggest that the learning that takes place in any facilitated group can be gained through observing the facilitator as well as through a 'learning by doing' involvement of group members. French and Bell (1990) suggest that one use of any organisation development (OD) intervention could be as a normative re-educative process, i.e. a *process of learning which has a normative intent*. This concurs with Baburoglu and Ravn's (1992) suggestions concerning the requirements of 'normative action research' (p.30). For our part, we suggest that an awareness by the facilitator that people may learn through observation of him or her, puts a requirement on that facilitator to become critically self-reflective, to be constantly appraising his or her own motives. In order to 'set an example' of instituting validity checks to increase communicative competence, the facilitator must be careful not to intervene only at points where the discussion is failing to accord with his or her own ideas. *The normative value of 'openness' underpins the process of appropriate intervention.*

In view of our discussion of facilitation approaches in Sections 2 and 3 above, and our conception of the problems arising from adopting a (purportedly) value-neutral stance, we confirm our original decision to adopt an explicitly confrontational, interventionist stance (coupled with an orientation to critical self-reflection). Let us now turn to a more detailed discussion of the way in which we tried to instantiate our facilitation approach (however, we do not claim that in all aspects of our workshop facilitation we would necessarily be interpreted, by ourselves or others, as fully instantiating our intentions). The discussion below should be treated as offering an indication of some of the possibilities for drawing on, and drawing out, our approach.

8.4 The Aims of the Facilitated Workshops

We conducted our research into standard setting in the chosen area of the NHS by attempting to set up conditions for discussion involving six cross-organisational groups, within both peer and multi-agency group settings. Our aim was to enhance communication channels between participants involved in sectors connected with hospital services in the area. The hope was to increase our own awareness, as well as that of the (other) participants, of the ways in which the various relevant issues may be perceived and the ways in which 'negotiations' between various perceptions and options for action may be generated.

We made explicit provision for dialogue to include an awareness of the differences that exist between perceptions. The aim of the dialogue, as we saw it, was to render people more aware of the areas of their starting disagreements, so that through the discussion their understanding of one another's views could be increased, and their understanding of their own position could shift as they encountered the challenges levelled by others. We believe that such shifts are necessary in order that negotiated options for action can be generated. This also means that, when any of the participants engage in any form of 'action', they bear in mind the viewpoints of others and act (accountably) in the light of this (Romm, 1994, explicitly explores this notion in relation to Ulrich's, 1983, critical systemic argument).

The facilitator's task on each occasion was to contribute to what we conceived to be a critical appreciation of the ways in which statements could be challenged or reviewed through the introduction and discussion of alternative perspectives on the various statements. See Gregory (1992, 2000) for her account of what is involved in critical appreciation.

As indicated earlier, we were aware of and informed by (while also modifying and extending) Habermas's notion of discursive encounter. Habermas (1984) argues that what he calls 'communicative rationality' involves an 'intersubjective relation that speaking and acting subjects take up when they come to an understanding with one another about something' (p.392). We drew on this argument of Habermas, but extended/modified his position by not insisting or even implying that the aim in discussion is to arrive at a consensus based on the 'force of the better argument' (Habermas, 1987, p.145). This meant that in 'debriefing' our 'subjects' (participants in the groups) we indicated that at any point there may continue to be disagreement, which should not be taken as a sign of failed communication. We indicated that all we expected was that speakers (and facilitators who could also 'speak') would be prepared to challenge one another's views with reference to the following four questions which could be used as an aid to the discourse:

- Do you understand what is being said?
- Is the speaker being sincere?
- Is the speaker's point acceptable to you?
- Do you agree with the speaker's use of information and/or experiences?

We indicated that if anyone felt their answer was 'no' to any of the questions, they should ask for clarification.

It should be borne in mind that by phrasing the questions in this way, we wanted to avoid the impression that finally there was a 'right', 'acceptable', 'true' way of seeing things that could somehow be 'reached' through discussion. We were at pains to indicate (through practice) that increased mutual insight could be gained from participating in a dialogue where such questions were asked, and that this (rather than some definite consensus) was the point of the dialogue. We were aware that we were using a modified version of Habermas's argument (1984, 1987) concerning the inter-subjective character of discussion and negotiation; and this was at the back of our minds as the theoretical framework for our involvement with the groups. Our design of the four questions and our debriefings at the start of each of the various sessions were meant to pragmatise (in modified form) the theoretical argument of Habermas.

Besides drawing on Habermas, we were also guided by authors emphasising that the point of systemic inquiry in the social world is not to 'find' data, but rather to generate greater 'comprehensiveness' in the ongoing search for understanding. Churchman (1979, p.116) proposes that comprehensiveness can be increased by enhancing the potential for opposition in the arena of debate. Our view that participants (including the researcher-facilitator as participant) should provide opposition to one another's viewpoints, was a concretisation of Churchman's suggestion in this respect.

Furthermore, our facilitation approach drew on Gouldner's account of what is involved in reflexive sociological research. Gouldner (1975) argues that: 'Fundamentally the task of the reflexive theorist is to help people to remain critical and sceptical of *good* news ... and correspondingly to help people to accept bad news and to *remember* it' (p.10). Gouldner's proposal ties in with the requirement to subject ideological language and thought to processes of critique: styles of thinking which seem to occlude invitations for critique and which incorporate a refusal to encounter 'bad news' for the position, may be subjected to ideology-critique. See Romm (1990) for an account of some epistemological implications of this argument.

Nevertheless, the question still arises as to how one can mitigate against researchers becoming unduly directive by contesting particular ideas and hence leading the discussion in particular directions. Put differently, how can one ensure that the researcher's own attitude remains sufficiently open, so that the direction of the discussion is not overly dictated by his or her particular participation? This is the subject of the next section.

8.5 Fair Intervention

In our view, facilitators trying to make a contribution in aiding group discussion should not attempt to extricate their own 'personhood' from the process. Rather, their fairness (or what might be conceived as a kind of 'objectivity') should be defined in terms of their expression of a commitment to aiding the highest possible level of discursive checking of claims (as made by others as well as by him/herself). To become 'fair' in this sense, s/he has to become a participant in the process of group discussion. S/he thus becomes—if only temporarily—a 'member' of the group. However we contend that, although there may be dangers in assuming the role of active participant, these can be tempered, as long as the researcher is aware of them. We wish to point out that even when facilitators attempt to remain 'apart' from the group, they cannot (legitimately) avoid decisions such as when to intervene to, for instance, protect minority viewpoints in the group against positions which appear to be too powerfully expressed. In other words, facilitators still need to make certain decisions regarding whether a particular position seems to 'need support' in the face of powerful/persuasive participants smothering the input of others. Whether (if at all) facilitators act to allow for viewpoints to be expressed, and to help their expression, already involves some form of 'direction'; even *not* acting implies that one allows the discussion to proceed in a particular direction. As Hanson (2001, p.548) notes, supposed 'non-action' can have consequences for how systems develop, and therefore itself may need to be accounted for.

Our argument is that it is preferable to recognise that one has to take decisions at every point in a facilitation process, coupled with an awareness of the problem of 'bias'/undue partiality on the part of the facilitator. This means (following Gouldner, 1975) that one is aware of the need to be open to an acceptance of 'bad news' for one's own position. This awareness is the best 'guarantee' that we have for facilitator fairness. What does this amount to? It means that the researcher has to adopt a stance of continued critical self-reflection in regard to the way in which s/he is directing the process. S/he has to be aware that s/he should not 'intervene' only to challenge statements that fail to accord with his or her own initial position. The facilitator himself/herself should be sufficiently open to recognise that his/her own position itself may be challenged. Participants should be advised that if they feel the facilitator always seems to challenge one particular type of viewpoint, they may bring this up as part of the effort to enhance the dialogical character of the group discussion. Of course, the facilitator also has to be conscious of the possibility that it may be powerful/persuasive group members who reject the 'intervention' of the researcher and hence are critical of it! Bearing this in mind, though, the self-reflective facilitator will take criticisms as an opportunity to re-consider the way in which s/he is conducting the facilitation.

There is, however, another point worth considering. It may not only be persuasive participants in the group who are particularly resistant to explicit intervention on the part of the facilitator. Any person may be resistant to such intervention when it is explicitly confrontational: when it confronts people with the 'bad news' of their favoured position. This means that the facilitator may be faced with the task of gaining 'rapport' with a group of people, some of whom are resistant to the style of facilitation which is being adopted. The facilitator has to take care to continue with the process, while not alienating people to the extent that they fail to trust it. It is of course hoped that through experiencing the possibility of hearing 'bad news' (as spoken by other participants and by the facilitator), having the opportunity to take part in the dialogue, and experiencing the enrichment that may spring from the process, participants will begin to trust the facilitation style. If they (or some of them) still show a measure of distrust, the facilitator has to take this into account. The problem is that s/he cannot then try to adopt a stance of seemingly 'neutral' outsider, because this would be at odds with the requirement to enhance dialogical capacity. So the tightrope walk between interventionist confrontation, and the development of a process of trust building within the group (with the facilitator being part of the group), cannot be side-stepped.

We present below a number of examples of the ways in which we intervened to challenge the statements made by some participants, in terms of the four questions mentioned above. It must be remembered that the point of our intervention was, firstly, to provide a concrete indication of what we meant by challenging statements with reference to the four questions, and, secondly, to evoke responses so that the discussion could proceed on a 'higher' (more enriched) level. Other discussions of these examples are provided in Gregory and Romm (1996, 2001).

8.5.1 *Some Examples of Facilitator Interjection*

In this section, we indicate four examples of occasions in which the facilitator injected an alternative position to the one that had been espoused by a participant:

1. Related to the question 'Do you understand what is being said?', the facilitator commented that she did not understand when a speaker indicated that there were gaps in the current way in which the system is designed to serve patients. This led to further discussion about the extent of the gaps—a discussion that highlighted a disagreement between participants concerning the way in which the introduction of reforms was impacting on patients. A discussion on ways in which patients may or may not be able to 'vote with their feet' also ensued, and this issue became a theme in subsequent peer and multi-agency meetings.

2. Related to the question 'Is the speaker being sincere?', the facilitator implicitly challenged a speaker's sincerity in regard to a statement that nursing staff can take a part in negotiating the contract with purchasers. The facilitator asked 'Do you really think so?', urging the speaker (and others) to reconsider the statement. It appeared that the speaker wanted to upgrade the status of nurses, and therefore portrayed them as impacting on the provider-purchaser contracts. On reconsideration the speaker suggested that she did not really think that at present the impact was significant. This, however, opened the door to discussions in subsequent meetings about the way the contracts were set up and the criteria for developing standards.

3. Related to the question, 'Is the speaker's point acceptable to you?', the facilitator interjected when a speaker stated (with disfavour) that nurses sit around having coffee in front of patients. The facilitator intervened by expanding on a point that had been made earlier by another participant, concerning the notion that nurses were often under pressure in their work. The facilitator found the complaint about coffee drinking 'unacceptable' in the way in which it had been put, and asked whether taking a coffee break could perhaps be seen as a response to pressure. This intervention roused the speaker to comment that the facilitator was overstepping her role by 'taking sides'. The facilitator responded that she was offering an alternative viewpoint, and allowing the issue to be seen from a different perspective. She also indicated that the speaker could still disagree in response, as part of continued discussion on the issue. From the facilitator's perspective the intervention in this instance was meant to help the expression of a sympathetic viewpoint towards nursing staff so that the issue of pressure, which had been raised earlier, could be highlighted and subjected to further consideration. In subsequent meetings this issue was further addressed and the task of management in dealing with staff workloads was explored.

4. Related to the question 'Do you agree with the speaker's use of information and/or experiences?', the facilitator commented on a speaker's statement that there were insufficient resources in her ward to introduce a buzzer system which would enable patients to call if they needed to go to the toilet. The facilitator intervened by suggesting that it may be possible to re-investigate the budgetary implications: though it may be a short-term cost, in the long run it may save money if one takes into consideration laundry costs, etc. Here the facilitator intervened by attempting to cast a different light on the experience of the 'insufficiency of resources'. The intervention also put onto the agenda the issue of control of the ward budget—something that was to be further explored in later meetings. Specifically, we regarded it as significant that some of the nurses, who initially had been reluctant to regard themselves as proficient in budgetary concerns, began to consider that they may be able to contribute—especially if their contribution could be regarded as increasing over time, rather than being based on an expectation of immediate full participation.

To summarise, we wish to reiterate that our facilitation inputs (of which the above were some examples) were intended to aid the discussion of various issues and to raise issues which otherwise may not have been put on the agenda. We accept that the way in which the facilitator intervenes will reflect some of his or her own beliefs about how to 'balance' a discussion so as to render it 'enriched'. In this sense, our mode of facilitation does not exclude facilitator-directedness. But it does hope to temper undue partiality (taking a specific side that happens to coincide with the initial viewpoint of the facilitator) by requiring the facilitator to reflect critically on his or her mode of intervening and to be aware of the possible impact of the intervention. The tempering of partiality does not lie in pretended non-involvement, but rather in a continued self-reflection to ensure (as far as possible) that the researcher's inputs do not spring from his or her wish to cling to his or her own existing perspective. The input and involvement in the discussion should be made in an effort to prevent others (including oneself) from clinging in this way. The aim of the facilitator in the process is to encourage people to consider options for thinking-and-acting which exceed the boundaries of their initial perspectives.

8.6 Conclusion

The discussion in this chapter has focused on exploring the justification for facilitators to intervene on questions relating to the truth, sincerity, intelligibility and moral acceptability of statements made in communities, in order to enhance their discursive character. We have pointed to a context in which we, as facilitators, were involved in exploring options for the redesign of NHS approaches to standard setting. We created the space for participants to develop their individual communicative skills, both through using the validity-claims and by observing their use in practice by a facilitator. In this way we believe we were able to work with the participants in aiding them to construct and structure their visions of possibilities through a dialogical process. We did not present ourselves as being 'experts' in making problems amenable to solution via the application of quantitative techniques. Instead we posed as being fairly adept in working with people to enhance the potential to develop a dialogical process to examine perceived-to-be problematic issues (also see Jones and Eden, 1981, for a discussion of this kind of orientation in Community OR).

In the project our agenda was informed by the possibility of suggesting directly to participants that they could (and should) raise questions in relation to the four (Habermasian-oriented) validity checks. We wanted to instil in participants a sense that any participant had the opportunity and the know-how to both contribute their own views and interrogate others' views, as the basis for discursive encounter. We also wanted to instil a sense that, in all areas of validity checking, there is no 'expertise' to which people may resort. No matter what supposedly 'expert' opinions or experiences may be brought to bear, participants all have their own experiences/ideas/value-beliefs that can be engaged as part of the discussion.

This is not to say that in principle we preclude the 'expertise' of those who wish to bring quantitative modelling tools to bear on the discussion in whatever contexts they are working. But we concur with Lane (1991, pp.66–69) that 'hard' modelling, as presented by any model-creation 'expert', is inappropriate to the extent that it is presented as a non-negotiable claim about the relevant variables, or indeed about the process of relevant modelling. We believe that it is both possible and desirable for operational researchers using hard modelling (or allowing others to use it) to attempt to create the space to define (and continually redefine) its 'validity' through discursive checking, in which concerned stakeholders can bring their views to bear on the construction and relevance of the model. Because of our approach to quantified information as itself requiring openness to dialogue (where others can challenge it on various grounds), we can ultimately be seen as oriented towards what could be described as a '*critical*' OR practice (also see Mingers, 1992). This approach allows facilitators, *inter alia*, to temper what they take to be asymmetries in a debate by offering a legitimate forum for the challenging of normally 'dominant' conceptions. But it also requires facilitators to try to be fair in the ways in which they add their own input into the debate and, in our terms, to be specifically wary of the temptation to direct the process primarily in accordance with their initial standpoint. They are required to critically reflect on the possible impact that pre-favoured stances might have on the course of the discussion.

It remains to consider the question of how we regard the status of the approach to facilitation (as fair intervention) that we have outlined in this chapter. Our answer is that we see our argument as an input into the discussion on facilitation. Like all inputs, its relevance for OR practitioners can be evaluated only as part of an ongoing conversation. We do not hope to have finalised the discussion on facilitation. We have concentrated in this chapter on sharing our reflections on what it might mean to employ an approach to facilitation that does not shy away from offering points of 'content', with the intent to enrich processes of debate in group contexts. We have offered some guidelines that may be used by practitioners who in turn will make sense of our approach in terms of their own judgements as to its possibilities.

References

Ackoff, R.L. (1979a). The future of operational research is past. *Journal of the Operational Research Society*, **30**, 93–104.

Ackoff, R.L. (1979b). Resurrecting the future of operational research. *Journal of the Operational Research Society*, **30**, 189–199.

Baburoglu, O.N. and Ravn, I. (1992). Normative action research. *Organization Studies*, **13**, 19–34.

Bernstein, R.J. (1991). *The New Constellation: The Ethical-political Horizons of Modernity/Postmodernity*. Polity Press, Cambridge.

Berry, M. (1993). Changing perspectives on facilitation skills development. *Journal of European Industrial Training*, **17**, 23–32.

Bleicher, J. (1982). *The Hermeneutic Imagination: Outline of a Positive Critique of Scientism and Sociology*. Routledge & Kegan Paul, London.

Burke, W.W. (1987). *Organization Development: A Normative View*. Addison-Wesley, Reading, MA.

Chensult, J. (1989). Training for changing consultant roles. *Journal of Management Consulting* **5**, 48–55.

Churchman, C.W. (1979). *The Systems Approach and Its Enemies*. Basic Books, New York.

French, W.L. and Bell, C.H. Jr. (1990). *Organization Development: Behavioral Science Interventions for Organization Improvement*, 4th edn. Prentice Hall, Englewood Cliffs, New Jersey.

Gouldner, A.W. (1973). *The Coming Crisis of Western Sociology*. Basic Books, New York.

Gouldner, A.W. (1975). *The Dark Side of the Dialectic: Toward a New Objectivity*. Sociology Institute of Amsterdam.

Gregory, W.J. (1992). *Critical Systems Thinking and Pluralism: A New Constellation*. PhD Thesis, The City University.

Gregory, W.J. (1993). Designing educational systems: A critical systems approach. *Systems Practice*, **6**, 199–210.
Gregory, W.J. (1996). Dealing with diversity. In: *Critical Systems Thinking: Current Research and Practice*, pp. 37–61. Flood, R.L. and Romm, N.R.A. (eds.). Plenum, New York.
Gregory, W.J. (2000). Transforming self and society: A 'critical appreciation' model. *Systemic Practice and Action Research*, **13**, 475–501.
Gregory, W.J. and Romm, N.R.A. (1996). Towards multi-agency dialogue: Facilitation as fair education. In: *Praxiology*, vol. 4. pp. 323–351. Gasparski, W.W., Mlicki, M.K., and Banathy, B.H. (eds.). Transaction, New Brunswick.
Gregory, W.J. and Romm, N.R.A. (2001). Critical facilitation: Learning through intervention in group processes. *Management Learning*, **32**, 453–467.
Gregory, W.J. and Walsh, M.P. (1993). Quality, ideology, and consumer choice: Health care standards and stakeholder participation. In: *Managerial Issues in the Reformed NHS*, pp. 169–181. Malek, M., Vacani, P., Pasquinha, J., and Davey, P. (eds.). John Wiley & Sons Ltd, Chichester.
Habermas, J. (1984). *The Theory of Communicative Action*, vol. 1. Beacon Press, Boston.
Habermas, J. (1987). *The Theory of Communicative Action*, vol. 2. Beacon Press, Boston.
Hanson, B. (2001). Systems theory and the spirit of feminism. *Systems Research and Behavioral Science*, **18**, 545–566.
Jackson, M.C. (2000). *Systems Approaches to Management*. Kluwer Academic/Plenum Publishers, New York.
Jones, S. and Eden, C. (1981). OR in the community. *Journal of the Operational Research Society*, **32**, 335–345.
Lane, D.C. (1991). Modelling as learning: A consultancy methodology for enhancing learning in management teams. *European Journal of Operational Research*, **59**, 64–84.
McKay, V.I. and Romm, N.R.A. (1992). *People's Education in Theoretical Perspective*. Maskew Miller Longman, Cape Town.
Midgley, G. (1992). The sacred and the profane in critical systems thinking. *Systems Practice*, **5**, 5–16.
Midgley, G. (2000). *Systemic Intervention: Philosophy, Methodology, and Practice*. Kluwer Academic/Plenum Publishers, New York.
Midgley, G. and Ochoa-Arias, A.E. (1999). Visions of community OR. *Omega*, **27**, 259–274.
Mingers, J.C. (1992). Recent developments in critical management science. *Journal of the Operational Research Society*, **43**, 1–10.
Mitroff, I.I. and Emshoff, J.R. (1979). On strategic assumption-making: A dialectical approach to policy and planning. *Academy of Management Review*, **4**, 1–12.
Parry, R. and Mingers, J. (1991). Community operational research: Its context and future. *Omega*, **19**, 577–586.
Pasmore, W.A. (1993). Designing work systems for knowledge workers. *Journal for Quality and Participation*, **16**, 78–84.
Rittel, H.W.J. (1972). Second-generation design methods. Interview by Grant, D.P. and Protzen J.-P. The Design Methods Group 5th Anniversary Report: DMG Occasional Paper No. 1, 5-10.
Romm, N.R.A. (1990). Gouldner's reflexive methodological approach. In: *Sociology and Society*, pp. 13–22. Alant, C.J. (ed.). Southern Book Publishers, Johannesburg.
Romm, N.R.A. (1991). *The Methodologies of Positivism and Marxism*. Macmillan, London.
Romm, N.R.A. (1994). Continuing tensions between Soft Systems Methodology and Critical Systems Heuristics. Research Memorandum 5. Centre for Systems Studies, University of Hull.

Romm, N.R.A. (1996a). Inquiry-and-intervention in systems planning: Probing methodological rationalities. *World Futures*, **47**, 25–36.

Romm, N.R.A. (1996b). Systems methodologies and intervention: The issue of researcher responsibility. In: *Critical Systems Thinking: Current Research and Practice*, pp. 179–193. Flood, R.L. and Romm, N.R.A. (eds.). Plenum, New York.

Romm, N.R.A. (1997). Becoming more accountable: A comment on Hammersley and Gomm. *Sociological Research Online, 2,* http://www.socresonline.org.uk/socresonline/2/3/2.html.

Romm, N.R.A. (2001a). *Accountability in Social Research: Issues and Debates*. Kluwer Academic/Plenum Publishers, New York.

Romm, N.R.A. (2001b). Organising a dialogical intervention strategy for development. In: *Development: Theory, Policy and Practice*, pp. 511–523. Coetzee, J.K., Graaff, J., Hendricks, F. and Wood, G. (eds.). Oxford University Press, Cape Town.

Schein, E.H. (1987). *Process Consultation: Lessons for Managers and Consultants*, vol. II. Addison-Wesley OD Series. Addison-Wesley, Reading, MA.

Ulrich, W. (1983). *Critical Heuristics of Social Planning*. Paul Haupt, Berne.

White, L.A. and Taket, A.R. (1993). Community OR—doing what feels good. *OR Insight*, **6**, 20–23.

White, L.A. and Taket, A.R. (1994). The death of the expert. *Journal of the Operational Research Society*, **45**, 733–748.

9

Perspectives of Engagement in Community Operational Research

John Friend

9.1 Dilemmas of OR and Action Research

The very phrase *operational research* implies a delicate balancing act. For it raises the question of how far a stance of scientific detachment can realistically be reconciled with a stance of direct *engagement* of the researchers with other people—most of whom would be more inclined to describe themselves as managers than researchers—as they go about their working lives. For these people will usually expect to gain something from whatever investment they may make in their engagement with the researchers—whether that investment takes the form of financial outlay, or sacrifice of time, or simply the granting of privileged access to their places of work.

A similar form of balancing act is implied by the concept of *action research* as developed by social scientists over a similar historical period (Lewin, 1946; Clark, 1976; Cunningham, 1993). Meanwhile, some of the more reflective practitioners of operational research (Ackoff, 1970; Boothroyd, 1978; Ormerod, 2002) and of systems science (Flood and Jackson, 1991) have developed conceptual frameworks of their own designed to guide working relationships between researchers and those in whose worlds their research projects are grounded.

Accepted for publication July 2002.

John Friend • Stradspan Limited, 17 Birks Wood Drive, Oughtibridge, Sheffield S35 0HY, UK.
Community Operational Research: OR and Systems Thinking for Community Development, edited by G. Midgley and A.E. Ochoa-Arias, Kluwer Academic/Plenum Publishers, New York, 2004

For many years, models of engagement between the researcher and the 'researched' drew their inspiration primarily from experiences of the world of the large manufacturing or mining organisation, as this was the world that provided the setting for many of the classical accounts of both action research and OR projects. In this context, relationships were typically expressed in terms of engagement between the individual researcher—or consultant or change agent—and his or her *client*. Gradually, however, the term *client system* has gained in favour, because of a growing recognition that the client role will often extend beyond the responsibilities of any one person—especially in the case of any relatively ambitious project conducted within a large organisation with a differentiated management structure.

9.2 Engaging in Community Development Projects

Since these concepts started to gain currency, researchers have been moving into other less sheltered organisational contexts that can be considered more *pluralistic*, in so far as there may be multiple sources of authority that are not related to each other in terms of any defined management hierarchy. One such context is that of *community development*, which can be broadly described as involving work with people in the communities in which they live in order to help them to enhance their communal resources to cope with the problems and opportunities around them.

The field of community development is one that presents especially complex challenges to the researcher or consultant. It is also a field that has been steadily attracting increasing programmatic support in both developed and developing countries, with significant flows of resources from the developed nations of the north towards the less wealthy countries to their south.

In general, the field of community development is one in which opportunities to design and conduct any substantial research project tend to depend on the provision of resources by external *sponsors* who are organisationally detached from the specific local milieu in which the work is to be done. Any such external sponsors may be governmental, charitable or commercial; indeed there are many instances where their structure of accountability takes on a more complex hybrid form.

Characteristically, the distinctive interest of such an external sponsor in investing in a community development project is typically in the broader lessons that can be drawn and transferred to other comparable contexts. Indeed the presence of this extra *sponsor* role, in addition to the more direct project *agent* and *host* roles, is often the key feature that turns a self-contained consulting project into a wider action research project.

For now the concern to generate wider learning outcomes becomes added to the immediate concern to generate action and learning outcomes for the primary project hosts.

9.3 The Roots of the Concept of Negotiated Project Engagement

The roots of the concept of *negotiated project engagement* differ significantly from those of other comparable models of engagement such as the *systems intervention* model as articulated by leading systems scientists (e.g., Flood and Jackson, 1991).

Its immediate roots can be found in the work of operational research scientists working alongside social scientists in the Tavistock Institute of Human Relations from the mid-sixties onwards[1] (Friend *et al*, 1988). The key phrase *project engagement* emerged from the work of a Tavistock team which, in 1980, was awarded a small grant by the Social Science Research Council[2] for a six-month study to review the experience of a cross-section of completed Tavistock projects involving field collaboration between OR and social scientists. This short study became known as the OROC project[3] (Friend *et al*, 1998). Its primary aim was to generate insights into factors contributing to success in bringing together the approaches of OR and the social sciences in field research engagements.

The OROC project could not itself be described as an action research project; for there were no direct hosts, despite the existence of a clear external sponsor. Yet the insights generated through the project have been further developed since that time, drawing on the subsequent experiences of the members of the OROC team in further action research projects in several parts of the world.

One of the most salient differences between the project engagement perspective and those perspectives based on concepts of systems intervention lies in the lack of any initial assumption that the field of

[1] The critical event was the formation in 1963 of a new *Institute for Operational Research* (IOR) within the organisational framework of the Tavistock Institute, as a result of discussions between the Councils of the Tavistock Institute and the UK Operational Research Society, both voluntary associations with charitable status.

[2] Since renamed the Economic and Social Research Council, this is an official agency within the UK government's structure of research councils through which grants for different forms of scientific research are awarded to universities and independent research institutions.

[3] Standing for Operational Research and Organisational Change.

enquiry can be viewed as a coherent *organisation* to which at least some systemic properties can be ascribed. The implications of this important difference will be developed further in later sections.

9.4 The Multi-Organisational Focus

The earliest project experiences of IOR were set in such disparate fields as the construction industry, local government and hospital management. Some of the most significant insights from these early projects were presented by Stringer in a seminal paper in which he set out to articulate his views on the general challenges of 'OR for Multi-organisations' (Stringer, 1967). He coined the term 'multi-organisation' to describe a (possibly ad hoc) *union of parts of several organisations, each part being a subset of the interests of its own organisation*. He saw a multi-organisation as *defined by the performance of a particular task (which may be a continuing one) through the interaction between individuals*.

The ideas proposed by Stringer were further developed a few years later in the course of an IOR research project, which set out to explore the inter-organisational relations that surrounded the planned expansion of the small town of Droitwich in the English Midlands[4] (Friend *et al*, 1974). The impetus for this expansion came from a government strategy to disperse population and employment from the West Midlands conurbation centred on Birmingham, and the chosen agency was a partnership structure bringing together two elected Councils representing very different levels of local government. The research team arrived in Droitwich about half way through the course of the 15-year programme of planned development. By that time, relations between the two partners, and also with other agents of development, were in some respects well developed, yet in other respects still fluid and indeed politically unstable and problematic.

The researchers introduced a set of quite simple graphical conventions through which to map the changing patterns of inter-organisational arrangements that developed through the interactions of individuals engaged in different aspects of their development task. One of the most crucial distinctions that the research team introduced in this mapping process was between relatively durable *policy systems* and more transient *decision networks* of a part formal, part informal nature.

Oven the ten years after the conclusion of this project, these ideas were extended through a succession of further projects conducted by IOR

[4] This project was supported by a three-year grant from the Social Science Research Council—see footnote 2.

staff to explore inter-agency relations in several domains of public policy, conducted under the auspices of government departments, research councils and other sponsoring agencies. Then later projects in the UK and other countries were to modify this view still further, towards a quite fluid view of planning and negotiation through shifting configurations of relationships in which few, if any, systemic features could sometimes be identified.

At this stage, a general hypothesis will be introduced, that is supported by many recent experiences of the author and those with whom he has collaborated in successive action research projects (Friend, 2001). This can be stated as follows:

> **Hypothesis A:** The more ambitious an action research project, the more likely it becomes that the configurations of relationships to be explored will be strongly influenced by transient or adventitious linkages that cannot realistically be understood in terms of systemic perspectives of the management of complexity.

9.5 *The Distinction Between Agent, Host and Sponsor Roles in Projects*

It has already been suggested that a tripartite distinction between *agent*, *host* and *sponsor* roles can offer a more helpful starting point in exploring relationships within action research projects than a more conventional dichotomy between consultants or researchers and their clients. This tripartite distinction is illustrated schematically in Figure 9.1.

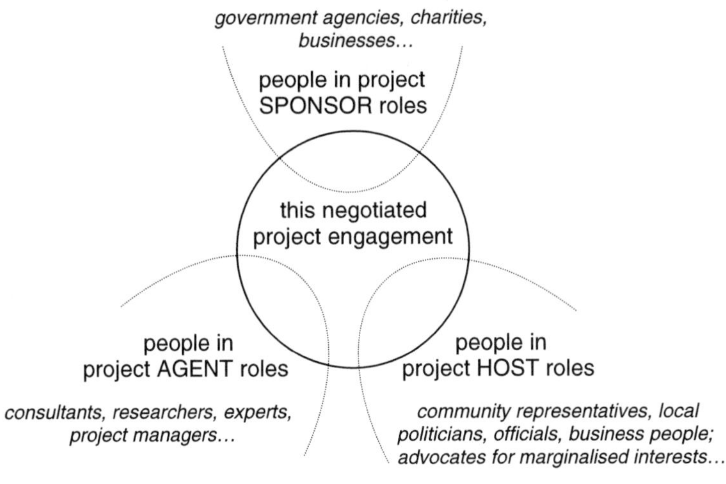

FIGURE 9.1 Types of role in action research projects with typical examples of stakeholders in community development projects

The essential transience of relationships between project hosts and agents of intervention is recognised by advocates of some of the more flexible and problem-focused forms of systems modelling such as Checkland (1981, 2001) or Flood and Jackson (1991). Yet, in the case of many of the more challenging action research projects reviewed in the course of the OROC project, the case became apparent for exploring elements of transient complexity impinging on those engaging in host roles. Indeed, there can be similar elements of transient complexity impinging on those in agent and sponsor roles, as indicated below:

- The configurations of relations among people engaging in a project in **Host** roles may be governed by transient as well as systemic influences. Such forms of *adventitious* complexity are especially likely to be found in fields such as that of community development where responsibilities for significant areas of decision-making may not only be diffuse but also *inchoate* or unformed;
- The configurations of relations among people engaging in **Agent** roles may also exhibit transient as well as systemic characteristics. Such complexity is often evident in ambitious projects guided by aspirations towards introducing an *interdisciplinary* approach. Often, this leads to the formation of an ad hoc project team, bringing together individuals who may not have worked with each other before and who in some cases indeed owe allegiances to different employers;
- The configurations of relations among people engaging in a project in **Sponsor** roles may also sometimes exhibit transient as well as systemic characteristics. This applies in particular to community development projects in developing countries with partial sponsorship from transnational development banks.

Stringer (1967) was among the first to note the conceptual challenges of this kind of multi-organisational sponsorship structure, in discussing the experience of the IOR building communications project. This was sponsored by a consortium of national professional bodies that turned out to be less stable than had at first been assumed.

9.6 *A Comparison of Two Community OR Projects*

At this point, it will be helpful to illustrate these points about the influence of adventitious as well as systemic complexity by drawing on two community operational research projects within the author's own experience. One of these projects concerned community health in inner London

***TABLE 9.1** Comparison of the Tower Hamlets and Thurnscoe Projects: Part I Project Context?*

Tower Hamlets health strategy 1987	PROJECT	Thurnscoe Tenants Housing Co-operative 1988
John Friend (JKF) as consultant working with a professor at a London School of OR	AGENTS?	JKF as consultant working with a senior associate from the OR department at Sheffeld City Polytechnic[5]
Three local community health groups concerned about local health services in this deprived inner London borough with a multicultural population of c. 180k	HOSTS?	A tenants' co-operative recently formed to manage a stock if about 300 houses acquired from the National Coal Board in a South Yorkshire mining village
Consulting fees contributed from research funds already allocated to hosts by a leading London health charity	SPONSORS?	Consulting fees met from OR Society seed funding to Community OR Unit
Shared concern at gradual erosion of local health services, heightened by imminent plan to close an accident and emergency unit	PROJECT IMPETUS?	Recent purchase of estate by tenants, with mortgage funds and professional management support, created need for campaigners to become managers
Local health planner was a recent part-time student at the OR school; professor already familiar with consultant's working methods	PRIOR LINKS?	Director of CORU had worked with Thurnscoe Co-op through education programmes at Northern College, and was aware of JKF's role in Tower Hamlets

and the other concerned co-operative housing management in the South Yorkshire coalfield. Both engagements were quite short, and were carried out in the late 1980s, when the Council of the UK Operational Research Society was in the process of launching a national initiative to expand the use of OR in support of local community groups (Rosenhead, 1986).

In Table 9.1, the two projects are compared in terms of their contexts. After reviewing the participants in agent, host and sponsor roles, the Table summarises the impetus behind each project and also the extent of prior links among the people and institutions involved. Tables 9.2 and 9.3 to follow will then review key dimensions of the *course* of each project and its *aftermath*. In each case, there were to be further engagements later, either

[5] Now Sheffield Hallam University.

with or without the involvement of the author; the analysis here refers to the engagement of the author as consultant as it was first negotiated.

9.7 *Similarities of Context*

Before commenting further on the differences between the two projects, it is worth mentioning that there also were significant similarities. In each case, the author was approached to act as a consultant by an acquaintance who was active in the early stages of the community OR movement in the UK. In the first case the approach came from Jonathan Rosenhead, Professor of OR at the London School of Economics and recently President of the OR Society. In the second case it came from Colin Thunhurst, recently appointed as the first Director of the Community OR Unit at Northern College in South Yorkshire.

It is significant that both invitations came at a stage of the author's career when he had recently left the Tavistock Institute to work as a free-lance consultant. This meant that it was not easy for him to work without a consulting fee; yet he was no longer bound by institutional expectations as to what minimum level of fee was acceptable. So both engagements involved contracts for him to spend a few days working at what most consultants would regard as a sub-commercial rate. There is a significant contrast here with the many community OR projects carried out primarily by university staff or students, where payment for working time may be less essential (Ritchie *et al*, 1994; Bowen, 1995).

In each case, there was a clear expectation that the main task of the consultant would be to facilitate one or more action planning workshops using the *strategic choice approach*. This is an interactive and relatively 'soft' OR approach that he had previously developed alongside colleagues in the Tavistock Institute (Friend and Hickling, 1997). Typically, such a workshop is concentrated within a single day, though the terms of engagement typically allow also for one or more informal meetings before that day when expectations can be negotiated on procedural matters, on the desired outcomes, and on the size and composition of the workshop group. In each case, it was agreed that a senior member of an academic OR group based in the host region should assist the author as co-facilitator of the workshop.

9.8 *Differences of Context*

As Table 9.1 indicates, there were more striking differences between the projects in the host roles than in the agent roles. Whereas the Thurnscoe

Tenants Housing Co-operative could be seen as a relatively cohesive voluntary organisation, with a single management committee, there were three different host organisations in the case of the Tower Hamlets strategy project. One was the Tower Hamlets Community Health Council, a statutory body with responsibility to represent local community interests to the agencies responsible for service provision. Another was the Tower Hamlets Health Campaign, with representation of local health service workers and a long history of campaigning against proposed reductions in levels of service. The third was the Tower Hamlets Health Strategy Group, recently formed to follow through the recommendations of an independently commissioned strategy study, which had recently mounted a critique of official plans for the area.

Both projects had an element of external sponsorship, which was quite modest compared to more ambitious and longer term community development projects, but which played a significant role in enabling consulting fees to be offered to the author. In the Tower Hamlets case, up to £1,000 was offered from an existing research budget allocated by the Kings Fund charity for health services in London. In the Thurnscoe case, a budget of up to £3,000 was offered though the Operational Research Society's pump-priming grant to the Community Operational Research Unit. In that case, the Unit's host institution, Northern College, already had well-developed prior links with the Thurnscoe tenants through established adult education programmes. In the Tower Hamlets case, the links that led to the project engagement were of a more adventitious nature.

9.9 *Comparisons of the Course of Each Project*

An enquiring framework that was first presented in the report of IOR's OROC project (Friend *et al*, 1998) will now be adapted as a means of comparing key aspects of the *course* followed by each project in practice, as shown in Table 9.2. The form of the table draws on some concepts introduced by a Tavistock colleague, Peter Spink, who played an advisory role to the OROC project team. Spink suggested describing the course of any project in terms of three complementary aspects: the mode of *enquiry*; the *relations* among the participants; and the strategy for *producing* desired outcomes.

A common feature of both project engagements was the agreement to hold an early strategic choice workshop as a means of helping the community hosts to deal collectively with concerns that were high on their current action agendas. In the case of Tower Hamlets, this represented a

TABLE 9.2 Comparison of the Tower Hamlets and Thurnscoe Projects: Part II Course of Project?

Tower Hamlets health strategy 1987	PROJECT	Thurnscoe Tenants Housing Co-operative 1988
Exploratory meeting led to agreement for JKF to facilitate an early action planning workshop, with follow up options open	TERMS?	Three month contract agreed to provide flexible workshop-based support to tenants in their management decisions
Interactive approach with focus on half-day workshop session to build a shared picture of issues and action strategy	ENQUIRY?	One or more strategic choice workshop sessions to explore key management issues, plus other advice as required
Consultant, with Professor as co-facilitator, working with a single group of about a dozen representatives of the three host organisations	RELATIONS?	Consultant, with Sheffield associate as co-facilitator, working interactively with available members of tenants' co-operative as a single working group
Progress during workshop in building strategy recorded on flipcharts; content recorded in photos, short report issued with progress summary chart	PRODUCING?	Progress towards decisions on house maintenance policy recorded on flipcharts, photo record used as basis for brief report back to hosts afterwards
Between initial planning session and workshop date, imminent horizon for official commitment to closure of A&E Unit created a sharper workshop focus	CHANGES OF COURSE?	Recent purchase of estate by tenants, with mortgage funds and professional management support, created need for campaigners to become managers
(Friend 1994; Rosenhead 1996) *Also workshop notes and photographs*	INFORMATION SOURCES?	(Thunhurst *et al*, 1992; Thunhurst and Ritchie, 1992) *Also workshop notes and photographs*

change from the terms of the original invitation to the professor of OR; for this had been merely to propose an alternative planning *approach* to that used by the Health Authority, based on a perspective of local community need. His analysis would then be offered as a critique of the logic of the Health Authority's present proposals. It was only during the first exploratory meeting between hosts and agents that the suggestion of a more interactive workshop approach was proposed, and readily approved.

In each case, it was encouraging that relations between agents and hosts developed in an informal yet constructive way during the course of the initial one-day workshop, despite significant differences in the composition of the two working groups. In the Thurnscoe case, all members of the Co-operative were invited; on the day, nine travelled by bus to Northern College, of whom eight were members of the management committee. The size of this group was ideal for interactive working, and the College surroundings were congenial for such an informal event.

In the Tower Hamlets case, there were three host organisations, two of which themselves brought together the interests of multiple stakeholder groups. So there was an extended prior discussion about which people should participate in the workshop, keeping in view the case for ensuring that the overall size of the group was modest enough for interactive working. The group of thirteen plus two guest facilitators that met at Oxford House in Bethnal Green was not unduly large in terms of group dynamics. However the rather small meeting room that had been reserved made it difficult for people to move around freely while the facilitators were building up a record of progress on flipcharts displayed around the walls, as is characteristic of this particular type of workshop.

Another consequence of this more differentiated host structure at Tower Hamlets was that many of the workshop participants were professional people who had taken on leading campaigning roles, even though they were not very representative of the impoverished multi-cultural population of the Borough. In contrast the participants in the Thurnscoe workshop were wives or widows of coal miners, retired miners, or unemployed people—all themselves living in the houses on the estate for which they were now collectively responsible.

The longer the duration of a project engagement, the more important it can become to explore significant changes of course during the project period. The very brevity of the two engagements compared here reduces the significance of this factor. However, in both cases there were significant changes afoot at the time of the workshop. In the case of Tower Hamlets, news had come out since the initial exploratory meeting which was to give the workshop fresh urgency. For the following week the Health Authority was now due to confirm its intention to close the Accident and Emergency Unit at the Mile End Hospital at the heart of the Borough.

There were elements of urgency too at the time of the Thurnscoe workshop, for it was only the previous week that the Co-operative had taken over direct responsibility for the management of the estate. So there were some pressing issues now to be resolved. Among them was the issue of whether the co-operative should begin to set up its own workforce to undertake medium scale repairs of the properties, or whether it should make arrangements for others to take on this work. It was agreed that this issue should provide the focus for the initial strategic choice workshop.

More broadly, it became apparent to the facilitators that a culture change was becoming necessary within the committee. The group had been set up to campaign for collective ownership of the houses in which they lived, and had met with widely acclaimed success in their campaign. Yet they were taking on direct responsibility for management of the properties now in their ownership; and this might mean facing invidious policy distinctions between different tenant categories.

Some further discussion of the course of the two project engagements will be found in published accounts as indicated in the final row of Table 9.2 (Thunhurst and Ritchie, 1992; Thunhurst *et al*, 1992; Friend, 1994; Rosenhead, 1996). The two accounts of the Tower Hamlets project bring out some interesting differences of perspective between the two facilitators, the second of whom had a deeper prior awareness of the power imbalances between medical consultants and community interests in inner London.

Further details of each workshop are recorded in informal notes made by members of the facilitation team immediately after the event; and also by photographs showing the contents of the flipcharts. These photographs were taken with a view to subsequent compilation as a 'photo-record' for discussion with the participants next time they met (Friend and Hickling, 1997). In some strategic choice workshops, photographs are also taken of the participants at work so as to record impressions of the interaction between them and thus bring the record to life. Yet this was not done in either of the cases reported here, because of a judgement that such an approach might be seen as intrusive in the early stages of work with a community group.

9.10 Comparisons of Project Outcomes

In reflecting on the outcomes of any negotiated project engagement, it is important to consider not just immediate outcomes but also any longer term influences, including possible learning outcomes for any of those involved (not only in host roles but also potentially in agent or sponsor roles). Judgements about such longer-term outcomes will of course depend on the time that has passed between the engagement and the reflection—some 14 or 15 years in the case of the two engagements that are compared here. This is a long enough period for some details to have faded from memory, yet also for more mature reflections to be offered.

The aftermaths of the two engagements are compared in Table 9.3. The more immediate outcomes of the two workshops were recorded

TABLE 9.3 *Comparison of the Tower Hamlets and Thurnscoe Projects: Part III Project Aftermath?*

Tower Hamlets health strategy1987	PROJECT	Thurnscoe Tenants Housing Co-operative 1988
Agreement on immediate campaign to stop early closure plan, including press publicity and lobbying, to be followed by further workshop with health managers	IMMEDIATE OUTCOMES?	Programme of investigations agreed into the implications of crash course compared to more gradual alternative
Two further workshops held, with focus shifting first to services in the community and then to child health services, and consequent changes in involvement	CONTINUING RELATIONS?	One further workshop held within initial contract period using different methods. Longer-term involvement by Community OR Unit, with further workshop in 1989
Some influence on later collaborations between the two facilitators, in London and elsewhere. One specific innovation in new visual format for progress report	INFLUENCES ON AGENT PROGRAMMES?	Contributed to build up of momentum of Community OR Unit, including transfer of responsibility to new full time research officer, later Director
Limited insights only into any impacts on later patterns of community engagement with statutory agencies in Tower Hamlets (probably marginal)	INFLUENCES ON HOST PROGRAMMES?	Contributions to early capacity-building within the committee, in particular through confronting a forceful member with alternatives to his proposals
Probably limited, as resource inputs from the London health charity were modest and probably not directly monitored	INFLUENCES ON SPONSOR PROGRAMMES?	This case included in the OR Society's book of 26 cases in Community OR (Ritchie *et al*, 1994); lessons of wider Thurnscoe work reported in the OR Society's Journal (Thunhurst *et al*, 1992; Thunhurst and Ritchie, 1992)
Through publications; also one-day conference on OR for community health at Northern College in June 1989	DIFFUSION OF LEARNING?	Further engagement of staff of Community OR Unit with other tenant housing co-operatives in Yorkshire
Introduction of managers as well as campaigners in a workshop can reduce momentum and commitment to actions	HYPOTHESES RE PROJECT ENGAGEMENT?	(1) Same as Tower Hamlets; (2) the facilitator can usefully play devil's advocate to counteract 'group think'

using the *progress package* format that is normally used to organise outcomes in the later stages of a strategic choice workshop[6]. In the case of the Tower Hamlets workshop, there was an urgency to agree a set of short term lobbying and publicity actions that might persuade the Tower Hamlets Health Authority to postpone its imminent proposals re closure of hospital services.

In the Thurnscoe case, it was agreed that the Yorkshire Metropolitan Housing Authority, as professional advisers to the Committee, was the most important agency to be consulted. The consequences of approaching different office holders was considered, in that some were influential in the management hierarchy while others—who were closer to the specific challenges faced by the Thurnscoe Co-operative—were thought to be more sympathetic to its aims. This kind of discussion of the internal political configurations of other associated organisations has been found to be a significant feature in the management of inter-agency relations in other contexts (Friend, 1990).

Relationships between agents and hosts were to become more continuous in the case of Thurnscoe than in Tower Hamlets, for the Community OR Unit (which was sponsored during its early years by the Operational Research Society) was based not far from Thurnscoe. It was therefore able to develop a close working relationship with the Housing Co-operative that was to endure for several years (Thunhurst and Ritchie, 1992). In Tower Hamlets, there were more limited opportunities for continuing contact, in that one member of the host group (Ann Taket) was an OR scientist who became active in community OR, especially in the field of health (Ritchie *et al*, 1994).

The longer term reflections offered here reflect the personal perspective of the author, who was invited to contribute to the initial engagement in each case as a project *agent*, working in that role with associates with whom he continued to have contacts through the developing community OR initiative. It is of course not easy for any person involved briefly in such an agent role to trace how far the engagement might have had longer-term influences among any of those who became involved in host or indeed sponsor roles. Anyone who operates briefly as a project consultant will tend soon to move on to other project engagements with different hosts and different sponsors. Meanwhile, at least some of the individuals

[6] In earlier writings (e.g. Friend and Hickling, 1997) these were referred to as a *commitment package*. The progress package takes the form of a grid within which a balance can be struck between commitments *now* and those deferred to the *future*; and also between responses to specific *areas of choice* and *areas of uncertainty* in addressing those choices.

in the host and sponsoring organisations may in the course of time move on to other responsibilities. Furthermore some of the configurations of contractual and other arrangements among the host organisations may have been relatively short-lived.

9.11 Developing a Framework for Interactive Learning

In developing a framework through which to trace such possibly diffuse influences of a negotiated project engagement, the members of the Tavistock Institute's OROC team (Friend *et al*, 1998) drew on the work of Hylton Boothroyd, a leading operational research scientist at the University of Warwick.

Shortly before the period of the OROC project, Boothroyd (1978) had presented a model of *articulate intervention* in which those engaged in projects were seen as influenced by partially intertwined programmes of action and of research. Boothroyd explicitly expressed a preference for use of the term *programme* rather than *system* in this context, acknowledging the influence on his thinking of the ideas about the evolution of scientific programmes propounded by modern philosophers of science such as Kuhn (1962) and in particular Lakatos (1972).

In telling the story of the development of operational research, Boothroyd discussed the subtle dynamic interrelationships of the evolving programmes of practising OR groups within specific commercial, industrial or other organisations; of academic schools within universities; and of professional associations such as the UK Operational Research Society. This led the OROC team to draw a wider distinction between what could be termed *institutional*, *disciplinary* and *personal* programmes. Such programmes could be seen as continually evolving and adapting through interactions not only among those acting as agents in successive project engagements, but also among those acting in project host and sponsor roles, as indicated in Table 9.4.

The comparisons offered in the third, fourth and fifth rows of Table 9.3 reflect personal impressions, any of which could be tested and possibly refined through further enquiries with others involved in either agent, host or sponsor roles. For project agents, this kind of enquiry tends to become more difficult to pursue in the case of project hosts, in the face of progressive loss of contact over the years.

The opportunities for project agents to pursue strands of further enquiry in relation to sponsors can be variable. In the case of the two projects compared here, there is more scope for further enquiry in relation to

TABLE 9.4 Interwoven Programme Strands in a Negotiated Project Engagement

	Project Agent roles	Project Sponsor roles	Project Host roles
Institutional programmes	*What programme strands are most influential?*	*What programme strands are most influential?*	*What programme strands are most influential?*
Disciplinary programmes	*What programme strands are most influential?*	*What programme strands are most influential?*	*What programme strands are most influential?*
Personal programmes	*What programme strands are most influential?*	*What programme strands are most influential?*	*What programme strands are most influential?*

the Thurnscoe project than in relation to Tower Hamlets. This is because the Operational Research Society, as the original external sponsor of the Community OR Unit, was to play a subsequent role in publishing further reflective accounts of community OR practice (Thunhurst and Ritchie, 1992; Ritchie *et al*, 1994). Also the author continued his association with the Society in other ways. In the case of Tower Hamlets, the scope for further enquiry was more limited, as the role of the sponsor (the Kings Fund) in the project was quite minor and indirect.

In general, the more ambitious and protracted a project engagement, the more scope there is likely to be for enquiry into the interwoven influences of the institutional, disciplinary and personal programmes that come to bear on the various people involved in host, agent and sponsor roles. Also, the more likely it becomes that the project will have brought together as agents a disparate group of individuals, including some with quite specialised skills. Some of these people may not only bring different disciplinary traditions into the project, but may have little or no prior personal experience of working with each other—and indeed may sometimes owe allegiance to different employing organisations. The virtues of inter-disciplinary working on important projects are often proclaimed; but the challenges associated with surmounting barriers of culture and accountability must also be recognised.

There may indeed be similar threads of complexity to disentangle among those engaged in host roles—not least in community development projects—and in external sponsor roles. Within any of the cells of Table 9.4, there may be many such strands of complexity to explore. Some of these may be amenable to investigation using a systems perspective, yet others may be of a more adventitious nature.

The OROC project team developed a sequence of three schematic diagrams to illustrate the *dynamics* of any negotiated project engagement; this builds further on the static representation presented here in Figure 9.1 (Friend *et al*, 1998). First, they presented a variant of Figure 9.1 in which a time dimension was introduced to illustrate the successive stages of engagement in a project before it starts, during its course and after its completion. The inclusion of a third stage of *disengagement* recognises that some projects may come to a quite abrupt and final end—not least in the case of agents who have been working with hosts in another part of the world—yet in other cases it may be more gradual. Indeed it may lead to further engagements among some or all of the parties.

In a further elaboration of this conceptual diagram, the OROC team introduced a picture, reproduced here in Figure 9.2, to show how the various types of strand distinguished in Table 9.4 might converge in the

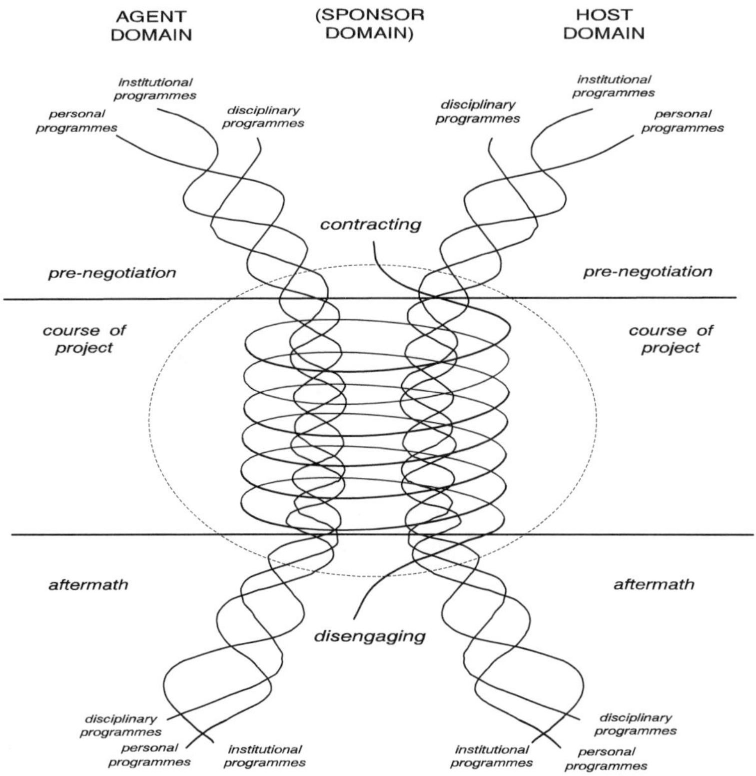

FIGURE 9.2 Strands of programme influence in a negotiated project engagement (source: Friend *et al*, 1998)

pre-negotiation stage, then become more tightly intermeshed while the project was under way. After the completion of the project, the strands would then tend to separate out again to a greater or lesser extent, influenced to varying degrees by any further engagements that might be negotiated among some or all of the participants.

This conceptual model of intertwined programme influences in turn led the OROC team to offer a skeleton framework to guide reflection and interactive learning among any of those who had participated in a project. This took the form of a structured checklist intended primarily to help people to learn lessons from a project engagement after its completion. It was however envisaged that the same framework might also be used to guide discussion of any relatively extended project while it was still under way.

This structured checklist has not been further developed since the time of the OROC project, though opportunities to apply it have arisen intermittently and there have been discussions of the scope for further development and use by consultants, research students and others (Friend *et al*, 1998). It is this enquiring framework that has been adapted here to guide the three phases of comparison between the Tower Hamlets and Thurnscoe projects in Tables 9.1–9.3.

As Table 9.3 suggests, it can be far from easy for anyone engaging in a project in an agent role to trace in retrospect how far that specific project might have influenced subsequent developments in the worlds of any of those participating in host or sponsor roles. For example, it can be asked whether there were—as had originally been intended—any significant influences on the planning processes of Tower Hamlets Health Authority. Then could the two later workshops that focused on services in the community and on child health services respectively have had any marginal impacts on the disciplinary outlooks of any of those from the medical, nursing, accountancy or other professions who participated at these later stages?

In the case of the Thurnscoe project, were there any significant influences on the broader institutional programme of the Yorkshire Metropolitan Housing Association, which went on to take responsibility for supporting and advising other co-operative housing projects within the region? And what can be asked about any influences on the personal outlooks and behaviours of any of the key participants?

To attempt to follow such multiple and partially interwoven strands in any comprehensive way would be so time-consuming as to become quite unrealistic, recognising the multiple pressures that impinge on the lives of most participants in a project, whether engaged in an agent, host or sponsor role. The initial assumption must therefore be that the influence of any one engagement on any of the contributing programmes will be so slight as to be negligible, unless there are circumstances to suggest otherwise. Then it will at least become possible to follow through a few of the more interesting strands in a more selective way.

9.12 Asking Wider Questions

Following the OROC framework of enquiry into the *aftermath* of a project, the sequence of questions in Table 9.3 concludes with the two headings of *diffusion of learning* and *hypotheses re project engagement*. The former question concerns the extent to which the experience of any particular project can be seen as coming to exercise wider influence on other people who were not specifically involved as agents, hosts or sponsors in that particular engagement. Such diffusion can occur either through formal channels such as publications and conference presentations, or through more informal communication networks. Table 9.3 indicates some of the principal channels of diffusion in the case of the two projects compared here, raising the question of whether the extent of any further influences might be explored in either case.

The final question in Table 9.3 raises the question of how far any particular project engagement might be viewed as having contributed to the formation, refutation or modification of wider hypotheses about the management of negotiated project engagements. Examples of two such hypotheses are given in Table 9.3.

The first hypothesis relates to the effect of extending the host involvement in an informal planning workshop that was originally arranged with a group of campaigners to engage other people with operational or policy responsibilities for the matters concerned. In Tower Hamlets, it seemed to make good sense after the initial workshop for the three community groups to invite senior managers of the Health authority to a follow-up workshop. However, the second workshop was judged to be less successful, partly because the urgency of the threatened closure of the hospital unit had receded and the action focus had become less clear.

Hypothetically, another reason for the disappointing results of the second workshop was that the invited managers, while appearing to participate in a co-operative spirit, saw this event as peripheral to the continuing web of policy meetings through which important investment decisions are developed. This leads to the formulation of a more general hypothesis as follows:

> **Hypothesis B:** Against the potential benefits from inviting more powerful stakeholders to participate in an interactive project engagement with representatives of relatively disadvantaged community groups must be set the

potential disbenefits of restricting the level of direct action outcomes that it may be possible to agree in such a setting.

This same hypothesis gained further support from the experience of the Thurnscoe project. Here again it was agreed to widen participation in a second workshop, by inviting a manager from the Yorkshire Metropolitan Housing Association to be present for part of the time.

A further hypothesis originating from the initial workshop in Thurnscoe relates to the ways in which a facilitator can deal with a dominant member of the group who, by force of argument and an air of superior knowledge, seeks to impose a particular strategy of action on the group. In that workshop, one participant with considerable experience of building maintenance proposed a crash strategy for training local unemployed people. This proposal appealed to the other members of the co-operative in that it seemed to relate to their wish to exercise full control over housing management as early as possible.

The author as facilitator now constructed on a flipchart a balance sheet to compare the consequences of this proposal against those of a more gradual approach, in terms of a range of criteria that had come out in discussion. This comparison at first suggested that all the advantages lay on the side of the proposed crash course. To test the realism of this, the author then asked whether there were any arguments at all to be considered for a more gradual approach, which might perhaps appeal to other parties. Examples might include the lending bank that had provided finance for the acquisition, or the Housing Association which, at the bank's insistence, had been appointed as management advisers over an initial transitional period.

At the workshop, there followed further discussion of advantages and disadvantages in which further criteria relevant to the concerns of other stakeholders were gradually introduced. The facilitator was here able to introduce some of the principles relating to pair comparison and the analysis of diverse sources of uncertainty—political as well as technical—that have come to characterise the strategic choice approach.

The result was that the group agreed not to come down conclusively in favour of their preferred strategy quite yet. Instead, they identified some of the key exploratory actions—including investigations and consultations with other people—that they might be wise to initiate before embracing that or any other strategy with confidence. In effect, the facilitator had acted as a devil's advocate in suggesting, in the face of a state of apparent consensus or 'group think' within the room, that there might be opposing points of view to consider if a balanced management decision was to be made.

A tentative general statement of such a hypothesis might be:

Hypothesis C: The conventional advice that a workshop facilitator should remain detached from the issues under discussion in certain situations may be modified in any situation where there appears to be consensus on preferred action among those present, yet other stakeholders may be influenced by different criteria. In that case, the facilitator can play a helpful role by asking questions designed to uncover these external interests.

9.13 *Generating Hypotheses to Guide the Future Practice of Community OR*

The cumulative experience of community OR so far has now become quite extensive yet can hardly be described as massive. Of particular interest in the context of this book is the question of whether a searching review of this experience can lead to a clearer expression of the defining characteristics of this relatively young field. In particular, can the practice of community operational research be said to *add value* to the existing body of expertise and wisdom in the wider realm of community development?

This is a question that it is important to address not only in relation to the circumstances of the more disadvantaged local communities, such as Tower Hamlets or Thurnscoe, to be found in parts of a developed country such as Britain. The question becomes all the more important in the wider circumstances of deprivation and disempowerment encountered in both urban and rural communities throughout the so-called developing world.

In starting to address questions about the added value that OR can contribute to the field of community development, it is worth referring back to the second of the two hypotheses presented in the preceding section. This put forward a conjecture about the circumstances in which a facilitator can usefully play a devil's advocate role, so as to bring out the concerns of stakeholders not present in the room. In the specific situation encountered in Thurnscoe, the author as facilitator drew on his repertoire of interactive OR methods to introduce a graphical method for developing comparisons between paired alternatives using multiple dimensions of comparison.

He introduced this method, not in the esoteric mathematical language favoured by some academic management scientists, but in the transparent visual terms that are characteristic of the 'soft' OR repertoire (Rosenhead and Mingers, 2001), allowing for expressions of disagreement or uncertainty within the group. However, the element of OR logic is there; and that is something that can be introduced whether the facilitator in person has a specialised OR background, or has been exposed to the relevant concepts and skills by training, apprenticeship or other indirect means.

The two cases of community OR compared here—as well as others in the author's personal experience—suggest that it can be restrictive to approach community development projects through a model of intervention that assumes that the object of intervention should be construed as an *organisation* with identifiable systemic properties. The tenants' housing co-operative at Thurnscoe could indeed be viewed as a purposeful open system, despite a significant change of purpose—from campaigning to management—at the time of the project engagement here discussed. Yet the project hosts at Tower Hamlets were more of a transient coalition of groups some of which were themselves multi-organisational in form. So it made sense—especially in Tower Hamlets, though to a large extent also in Thurnscoe at that time—that the problems to be addressed should be viewed as transient rather than systemic in nature. This leads to the suggestion of the following specific hypothesis to guide the practice of community OR:

> **Hypothesis D:** In the practice of community OR, there is value in drawing on a repertoire of methods not all of which start from the premise that project hosts can be represented in terms of systems concepts, or that their problems can be construed in terms of exploring opportunities for change in their systemic characteristics.

Some systems thinkers such as Checkland (1981) have succeeded in breaking away from the restrictive assumption that there is one recommended way in which a system of management should be modelled, by encouraging the articulation of more than one *root definition* of system purposes and functions. Other systems thinkers have gone further by encouraging the use of different *metaphors* for organisations as an instrument to inform choice of the most appropriate systems methodology to guide intervention in any situation (Jackson and Keys, 1984; Flood and Jackson, 1991). Also, Ulrich (1983), through his *Critical Systems Heuristics*, has introduced methods for asking normative as well as descriptive questions about the properties of socio-technical systems.

However, there are more fundamental differences between the worldviews of scientists who claim to be systems thinkers and those, such as the author, who are reluctant to make such claims. One of the most significant differences can be found in the way in which *problems* are perceived, and possible lines of response are formulated. The systems scientist normally sets out to explore relationships within a purposeful organisation, viewed as an open system, in terms that help highlight weaknesses in the arrangements through which its ongoing patterns of activity are managed. These weaknesses are seen as presenting opportunities for creative systems reconfiguration or redesign. In some systems approaches, several nested levels of system guidance are distinguished (Beer, 1985).

Other interactive approaches in the repertoires of OR scientists focus not so much on the understanding of systems as on *areas of choice* or *options* that may be identified in a here-and-now situation, and these may then dissolve once responsive actions are taken and new situations arise. The strategic choice approach (Friend and Hickling, 1997) is one such example; another is the approach of Bennett *et al* (2001), drawing on the metaphors of game theory and drama theory.

A useful way of highlighting this basic distinction between systems and other approaches is through consideration of the two distinct translations for the English verb *to be* in Spanish, Portuguese and certain other European languages. Does a problem arise from a current *situation* (Spanish verb *estar*) that can be dealt with through here-and-now decisions? Or does it exist in *essence* (Spanish verb *ser*), indicating the potential for dealing with it through some form of redesign of systemic relationships? The *ser/estar* distinction can be seen as a key indicator of whether a problem is constructed in systemic or more transient terms (Friend, 2001).

> **Hypothesis E:** There is a case for the development of more explicit diagnostic tools to help practitioners of community OR in appreciating whether they should construe the key issues to be addressed in a negotiated project engagement in terms of situational elements or elements with more systemic properties.

It is of the nature of hypotheses such as those introduced so far that they can be criticised, modified and replaced at any time. Many further hypotheses relating specifically to community OR could be advanced—for example, relating to the importance of distinguishing the influences on the course of a project of those who engage in agent, host or sponsor roles. Further hypotheses might address the availability to people in any of these roles of tools through which to probe the influences of the various disciplinary, institutional and personal programme strands that become intertwined before or during the course of the project.

9.14 *Wider Insights to Guide the Future Practice of OR*

How far can hypotheses suggested by experience in the field of community OR be seen as of potential value in guiding the practice of OR in other fields where configurations of host relationships may be complex, developmental or unstable in form? And can they help in other fields of project engagement where the distinction between host and sponsor roles may be a significant one?

One traditional definition of OR, which is no longer accepted as authoritative but which for many years was printed at the front of each issue of the *Journal of the Operational Research Society*, reads as follows:

> Operational Research is the application of the methods of science to complex problems arising in the direction and management of large systems of men, machines, materials and money in industry, business, government and defence. The distinctive approach is to develop a model of the system, incorporating measurement of factors such as chance and risk, with which to predict and compare the outcomes of alternative decisions, strategies and controls. The purpose is to help management determine its policy and actions scientifically.

While much of the language of this statement now reads as dated—from the restriction to 'large' systems to the lack of gender sensitivity—its general tenor still applies to much of the conventional practice of OR in larger corporate contexts. Yet it is a definition that scarcely reflects the developing experience of some OR scientists in introducing the so-called 'softer' OR methods in less sheltered organisational environments, such as those to be found in community development contexts. So can an alternative starting point be offered through which to approach the expression of a broader, more up-to-date view of the distinctive competence of OR?

It is possible to distinguish five broad aspects of difficulty in decision making that have demonstrated that they are amenable to recognised OR methods. As viewed by the decision-makers themselves, these difficulties can be expressed as follows:

- They experience difficulty in exploring structural links among problems they face;
- They experience difficulty in searching for feasible and acceptable solutions;
- They experience difficulty in reconciling potentially conflicting interests or goals;
- They experience difficulty in predicting the consequences of proposed actions;
- They experience difficulty in agreeing how to address sources of uncertainty.

Most if not all of the methods within the recognised repertoire of the OR scientist can be seen as intended means of response to at least one—and in many cases more than one—of these types of difficulty experienced by decision makers. There are few such methods that would claim to address all five of these areas of difficulty at the same time; yet such recognised OR methods as mathematical programming, system dynamics, multi-criteria decision analysis, simulation and risk analysis do aspire to address two or more of them together.

Seen as forms of useful service to decision-makers, the softer, more interactive OR methods, as well as the so-called 'harder', more mathematical methods, can be viewed as addressed to the same above five aspects of difficulty in decision-making. Certainly, this has been the author's personal experience in developing and applying the set of tools for shaping, designing, comparing and choosing that together comprise the strategic choice approach (Friend and Hickling, 1997).

The choice of preferred approach in any engagement—hard or soft, facilitative or analytical, systems centred or decision-centred—will be influenced by the personal repertoire and judgement of the agent or, if there are several people working together in agent roles, by some process of negotiation among them. This negotiation may sometimes also draw in other people in host or sponsor roles. This in turn will mean discussing the nature of the problems being faced: should they be treated as situational—Spanish *estar*—or problems of essence—Spanish *ser*?

From a host perspective, the practice of 'soft' OR can be seen as not so much an aberration from the 'hard' mainstream, but an equally valid adaptation of the five core capacities of OR to more fluid, developmental contexts of decision making. For the extensive history of development of soft OR methods has not merely involved a relaxation in the standards of scientific rigour that the more traditional OR scientists see as guiding their work. Rather, it has involved a continuing openness to insights from the social sciences, as evidenced by successive developments in the United States, in Britain and elsewhere (Jackson *et al*, 1989).

This supports the important point that the OR scientist who works with communities cannot assume that tools to manage the five types of difficulty listed here will be *sufficient* as a platform from which to engage in this field. It is vital too that the OR scientist learns to recognise the contribution of social scientists to the art of negotiated project engagement, and also the skills that experienced community development workers have crafted through their years of practice in the field.

References

Ackoff, R.L. (1970). On the ambiguity of the researcher and the researched. In: *The Place of Research in Social Choice: A Symposium Dedicated to the Memory of William Neil Jessop*, pp. 21–32. Tavistock Institute of Human Relations, London.

Ackoff, R.L. (1981). *Creating the Corporate Future*. Wiley, New York.
Beer, S. (1985). *Diagnosing the System for Organisations*. Wiley, Chichester.
Bennett, P., Bryant, J. and Howard, N. (2001). Drama theory and confrontation analysis. In: *Rational Analysis for a Problematic World Revisited*, pp. 225–248. Rosenhead, J.V. and Mingers, J. (eds.). Wiley, Chichester.
Boothroyd, H. (1978). *Articulate Intervention*. Taylor and Francis, London.
Bowen, K. (ed.) (1995). *In at the Deep End: MSc. Student Projects in Community Operational Research*. Operational Research Society, Birmingham.
Checkland, P.B. (1981). *Systems Thinking, Systems Practice*. Wiley, Chichester.
Checkland, P.B. (2001). Soft systems methodology. In: *Rational Analysis for a Problematic World Revisited*, pp. 61–89. Rosenhead, J.V. and Mingers, J. (eds.). Wiley, Chichester.
Clark, A.W. (ed.) (1976). *Experimenting with Organisational Life: The Action Research Approach*. Plenum, New York.
Cunningham, J.B. (1993). *Action Research and Organizational Development*. Praeger, Westport, CT.
Flood, R.L. and Jackson, M.C. (1991). *Creative Problem Solving: Total Systems Intervention*. Wiley, Chichester.
Friend, J.K. (1990). Handling organizational complexity in group decision support. In: *Tackling Strategic Problems: The Role of Group Decision Support*. Eden, C. and Radford, J. (eds.). Sage, London.
Friend, J.K. (1994). Community Involvement in health strategy in Tower Hamlets. In: *Community Works: 26 Case Studies Showing Community Operational Research in Action*, pp. 137–145. Ritchie, C., Taket. A., and Bryant, J. (eds.). Operational Research Society, Birmingham.
Friend, J.K. (2001). Engaging with transient complexity in development projects. In: *Understanding Complexity*. Ragsdell, G. and Wilby, J. (eds.). Kluwer/Plenum, London.
Friend, J.K., Bryant, D.T., Cunningham, J.B., and Luckman, J. (1998). Negotiated project engagements: Learning from experience. *Human Relations*, **12**, 1509–1542.
Friend, J.K. and Hickling, A. (1997). *Planning Under Pressure: The Strategic Choice Approach*, 2nd edn. Butterworth-Heinemann, Oxford.
Friend, J.K., Norris, M.E., and Stringer, J. (1988). The Institute for operational research: An initiative to extend the scope of OR. *Journal of the Operational Research Society*, **39**, 705–713.
Friend, J.K., Power, J.M., and Yewlett, C.J.L. (1974). *Public Planning: The Inter-Corporate Dimension*. Tavistock Publications, London. Reprinted Routledge 2001 in the International Behavioural and Social Sciences Library.
Jackson, M. and Keys, P. (1984). Towards a system of systems methodologies. *Journal of the Operational Research Society*, **35**, 473–486.
Jackson, M., Keys, P., and Cropper, S. (eds.) (1989). *Operational Research and the Social Sciences*. Plenum, New York.
Kuhn, T.S. (1962). *The Structure of Scientific Revolutions*. University of Chicago Press, Chicago.
Lakatos, I. (1972). Falsification and the methodology of scientific research programmes. In: *Criticism and the Growth of Knowledge*. Lakatos, I. and Musgrave, A. (eds.). Cambridge University Press, Cambridge.
Lewin, K. (1946). Action research and minority problems. *Journal of Social Issues*, **2**, 34–46.
Ormerod, R.J. (2002). On the nature of OR: Taking stock. *Journal of the Operational Research Society*, **53**, 475–491.
Ritchie, C., Taket. A., and Bryant, J. (eds.) (1994). *Community Works: 26 Case Studies Showing Community Operational Research in Action*. Operational Research Society, Birmingham.
Rosenhead, J.V. (1986). Custom and Practice. *Journal of the Operational Research Society*, **37**, 335–343.

Rosenhead, J.V. (ed.) (1989). *Rational Analysis for a Problematic World*. Wiley, Chichester.

Rosenhead, J.V. (1996). What's the problem? An introduction to problem structuring methods. *Interfaces*, **26**(6), 117–131.

Rosenhead, J.V. and Mingers, J. (eds.) (2001). *Rational Analysis for a Problematic World Revisited*. Wiley, Chichester.

Stringer, J. (1967). Operational research for multi-organisations. *Operational Research Quarterly*, **18**, 105–120.

Thunhurst, C., Ritchie, C., Friend, J., and Booker, P. (1992). Housing in the Dearne Valley: Doing community OR with the Thurnscoe Tenants Housing Co-operative. Part 1—The involvement of the community OR unit. *Journal of the Operational Research Society*, **43**, 81–94.

Thunhurst, C. and Ritchie, C. (1992). Housing in the Dearne Valley: Doing Community OR with the Thurnscoe Tenants Housing Co-operative. Part 2—An evaluation. *Journal of the Operational Research Society*, **43**, 677–690.

Ulrich, W. (1983). *Critical Heuristics of Social Planning: A New Approach to Practical Philosophy*. Haupt, Berne.

10

Systemic Intervention for Community OR: Developing Services with Young People (Under 16) Living on the Streets

Alan Boyd, Mandy Brown and Gerald Midgley

10.1 Introduction

For ten years, researchers in the Centre for Systems Studies at the University of Hull (where all the authors of this chapter were based until relatively recently[1]) have been participating in an on-going research programme to explore the value of using *systemic intervention* in Community OR. Approximately thirty Community OR projects conducted by Centre staff

[1] Alan Boyd and Gerald Midgley no longer work for the University of Hull, but retain Visiting Researcher status there.

Final version accepted for publication March 2002.

Alan Boyd • Manchester Centre for Healthcare Management, University of Manchester, Devonshire House, University Precinct Centre, Oxford Road, Manchester, M13 9PL, UK.
Mandy Brown • Centre for Systems Studies, Business School, University of Hull, Hull, HU6 7RX, UK.
Gerald Midgley • Institute of Environmental Science and Research (ESR) Ltd., Christchurch Science Centre, 27 Creyke Road, PO Box 29-181, Christchurch, New Zealand. Centre for Systems Studies, Business School, University of Hull, Hull, HU6 7RX, UK.
Community Operational Research: OR and Systems Thinking for Community Development, edited by G. Midgley and A.E. Ochoa-Arias, Kluwer Academic/Plenum Publishers, New York, 2004

have been a part of this research programme. We have therefore been able to evolve our practice in the light of methodological learning, and simultaneously we have developed our methodology through reflections on practice.

We begin this chapter by providing a summary of our systemic intervention methodology. A case study is then presented of a project to support the design of new services for young people under the age of 16 living on the streets of Manchester. This illustrates how our ideas can be translated into practice. Finally, we end with some reflections on the wider implications of our methodology for Community OR.

10.2 *Systemic Intervention*

It is very noticeable that many writings about the practice of Community OR take the form of first person narratives: they tell stories about *what* the practitioner did and *why*. The focus is often at a methodological level: why certain methods seemed appropriate, and how they were applied or adapted to take account of specific circumstances (see, for example, Ritchie *et al*, 1994).

When the actions of the practitioner are the focus of research, we suggest that the concept of *intervention* is particularly useful. Midgley (2000) defines intervention as 'purposeful action by an agent[2] to create change' (p.132).[3] The Community OR practitioner can be viewed as just such an intervening agent.

[2] Midgley (2000) suggests that an agent can be viewed as either a single human being, or an identifiable group of human beings in interaction (e.g., a family, team or organisation), that have purposes ascribed to them. In the case of a group, this definition does *not* assume that all participating individuals share the purpose of the whole (indeed, some sub-agents may act in opposition to the dominant purpose). However, a group can be called an agent when it (or its representatives) is perceived as acting to realise a dominant purpose at the group level regardless of the actions or views of sub-agents. The word 'dominant' here is crucial. It indicates that the group purpose is a function of whatever mechanisms of legitimation exist within and beyond the group that allow it to be perceived as moving in one particular direction, regardless of any counter-arguments being produced by internal opponents. Therefore, when a government minister declares war on behalf of a nation, it is generally accepted that the nation is at war even if half of its citizens wish to dissent.

[3] This is a deceptively simple statement. There is actually a great deal of explicit theory lying behind this understanding of agency and intervention, but there is insufficient space to present it all in this chapter. For further details, see Midgley (2000).

But why do we talk about *systemic* intervention? Our answer is that systems thinking, with its focus on 'holistic' understanding, can provide us with some essential insights. In the tradition of Churchman (1970), Ulrich (1983) and Midgley (2000), we argue that it is the *boundary* concept that lies at the heart of systems thinking. Because of the fact that everything in the Universe is directly or indirectly connected with everything else, where the boundaries are placed in any analysis becomes crucial. The 'cut-off point' for analysis will make some things visible and others invisible. Systems thinkers pursue an ideal of comprehensiveness, but know that this is unattainable. However, reflection on the boundaries of knowledge at least enables us to consider options for inclusion, exclusion and marginalisation. It also reminds us that all understandings are incomplete: there is a need for humility and openness to the perspectives of others (Ulrich, 1983).

If intervention is purposeful action by an agent to create change, then *systemic* intervention is purposeful action by an agent to create change *in relation to reflection on boundaries*. This statement embodies the core concern of the methodology of systemic intervention that we will be introducing over the coming pages.

At the bare minimum, we suggest that an adequate methodology for systemic intervention should be explicit about three things: boundary critique; theoretical and methodological pluralism; and action for improvement. These are discussed below.

10.3 Boundary Critique

There is a need for agents to reflect critically upon, and make choices between, boundaries (as mentioned previously). Boundaries define both *what issues* are to be included, excluded or marginalised in analyses, and *who* is to be consulted or involved (the two are obviously linked, as different agents will have different concerns). Because of the 'who' question, issues of power and participation take centre stage in systemic intervention (Churchman, 1979; Ulrich, 1983; Brown, 1996; Midgley, 1997a, 2000; Vega-Romero, 1999; Córdoba and Midgley, 2002).

An important aspect of our understanding of boundaries is that boundary judgements are intimately linked with *value* judgements (Churchman, 1979; Ulrich, 1983): the values adopted in any intervention will direct the drawing of boundaries that define the knowledge accepted as pertinent. Similarly, the inevitable process of drawing boundaries will constrain the ethical stance taken and the values pursued. Making decisions about boundaries is therefore an *ethical* business. It is also important to note that, regardless of how detailed the process of critical reflection on values and boundaries actually is, there may still be surprises as things excluded from view interact with whatever is the focus of attention. While boundary critique cannot altogether eliminate surprises, it can at least help us anticipate some effects of intervention that might otherwise have been ignored. Indeed, because things change over time, boundary judgements need to be regularly reviewed as part of a *learning process* (Ulrich, 1983; Brown and Packham, 1999).[4]

Of course, making boundary judgements is an *activity*. It is intervention to shape the agent's and others' understandings, which may in turn influence future action. Critical reflection upon boundary judgements is vital because it is only by way of boundary critique that the ethical consequences of different possible actions (and the ways of seeing they are based upon) can be subject to analysis.

Because boundary critique is an *ethical* practice, it is concerned with how things *ought* to be as well as how things currently are (Ulrich, 1983).[5] Therefore, we usually facilitate debate with people involved in, and/or potentially affected by, an intervention to support their participation in both defining the boundaries of social issues and planning how they should be addressed. Ulrich (1983) has developed a list of twelve questions for this purpose. These support people in exploring issues of purpose, control, expertise and legitimacy, and we have found them to be very useful in a number of interventions (e.g., Cohen and Midgley, 1994; Midgley *et al*, 1998; Midgley, 2000).

One further aspect of boundary critique should also be outlined. So as not to be naïve about issues of power and participation, it is important to be aware of what tends to happen when there is a conflict between different groups of people who have different ethics relating to the same issue, and thereby make different boundary judgements. Midgley (1992, 2000) argues that, if one group makes a narrow boundary judgement and another makes a wider one, there will be a *marginal* area between the two boundaries. This marginal area will contain elements that are excluded by the group making the narrow boundary judgement, but are included in the wider analysis undertaken by the second group. We can call the two

[4] There is a substantial body of literature on the theory, methodology and practice of boundary critique. See, for example, Churchman (1970, 1979); Ulrich (1983, 1987, 1994, 1996, 2000); Midgley (1992, 1994, 2000); Midgley *et al* (1998); Brown and Packham (1999); Vega-Romero (1999); Córdoba *et al* (2001); Yolles (2001); and Córdoba and Midgley (2002).

[5] Ulrich (1983) derives the distinction between 'is' and 'ought' from Kant (1787, 1788), as well as a variety of critical theorists (e.g., Horkheimer, 1937; Wellmer, 1970; Habermas, 1971; and Fay, 1975).

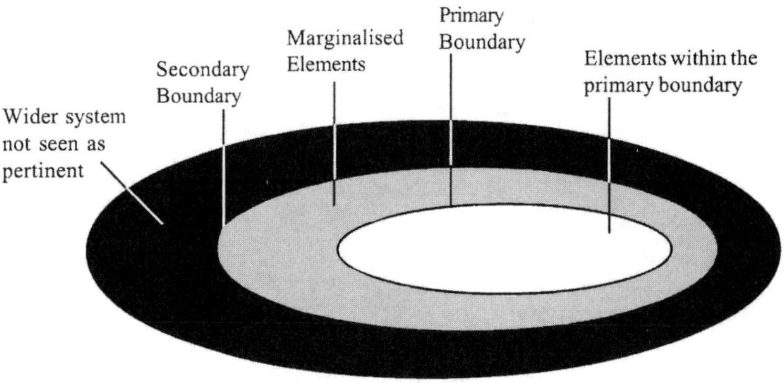

FIGURE 10.1 Primary and secondary boundaries

boundaries the *primary* and *secondary* boundaries (the primary boundary being the narrower one). This is represented visually in Figure 10.1.

When two ethical boundary judgements come into conflict, the situation tends to be stabilised by the imposition of either a *sacred* or a *profane* status on marginal elements. The words 'sacred' and 'profane' mean valued and devalued respectively. This terminology has been borrowed from the tradition of anthropology, exemplified by the work of Douglas (1966), and it should be stressed that it is not meant in an exclusively religious sense, but refers to the special status of a marginalised element. The imposition of either a sacred or a profane status on marginal elements stabilises a conflictual situation in the following manner. When marginal elements become profane, the primary boundary and its associated ethic is focused upon and reinforced as the main reference for decision making. People or issues relegated to the margins are disparaged, allowing the secondary boundary to be ignored. Conversely, when marginal elements are made sacred (and thereby assume a special importance), the secondary boundary and its associated ethic is focused upon and reinforced.

However, this is not the end of the story. Not only do ethical tensions give rise to sacredness and profanity, but this whole process comes to be overlaid with social ritual. Ritual is behaviour, in whatever context, that contains certain stereotypical elements that involve the symbolic expression of wider social concerns).[6] An observation of the presence of ritual can tell us

[6] Also see Douglas (1966), and Leach (1976) for further thoughts on the relationship between ritual, sacredness and profanity.

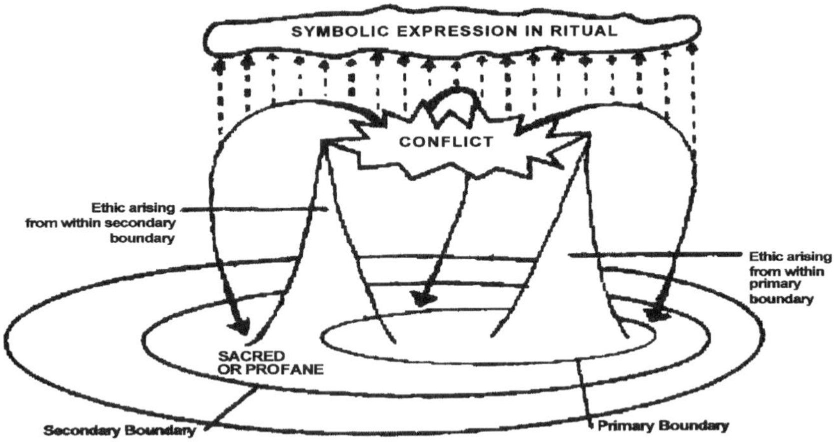

FIGURE 10.2 Marginalisation, ethical conflict, sacredness, profanity and ritual

where sacredness and profanity might lie, and hence where ethical conflicts related to marginalisation might be found. In order to make this clearer, the whole process has been represented diagrammatically in Figure 10.2.

To explain, in Figure 10.2 we see one ethic arising from within the primary boundary, and another from within the secondary one. These come into conflict—a conflict that can only be dealt with by making one or other of the two boundaries dominant. This dominance is achieved by making elements in the margin (between the primary and secondary boundaries) either sacred or profane. The whole process is symbolically expressed in ritual which, in turn, helps to support the total system.

Of course, the 'system' represented in Figure 10.2 is a model, and like all models it does not fully express the complexity of the many value and boundary judgements that interact dynamically in social situations. Some discussions of this complexity, together with practical examples that further clarify the theory, can be found in Midgley (1992, 1994) and Midgley *et al* (1998). One particularly important point about the complexity lying beyond the model should be borne in mind, however: this kind of 'system' does not exist in isolation. It is 'held in place', or granted integrity, by virtue of the fact that it expresses wider struggles between competing discourses (Midgley, 2000). These competing discourses can develop in families or small groups; at an organisational or community level; or they can be society-wide. Indeed, there may be multiple levels of mutually supportive discursive conflict, and a boundary judgement needs to be made about which level(s) of analysis will be most appropriate for the purposes being pursued.[7] We suggest that relatively local ethical conflicts are, generally speaking, a lot easier to address than those with wider societal connections: once ethical conflicts become embedded in influential institutions and cultures, they become quite resistant to change.

The model of marginalisation described above is not just an interesting sociological theory. It is useful for intervention, in that reflection upon the model in relation to local situations can deepen boundary critique by surfacing issues of power and participation that might not otherwise have been apparent (Midgley *et al*, 1998; Midgley, 2000).

10.4 Theoretical and Methodological Pluralism

The second aspect of a methodology for systemic intervention that should be made explicit is the need for agents to make choices between theories and methods to guide action, which requires a focus on *theoretical and*

[7] To make this clearer, we can provide an example of a society-wide discursive conflict that results in the marginalisation of people who are unemployed. There is a conflict in many Western societies between the liberal discourse of citizenship (where all people are seen as having equal value because of their status as rational beings) and the capitalist discourse of good employment practice (which limits the responsibility of organisations to their employees alone). This conflict is not stabilised by either the inclusion *or* exclusion of the unemployed, but by their marginalisation. If unemployed people were to be fully included along with employees in the primary boundary of industrial organisation, 'good employment practice' (indeed, the whole capitalist system of organisation) would become untenable. However, if they were fully excluded, the liberal ideal of equal citizenship would become untenable instead. Both the liberal and capitalist discourses have long histories in the West, and have come to be institutionalised throughout the economic and legal systems of our societies. While on the whole the two discourses are mutually supportive (Booth Fowler, 1991; Midgley and Ochoa-Arias, 1999), there are still significant tensions, and the phenomenon of unemployment points to one of them. The key to understanding the status of the unemployed is to realise that it is only possible to maintain the dual commitment to liberalism and capitalism if people who are unemployed are neither fully included nor excluded. People who are unemployed therefore become marginalised, but the conflict is finally stabilised when a sacred or profane status is imposed on them. When they are regarded as profane, it justifies thinking in terms of narrow organisational boundaries. When they are regarded as sacred, this justifies programmes to support social inclusion. There is rarely a consensus on whether a marginal group or issue should be viewed as sacred or profane, but there are dominant patterns of social action which come to be solidified in rituals. In the case of the unemployed, a typical example is 'signing on' as available for work, which many people view as an exercise in ritual humiliation.

methodological pluralism. These two forms of pluralism have meaning in terms of the focus on boundary judgements mentioned above. If understandings can be bounded in many different ways, then each of these boundaries may suggest the use of a different theory (and conversely, each theory implies a particular boundary judgement). *Methodological* pluralism then also becomes meaningful because methods and methodologies embody different theoretical assumptions: choices between boundaries and theories suggest which methods might be most appropriate (and conversely, choices between methods imply particular theoretical and boundary judgements).

Choice between theories and methods is also a form of action, in the same way as reflection on, and choice between, boundary judgements can be seen as action: it is intervention in the present to shape a strategy for future intervention.[8]

For the purposes of this chapter, we will provide further thoughts on *methodological* rather than theoretical pluralism. This is because choice between methods, and the practice of mixing them creatively in interventions, has been widely recognised as valuable in Community OR (e.g., Jackson, 1987a, 1988; Gregory and Jackson, 1992a,b; Ritchie *et al*, 1994; Midgley, 2000). Indeed, this is a theme of relevance in operational research more widely too (e.g., Jackson and Keys, 1984; Mingers and Gill, 1997; Taket and White, 2000).

It is important for the purposes of this discussion to differentiate between 'method' and 'methodology'. While 'methodology' refers to the theory that justifies the use of particular methods, a 'method' is a set of techniques operated in a sequence (or sometimes iteratively) to achieve a given purpose (Checkland, 1981; Jackson, 2000; Midgley, 2000). When we talk of methodological pluralism, we mean embracing the possibility of engagement at two levels. First is the level of *methodology*, where we can respect others' methodological ideas, thereby allowing their insights to inform our own understandings of how we should approach practice (either temporarily, during a particular intervention, or on a longer-term basis as continual reference points). Second, we need pluralism at the level of *method*, where we can use a wide range of methods in support of particular purposes. Figure 10.3 illustrates these two levels.

In Figure 10.3, we see that different methodologies (in the ellipses) are linked with different methods (in the rectangular boxes) by arrows that

[8] Many issues have been left unexplored by this short exposition, including paradigm incommensurability, standards for choice between theories, theoretical coherence/incoherence, how to develop the methodological knowledge base of agents, etc. All these issues are covered in Midgley (2000).

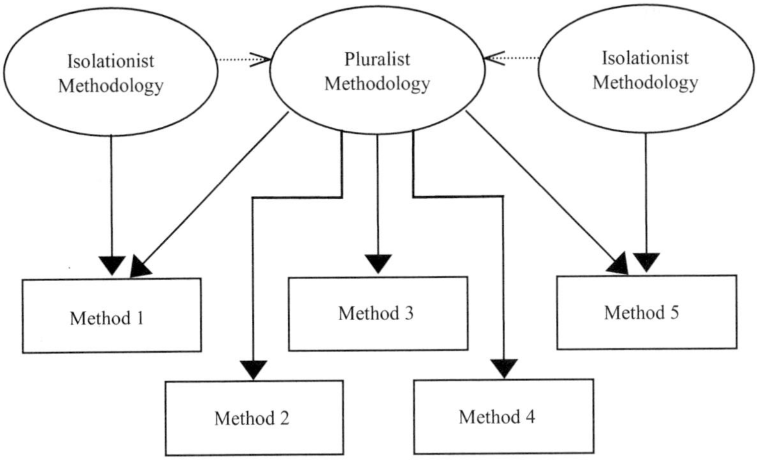

FIGURE 10.3 Relationships between methodologies and a variety of methods

represent the ascription of validity and/or legitimacy. In two cases (the left and right ellipses) this is a one-to-one linkage, with a particular methodology indicating the validity and/or legitimacy of just one method (or a narrow range of methods). When a methodology is proscriptive in this way, refusing validity or legitimacy to the majority of methods, it can be called 'isolationist' (Jackson, 1987b; Flood, 1989). Most methodologies produced during the 20th Century (whether for scientific study or intervention) are isolationist: they prescribe a 'one best way' of doing things. In contrast, a pluralist Community OR practitioner (using the methodology represented by the central ellipse in Figure 10.3) can use the full range of available methods, but they are seen through the theoretical lens of his or her own methodology. Also, they are made meaningful in local situations by the way they meet (or fail to meet) the purposes of the agent(s) engaged in intervention (and of course these purposes may be evolved through the intervention itself).[9] The pluralist practitioner defines his or her methodology in contrast

[9] The purposes of an agent engaging in intervention using a systems methodology are likely to be informed by the methodology itself, as well as his or her life experiences. Other agents involved in the intervention may have less or no knowledge of systems methodology, and for them life experiences are more likely to dominate. As the systems methodology described here requires the Community OR practitioner to open up to the perspectives of others involved in the intervention, the possible actions suggested by the methodology come to be seen as potential resources for all the agents involved and affected. Of course, when a methodology is used that has the exploration of boundary judgements at its heart, purposes are not pre-defined but can be developed through the process of intervention.

with others (most of which will be isolationist). Although pluralist methodologies are distinguished from isolationist ones, the pluralist will still be able to learn from the latter—hence the dotted arrows going between the pluralist methodology and the isolationist methodologies in Figure 10.3.[10] The idea of developing a pluralist methodology through reflection on the strengths and weaknesses of other methodologies (including isolationist ones) is expanded upon by Midgley (2000), who offers an ideal model of practitioner learning.

As there are different rationales for pluralism at the levels of methodology and method, they are dealt with separately below.

The essential value of being aware of, and learning from, a variety of methodological positions comes from the knowledge that no one theory, or set of theories—whether or not they have been codified into a methodology—can ever be comprehensive (e.g., Morgan, 1986; Francescato, 1992; Romm, 1996). Therefore, it is bound to be the case that others will have different insights to us. While we may disagree with some of their assumptions and want to challenge them, other insights and assumptions will be valuable to us—indeed, valuable enough for us to want to use them in a methodology of our own.

The purpose of learning from other methodologies is therefore that reflections on the similarities with, and differences from, one's own ideas can enable the continued evolution of one's own methodology in a manner that enhances the conceptual basis with which interventions are planned (Gregory, 1992; Romm, 1996). The key to this learning is to welcome the insights of others without taking on any idea to the exclusion of all others, thereby losing the possibility of pluralism. Therefore, to say that (for example) use of the scientific method can be valuable at times should not lead to the conclusion that only this one kind of method is valid. Those aspects of scientific methodology that invalidate other methods need to be challenged, but the scientific method (and indeed any other method that may originally have come from an isolationist perspective) can still be seen through the 'lens' of a pluralist methodology.

Having discussed methodology, we can now move on to methods. The value of pluralism at the level of methods comes from observations of what happens if only a very narrow set of methods are used—indeed, it has been known for some people to specialise in the use of just one.

[10] Because of the proscriptive nature of isolationist methodologies, it is less likely that learning will go in the other direction, from a pluralist to an isolationist methodology, but such learning is not impossible.

With an armoury of just a couple of methods, three significant inter-linked problems arise:

First, the practitioner is likely to be unresponsive to diverse understandings of issues arising out of the different perspectives of those affected by them. If the way of seeing the issue isn't supported by the chosen method, then dissatisfaction is likely to result. If those who are marginal to decision making experience the dissatisfaction, then the most likely outcome will be decisions which further marginalise stakeholders with a different point of view from the decision-makers. An example is Spash's (1997) observation that methods of cost-benefit analysis used in environmental management projects which involve stakeholders in making trade-offs between economic and environmental values automatically marginalise environmentalists who cannot accept the trade-off mentality embedded in the methods. In the longer term, this kind of marginalisation can lead to significant social problems (not to mention environmental ones in the above example) including, in the more extreme cases, civil unrest and violence.[11]

In contrast, if the dissatisfaction is experienced by the decision makers themselves, and the chosen method only supports the understanding of a marginalised group without initiating a dialogue with decision makers, then it is likely that the intervention will be ignored by the latter (Midgley and Milne, 1995; Liebl, 2002). After all, it is rare for somebody to listen to others who are not prepared to listen to them in return, and those who are already in control of decision making are simply able to turn their backs in such a situation (Gregory, 1992). The only two scenarios in which decision-makers are likely to listen to marginalised groups without expecting the same courtesy in return is when they are required to do so by a 'higher' authority, or when they are already wedded to an ethic of participation.[12]

[11] Pinzón and Midgley (2000) argue for boundary critique and methodological pluralism in the context of conflict resolution in Colombia because it is only by respecting the issues as they are seen from all sides that it becomes possible to start meaningful negotiations. The idea of trying to resolve conflict in the Colombian context without resorting to the use of a wide variety of methods for use in different local situations seems to us to be impossible.

[12] The issue of what may motivate decision-makers to listen to marginalised groups is a difficult one. Some other methodologies, such as Ulrich's (1983) Critical Systems Heuristics which (in many respects) is not naive with regard to power issues, have been criticised for failing to consider this in an adequate manner (Jackson, 1985; Flood and Jackson, 1991a,b; Ivanov, 1991; Mingers, 1992; Romm, 1995; Midgley, 1997b). While Gregory (1992) suggests that a guarantee of being listened to in return is the only thing that can motivate decision makers to listen to marginalised stakeholders, Foucault (1980) and Vega-Romero (1999)

The second significant problem with using a narrow range of methods is that, as intervention proceeds, the issues of relevance to those involved in, and affected by, the intervention may change as people's understandings develop. So what may have started out as, say, an issue of organisational restructuring to improve efficiency and effectiveness, might eventually come to be seen as an issue of fostering participative decision making to improve workforce satisfaction and commitment. If the practitioner can only handle methods for restructuring, then s/he will be unable to deal with this change in focus (unless the new issue can be 'forced' into the mould of the old one).[13]

The third problem associated with the use of a narrow range of methods is that practitioners may see all issues through the methodological 'lens' that these methods have traditionally been associated with. There is an old saying that rings true in this context: if the only tool you have is a hammer, then everything looks like a nail. Therefore, it is not just that the use of a narrow range of methods makes the practitioner unresponsive to a diversity of interpretations of issues, and unable to react appropriately to changing agendas, but also *s/he is unlikely to be aware of this*. This lack of insight, which comes from an unwillingness to explore the possibility that there may be other ways of seeing and doing, is a substantial obstacle to practising systemic intervention.[14]

both argue that dialogue tends to result from relations of power. Decision-makers will only listen to others if they will lose out from not doing so. It could be that, when decision-makers engage in dialogue with marginalised groups, they do so, not because they fear direct retribution from those groups, but because they fear a more general loss of public trust. However, they can get away with ignoring marginalised groups if the latter are the ones who refuse to engage in dialogue (there is a general moral expectation in Western culture that dialogue will be the first recourse in cases of conflict).

[13] This is actually an example from Ho (1997). He was asked by Tainan City Council in Taiwan to support their restructuring efforts, but very quickly found that issues of democratisation came to the fore. Had he not been able to show how these could be dealt with alongside the restructuring by using different methods from those originally envisaged by the leader of the Council, the restructuring would almost certainly have been thwarted by employee sabotage.

[14] This discussion of methodological pluralism has omitted the issue of how various barriers to pluralist practice can be addressed. Mingers and Brocklesby (1996) suggest that there are three barriers: *philosophical* (concerning issues of paradigm incommensurability), *cultural* (how academic and organisational cultures can get in the way) and *psychological* (how individuals can move between different methods requiring different skills). For a review of the debate surrounding these barriers, and some proposed solutions, see Midgley (2000).

10.5 Action for Improvement

Finally, an adequate methodology for systemic intervention should be explicit about taking *action for improvement*—action for the better, which cannot, of course, be defined in an absolutely objective manner. 'Improvement' needs to be understood temporarily and locally: as different agents may use different boundary judgements, what looks like an improvement through one pair of eyes may look like the very opposite through another (Churchman, 1970).[15] The word 'local' does not necessarily imply that the understanding is confined to an organisation or small group (although this may be the case)—a 'local' understanding may be very wide spread, even to the extent of providing the foundations for legal statutes and international agreements. However, the terms 'local' and 'temporary' remind us that it can be dangerous to assume that an 'improvement' will be universally regarded as such.

Also, even if there is widespread agreement between all those directly affected by an intervention that it constitutes an improvement, this agreement may not stretch to future generations. The temporary nature of all improvements makes the concept of *sustainable* improvement particularly important: while even sustainable improvements cannot last forever, gearing improvement to long-term stability is essential if future generations are to be accounted for. We can say that an *improvement* has been made when a desired consequence has been realised through intervention. In contrast, a *sustainable* improvement has been achieved when this seems like it will last into the indefinite future without the appearance of undesired consequences (or a redefinition of the original consequences as undesirable).

The notion of improvement is important because agents are restricted in the number of interventions they can undertake, and must

[15] An example is logging a stretch of rain forest. This may bring about an improvement in the eyes of the logging company's employees and those who consume the wood that is generated, but may be considered as damaging by tribal people who are displaced from their ancestral lands, and by conservationists concerned with the preservation of species diversity. As Churchman (1970) says, every improvement assumes boundaries defining what consequences of intervention are to be taken into account, and what are to be ignored or regarded as peripheral. In the above example, the logging will only be viewed as bringing about an improvement if the displacement of tribal people and the reduction of species diversity are excluded from the boundaries of analysis. Clearly, *what* is included in the boundaries of analysis and *who* conducts this analysis are both vital issues in defining improvement.

therefore make judgements about what they should and should not do. The extent to which various interventions look like they may or may not bring about improvements, or may bring about improvements that have greater or lesser priority, is a useful criterion for making these judgements.

Of course, we should say why we have used the term 'improvement' rather than, say, the creation of beauty, pleasure, knowledge, understanding, emancipation or spiritual enlightenment. The answer is that, if we value any of these things, the creation of these *represents* an improvement. The term 'improvement' is therefore general enough to have meaning in relation to almost any value system: it simply indicates the purposeful action of an agent to create a change for the better.[16]

10.6 Interrelating the Three Activities

These three activities—reflecting on value and boundary judgements; making choices concerning theories and methods; and taking action for improvement—are clearly inseparable. Doing one always implies doing the other two as well, although the focus of attention may shift from one to another aspect of this trinity so that none remain implicit and thereby escape critical analysis. The separation between the three is therefore analytical rather than factual: it ensures a proper consideration of a minimum set of three 'angles' on possible paths for intervention. Making all of them a specific focus of a methodology for systemic intervention guides the reflections of the agent, ensuring that boundaries, values, theories, methods, and action for improvement all receive explicit consideration. The three activities, diagrammed in relation to one another, are presented

[16] It should be noted that there is a counter-argument to this. According to Rorty (1989), using a term like improvement (or truth, legitimacy, ontology, morality, etc.) suggests a belief in absolute facts or values. Rorty believes that such words are tainted. To talk of improvement is to talk about the attainment of a state that *everybody* would agree is better. Rorty has launched a fierce critique of the apparent certainties of modernity. He offers a powerful argument, but why *abandon* words like truth, morality and improvement? If we are prepared to be critical about the business of making boundary judgements, there is no need to assume that understandings of improvement are universal. To abandon words like truth, morality and improvement is to risk slipping into negativity and inaction. To tear away the modernist certainties surrounding their use and clothe them with an awareness of the frailty of human understanding is to preserve the possibility of positive action while facing the complexities of this head on.

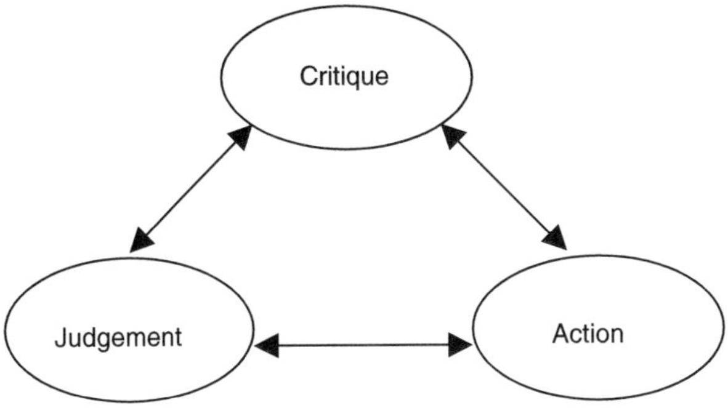

FIGURE 10.4 Three aspects of a methodology for systemic intervention

in Figure 10.4. *Critique* specifically means boundary critique (reflection on, and choice between, boundaries and associated values); *judgement* means judgement about which theories and methods might be most appropriate; and *action* means the implementation of methods to create improvement (however this is to be understood by different actors in the local context).

This brings us to the end of our discussion of methodology. Now it is time to put flesh on the bones by detailing a Community OR intervention that was informed by these ideas.

10.7 *Young People on the Streets*

Our intervention was commissioned by three voluntary organisations: Safe in the City (a Children's Society initiative), NCH Action for Children, and the Manchester & Salford Methodist Mission. They asked for some action research over a nine month period to:

- Investigate the situation faced in Central Manchester by 'detached' under-16 year olds (young people living on the streets[17]);
- Identify appropriate responses and services to meet the needs of this group; and
- Assess the feasibility of potential projects.

[17] Many young people in their teens resent being called 'children' and prefer the label 'young people'. As a mark of respect for this, we have called them 'young people' in this chapter, even though 'young people on the streets' doesn't have quite the same emotional impact as '*children* on the streets'.

Stein *et al* (1999) conducted a nation-wide survey of young people under 16 in the UK and concluded the following:

> Overall, we estimate that 11% of young people run away for one night or more on one or more occasions before the age of 16, amounting to around 77,000 young people running away for the first time each year.... At first glance, these findings may seem rather bland. However, they are of considerable importance in that they represent, for the first time, firm and reliable evidence that there is a significant prevalence of running away in all parts of the UK. The implication of this is clearly that services must be developed in all kinds of areas.... if there is to be an effective and inclusive response to the needs of young people who run away. Whilst many young people only run away once, others go on to run away repeatedly. We estimate that there are a total of around 129,000 incidents of running away over night per year in the UK. (Stein *et al*, 1999, p.38)

Some of these young people stay missing for a considerable time and become completely 'detached' from the parents, carers and the agencies that should be supporting them (English, 1973; Barter, 1996). As they have no source of income, they have to develop survival strategies that can expose them to risks of violence, involvement in crime, drug or alcohol misuse and sexual exploitation (Morgan-Klein, 1990). They do not have access to educational opportunities, primary health care or safe accommodation (Stein *et al*, 1994; Power *et al*, 1995), and their mental and physical health can be compromised (van der Ploeg and Scholte, 1997). In the UK, work with these young people is generally under-funded, unpopular and poorly co-ordinated: responsibilities tend to fall between different agencies, and service provision is piecemeal at best, failing to meet the full range of needs of young people on the streets (Boyd *et al*, 1999a).

During our intervention, the Police in Manchester estimated that they dealt with around 2,000 young people living on the streets of that city in the previous year alone. When we first heard this we were shocked, and most of the people we have talked with about it have the same reaction. It is largely a 'hidden' problem—something one expects in Latin America, where absolute poverty can be found in just about every major city, but not in the UK with its supposedly advanced welfare state. It seems that our society manages to keep absolute poverty (including that faced by young people under 16) largely invisible.

10.8 First Contacts

We first found out about the possibility of conducting this intervention when we were contacted by the three voluntary organisations. They had heard about our work at the Centre for Systems Studies from Roger

Adams, an internal consultant in the Children's Society. The consortium had short-listed us along with nine other research groups whom they thought might be appropriate to undertake the intervention. We were invited to tender for the work.

The request for tenders was quite explicit about the fact that the voluntary organisations wanted the work to be undertaken in an *action research* mode (i.e., it was quite clearly an intervention, not a pretence at observation). They believed that the exploration of values should be at the heart of the intervention; the researchers had to have experience of working with young people; and they had to demonstrate commitments to the participation of young people, equal opportunities and anti-oppressive practice. Given everything we have already said about our methodology, it should be quite clear that the values of the commissioning organisations and our own were very close (at least it seemed that way from the documentation, and this was later confirmed by our experience of working with the agency representatives). In addition, we were in agreement that this was an important project to put our time into, in the sense that anything we could contribute to improving the lives of young people on the streets (many of whom are in crisis and extremely vulnerable) would be very much worthwhile. We felt that we could make a positive difference. The obvious 'fit' between the consortium's expressed purposes and values and our own encouraged us to put some considerable work into producing our bid, and we were rewarded with the contract.

10.9 *Our Initial Proposal*

Our specification of methods in the tender document was quite general, proposing an initial period of interviews and workshops with young people on the streets, feeding into a series of multi-agency workshops (with participation from young people) to develop ideas for change. We proposed interviews followed by planning workshops for three reasons:

First, it is important for agencies to listen to the voices of young people so that the services they design make a genuine difference to their lives, in the young people's own terms rather than in terms set by a purely 'adult' agenda. A young person rarely runs away 'for fun': there is usually a serious problem that s/he is trying to escape, and for this to be addressed the young person must first be heard.

Second, young people still occupy a marginal position in British society (albeit less so than was once the case), in the sense that in many families they are seen as subordinates. Also, they are not regarded in law as capable of making 'adult' decisions, and they are often looked upon as less 'rational' than people over 16. This is reinforced by legislation that establishes minimum ages for driving, marrying, receiving welfare benefits, consenting to sex, buying cigarettes and alcohol, etc. While there are obvious reasons for much of this legislation (for instance, without an age of consent it would be much more difficult to prosecute child sex abusers), it nevertheless has the side effect of adding to the marginalisation of young people. Indeed, young people have no legal right to control over some aspects of their lives that we actually regard as essential.[18] We therefore wanted to be sure that young people's views were included in a meaningful manner, and we were conscious of the need to avoid replicating their marginalisation (as far as possible) in the design of our intervention. Here we see the first instance of boundary critique in our work.

Our third reason for interviewing young people before starting a series of planning workshops was strategic. We were aware that most workers in the 'caring' professions are graduates who have been socialised into an 'observational' mode of research: even if qualitative methods are accepted, there is usually still an assumption that these will be used to *find out about* the situation rather than change it. As the commissioning organisations had asked explicitly for *action* research, we knew that we would not encounter major resistance to intervention. However, we were still aware that some interpretations of action research are less explicitly interventionary than the one usually embraced by Community OR practitioners (applied social science, where researchers conduct a study and then write it up for others to use to make changes, is occasionally confused with action research). Therefore, we thought it would be wise to talk about finding out about the situation before launching into planning workshops.[19]

Although we set out this general direction for the intervention, we made it clear that local contingencies may require a change of direction,

[18] We came up against a good example of this in our intervention. When children are in care, it appears that parents have an automatic right to visit them—regardless of whether the child wants this or not. We met several young people who said that they did not want their weekly visit from a parent whom they regarded as an abuser, even though the meeting was supervised. One young woman said that she wanted to break off all contact with her mother, yet she was being prevented from doing so by the care home staff. However painful this is for the parent, we feel that young people should be able to choose not to have enforced social contact of this kind.

[19] In an evaluation session, conducted near the end of our intervention, Roger Adams (a Children's Society action researcher) expressed surprise that we had

and if this turned out to be the case it would be negotiated with our steering group (made up of representatives of the three funding organisations). We did indeed find it necessary to depart from our original proposal, and many things happened that we had not anticipated. Rather than dwell on the original proposal, we will provide an overview of our approach (next), and then a narrative of the course of events, the questions they raised, and how we answered them through a creative process of mixing methods.

10.10 *An Overview of the Intervention*

Altogether, there were four phases in the intervention. Phase One was about identifying key issues (asking 'what *is* the case?') through interviews with young people on the streets, thereby building up a picture of their situation from their own point of view. Phase Two was about identifying important goals for different stakeholders (specifying 'what *ought* to be the case?'), including young people aged under 16; young people over 16 with previous experiences of running away; carers; and all the various agencies involved in working with young people. This was done through a series of workshops. Phase Three involved workshops with individual agencies and multi-agency groups to plan how to realise the goals identified. Phase Four involved communicating about the project through a variety of publications aimed at different audiences.

Phases One to Three took nine months to complete, and then Phase Four was completed in the six months after the official end of the project. A narrative of our progress through the intervention is provided below.

10.11 *Establishing the Ground Rules*

We started the intervention with a series of meetings with our steering group to work out the details of how we were going to proceed. Because we were going to talk with vulnerable young people under 16, we were

not gone straight for planning workshops involving a partnership of young people and professionals. Interestingly though, the consensus amongst the agency representatives was that our initial investigative research based on interviews with young people was an essential aspect of our intervention. They said that it produced a powerful picture of the needs of young people that agency representatives felt they had to respond to by making positive changes.

asked to produce written documentation about our ethics, principles and practices; design a complaints procedure in case any young person had a grievance against us; and agree a confidentiality statement.[20] We also agreed that any quotations by young people reproduced in reports would be anonymised. It is a legal requirement in the UK that, whenever someone under 16 is interviewed, an 'appropriate adult' other than the interviewer is present to represent the young person's interests (see Sieber, 1992, for a discussion of the ethical issues associated with involving potentially vulnerable people in research). The Children's Society volunteered two of their street workers, a man and a woman, to act as appropriate adults and accompany us during interviews and workshops.

Another feature of these meetings was a series of discussions about how we were going to get access to young people and involve them in the intervention. We were talking about young people with no fixed abode, many of whom were actively avoiding any contact with the 'authorities' for fear of being returned to the abuse they had fled from. Again, the Children's Society street workers were the answer: they knew where the young people hung out, and volunteered to take us around in the evenings and at night to make contacts and conduct interviews. However, we were told that it would be impossible to arrange a participative workshop with young people: they tend to live hand-to-mouth, one day at a time, and would not be able to plan for attending such an event. An alternative, however, would be to hold a workshop with young people who had previously run away from children's homes, so this is what we agreed to do.

Dealing with the above issues involved boundary critique. Although they are all 'practical' issues (about addressing vulnerabilities and finding ways to make enough contacts with young people), to have neglected them would have resulted in the marginalisation of the very people the project was designed to help. We were very conscious of the need to guard against this kind of marginalisation.

[20] The confidentiality statement we ended up with was adapted from one used by the Children's Society, whose street workers go out at night to make contact with young people on the streets. It basically states that all discussions remain confidential unless a young person discloses harm being done to themselves or others. This caveat was necessary because of the possibility that a young person might allege abuse. Indeed, an allegation of physical abuse at a children's home was made during the course of our intervention, which the Children's Society followed up on our behalf.

10.12 *Phase One: Identifying Key Issues*

We considered a variety of options for identifying the key issues from the perspectives of young people—we did not automatically go for interviews. One idea was to produce a video or photographic exhibition in partnership with young people on the streets, as suggested by Dewdney *et al* (1994).[21] This seemed like an attractive option because it would present the issues in an emotionally powerful manner, making it hard for the agency representatives planning new services not to respond positively (and it would also be fun for the young people). However, we ended up rejecting it on the grounds of lack of time, the need for confidentiality, and lack of sufficient expertise to guarantee a polished product.

We therefore accepted that interviews were going to be the most effective option, but what kind would be best? We were aware of a technique for surfacing issues that involves the use of photographs to stimulate debate. The idea is that people pick photographs that have some meaning for their lives, and then discuss the issues they raise (Prosser, 1998; Weil, 1998). However, because the situation we were entering was relatively unusual, we were unsure exactly how the use of this method might pan out. We therefore decided to pilot it. We gathered together two sets of images. One was photographs, culled from magazines, that seemed (in our eyes) to be relevant to the situation of homeless young people; and the other was a set of cards with evocative paintings which had been specially designed for raising issues in the context of counselling. We went out for a couple of nights, using the cards with two young people; the photographs with another two; and ordinary questioning (without visual images) with a final two. We compared the outputs from the interviews and asked the young people themselves what they thought of the pictures, and the consensus seemed to be that the visual images did not add much value to the interviews. There was one exception to this, however, which was a young man with a learning disability who seemed to find the cards helpful. We therefore decided to go with ordinary semi-structured interviews, and hold the cards in reserve in case they were needed in individual cases.

[21] Maturana's (1988a,b) theory of autopoiesis, and Bilson's (1996, 1997) application of it in interventions, informed our thinking about this. Maturana (1988a,b) argues that people are able to switch to a new 'rational domain' (they can think in new ways) if an emotional reaction is triggered that makes the switch compelling. Bilson (1996, 1997) explains how emotions can be stirred through the use of personal anecdotes when presenting research data. A video, which allows personal appeals to be made by young people, would be even more compelling.

10.12.1 *Designing the Interviews*

In planning the interviews, consideration was given to producing an account of the situation of 'detached' young people that would engage and interest other stakeholders. We therefore wanted to elicit compelling stories about the young people's lives. We also wanted to strike a balance between our own wish to get these stories (which would require some imposed structuring from us), and the wishes of young people to use the interviews in other ways (if indeed they did). For example, we foresaw the possibility that some young people might want to use the interviews therapeutically. This might happen directly through the discussion of issues, and indirectly on account of information about their individual situation being heard by the appropriate adult who would be in a position to offer longer-term support. Here, using the language of boundary critique, we were trying to ensure the inclusion of multiple stakeholder purposes: it is very easy for researchers to marginalise the purposes of others, making them 'subjects' of research instead of participants (Heron, 1988).

We used some of the literature on interview methods[22] to inform the drawing up of guidelines (which were approved by our steering group), and these can be found in our final report on the intervention (Boyd *et al*, 1999a). We also used a pre-prepared list of questions to guide the interviews, but this was not intended to be overly prescriptive, and generally questions arose naturally out of a fairly open-ended discussion (see also Foddy, 1993). An 'expenses' payment of £5 was made at the end of each interview, allowing us to formally recognise the value of the young person's time. When researchers are paid for their time and other participants are not, this can sometimes alienate the unpaid volunteers who may view the researchers as relatively privileged compared to them. In terms of the theory of boundary critique, if we had expected young people to contribute without being paid, this may have established a marginalisation dynamic that would have prevented effective involvement.

There were, of course, inherent limitations regarding what could be achieved by way of involving 'detached' young people through interviews, and we decided that we simply had to accept these. Young people who had been away from home for a relatively brief length of time might either be in a state of crisis or on a high with the novelty of being away, and participation in our intervention might simply be seen as an unwelcome

[22] See, for example, Rich (1968); Bradburn and Sudman (1980); Powney and Watts (1987); Fine and Sandstrom (1988); Freire and Faundez (1989); Morrissette (1992); Weber *et al* (1994); Holstein and Gubrium (1995); Drever (1995); Kvale (1996); and Graue and Walsh (1998).

imposition. Also, these young people would quite likely move on within a few days, making them unavailable for more than one-off involvement. Furthermore, we envisaged that involving black young people would be more difficult than involving their white counterparts, since the former tend to make more use of networks of friends and family, and generally adopt a lower profile to avoid racism (Julienne, 1998).

We recognised, however, that it would be possible to contact other young people with relevant views and experiences, including older young homeless people who had been runaways in the past. van der Ploeg and Scholte's (1997) research suggests that a relatively high proportion of homeless adults started living on the streets as teenagers. It would also be possible to involve young people who had been missing in the past, but who were currently living at home or in care (as suggested by our steering group). More generally, we felt that young people in care could make a useful contribution, even if they had not gone missing themselves: this is because they are at a high risk of running, and most would be living with other young people who had gone missing in the past.

Our steering group suggested that some young people might prefer to be interviewed with just ourselves and the appropriate adult present, while others might prefer to be seen together with a group of friends (for support). They argued that we should respect the young people's wishes on this, not try to make them talk individually if they did not want to. When it actually came to it, most were happy to talk on their own, but four group interviews were undertaken. In these, we tried to encourage open, respectful but critical discussion amongst the participants.

At this early stage we also asked ourselves, in the spirit of boundary critique, should we involve other stakeholders? We thought of various 'pros' and 'cons'. An important 'pro' was the fact that involving others would enable us to build a richer picture of the situation, including a diversity of perspectives. A 'con', however, was the desire to ensure that young people were really heard by the agencies: mixing in other viewpoints could dilute their voices. We therefore decided not to involve any 'professionals' at this stage, but thought it worthwhile to consider the involvement of other non-professionals who might find it difficult to get their voices heard. We had parents and foster carers in mind. Our steering group advised us that it would not be easy to involve parents, partly because they would be difficult to contact (none of the organisations we were in touch with at that time kept records of their names and addresses). It was also unlikely that they would want to participate because of the stigma, and because of their emotional state if their child was missing at the time. Nevertheless, we thought we should try. We got the support of a local newspaper and radio station, and asked for members of the public with an interest in child homelessness, including parents of young people who had run away, to come forward. Unfortunately, none did. However, involving foster carers (whom we contacted via Social Services) was much easier, and several were interviewed.

10.12.2 Conducting the Interviews

When it came to interviewing the young people, we felt that it was important to have a preliminary discussion with the potential interviewee to assess his or her current situation, and for an informed decision to be made about whether the interview should go ahead. The appropriate adult made the first introduction, as in a number of cases they already knew the young person. We always tried to maintain a gender mix among the adults: if the interviewer was male, we asked for a female appropriate adult, and vice versa. Before beginning, we ran through the contents of a leaflet that we gave to the young person, explaining the main aims of the intervention. We then gave the young person a copy of the confidentiality statement, explained it, and asked if it would be OK to record the interview and take some notes (saying that these would help in putting across to the agencies what young people had to say). Potential interviewees were encouraged to ask questions, expressing any concerns that they might have about the interview, and to decide whether they wanted to go ahead with it, and on what terms.

In addition to participating in the discussions prior to the interview, the appropriate adult kept time and observed what appeared to make the process effective or ineffective for the young person. By and large, the roles of the interviewer and the appropriate adult were kept separate, but occasionally the appropriate adult did ask questions during the interview when they had encountered the young person previously and felt that they could help the interviewer to reveal relevant information.

For each interview, the date, time and location were noted, together with the age, gender and ethnic origin of the interviewee. Young people were not asked about their ethnic origin or sexual orientation, as it was felt that this might inhibit the establishment of a rapport, so ethnicity was a judgement by the interviewer. Some young people did, however, make their sexuality known.

The series of interviews ceased when we felt that no new information was being revealed. Twenty-three young people were involved in total. No young person refused to be interviewed or to have their interview tape recorded, and all except two of the recordings could be fully deciphered. Because the young people volunteered to be interviewed, and were keen to tell their stories, the presence of an appropriate adult and a tape recorder appeared to be relatively minor barriers. Our 'externality', being from a University and (in the case of two of us) from another city, may also have enabled interviewees to be open about their experiences.

At the end of the interviews, interviewees were asked how they had felt during the interview and how it could have been improved. Although they inevitably found aspects stressful, as confirmed on occasion by their body language, they did not find the experience negative overall, and some said that it had definitely been worthwhile from their point of view. The value of the process to the young people was indicated by the fact that, although the majority of them were happy to end their involvement with the intervention at this stage, four subsequently became involved in later phases of it.

On a number of occasions, the appropriate adults who sat in on the interviews said afterwards that they had learned new information about the young person, which would be relevant to their future work with him or her. Indeed, the Children's Society instigated a review of their approach to street work after experiencing our interview technique: they realised that the young people appreciated a supportive questioning approach. Before the review they had tended to avoid asking direct questions for fear of frightening off the young people. This change was the first positive result from our intervention.

The expenses payment proved to be an incentive for some young people to participate who would not have done so otherwise, this being particularly apparent for young people at a children's home who regularly went missing and claimed that they did so in order to acquire money through illegal activities. For most, however, the chief motivation appeared to be the opportunity to tell their story to someone who would listen, and might bring about change.

This first phase of the intervention lasted two months, and we worked a couple of nights per week on it. It was generally only possible to conduct one or two interviews per night, as most of the time was spent just looking for new interviewees to talk with.

10.13 *Phase Two: Evolving Stakeholder Goals*

Having explored what *is* the case (from the perspectives of young people, plus several foster carers) in the first two months of the intervention, we then wanted to move on to look at what *ought* to happen (from the points of view of a wide range of stakeholders). In putting together our original tender document, we had in mind that the interviews would be followed by a series of planning workshops, and these would be similar to the planning workshops that we had run in two previous interventions (Cohen and Midgley, 1994; Midgley *et al*, 1998). In both those interventions we created a synergy of methods from Critical Systems Heuristics (CSH) (Ulrich, 1983) and Interactive Planning (IP) (Ackoff, 1981). This involved asking Ulrich's (1983) twelve questions about purposes, control, expertise and legitimacy (mentioned earlier) in relation to what ought to be designed, taking account of only minimal constraints. These constraints are that people's proposals must be feasible, viable and adaptable (see Ackoff, 1981, for precise definitions). Thus, people are able to explore different possible boundaries and values while generating their plans, while simultaneously avoiding utopian visions that can never be operationalised. The result should be an 'ideal' plan (Ackoff, 1981): one that transcends the current limitations people experience, but is still possible to implement.

10.13.1 Designing the Workshops

As we entered the second phase of our intervention, we reviewed the idea of a CSH/IP synergy in discussion with our steering group. First of all we asked the question, who are the key stakeholders? In addition to young people and carers, we identified a wide range of agencies. These included Social Services (which runs several children's homes and has responsibility for child protection more generally); the Housing Department; the Police; the Education Department; a Methodist Housing Association; a training facility for unemployed young people; the Careers Service (which provides careers guidance, mostly in schools); a drop-in for homeless people; the Children's Society; a mental health crisis service; a mental health drop-in centre; the Health Authority; NCH Action for Children (a voluntary organisation running children's homes); and the Manchester & Salford Methodist Mission.

We discussed the fact that, at this stage, there was no need for a great deal of detail. The agencies did not already have a clear idea how to move forward, so it would be more important to communicate what the young people had to say, and to surface the key values which should inform the development of services. In addition, it was anticipated that time would be an issue for the agency representatives. It would be unlikely that they would be able to put aside more than half a day at a time, and then only on an occasional basis (none of the agencies other than the Children's Society had staff whose sole remit was homeless young people). We were told that there was some scepticism about multi-agency working in the locality: there was a history of conflict between a couple of the agencies, and some previous attempts at co-operation had been seen as 'all talk and no action'. It would therefore be important to make a positive impact first time, otherwise there was a risk that participation could fall away sharply. All of this confirmed our belief that the CSH/IP synergy was appropriate: it can indeed be carried out in half a day; it focuses on values; it does not assume that people already have a strategic direction in mind; and it tends to produce a striking consensus. In our experience the areas of disagreement that are surfaced are minor in comparison with the areas of agreement, so people almost always get a sense of positive achievement from its use.

We also discussed with our steering group the possible participation of young people in the workshops. We had already agreed that we should work with groups of ex-runaways in children's homes rather than young people who were living on the streets because of the logistical problems of getting the latter to attend workshops (see earlier). So the question was not 'who should be involved?', but 'how should the involvement be handled?' We determined that it would not be advisable for young people to attend the same workshops as adults, at least initially. Some young people may be embarrassed in front of professionals by poor reading and writing skills; they would be less used to participating in relatively large meetings; and they might prefer a more 'playful', creative approach. In addition, young people might be inhibited by the presence of staff from Social Services and the Police. Many regard these agencies as part of the problem they are trying to deal with: the Police and Social Services are required by law to return young people to where they have run from unless there is clear evidence of abuse.[23] Also, more generally, lower status tends to be accorded to children's views compared with the views of adults (see the earlier discussion of the marginalisation of young people), and we did not want to replicate this dynamic in our workshops. For all these reasons, we decided to work separately with professionals and young people. Indeed, working with young people in a separate group had the added advantage of enabling their viewpoints to be built up and thought through without being 'squashed' at an embryonic stage by adults.

Then we asked the question, would it be better for all the agencies to meet together in a multi-agency forum, or would it be preferable to keep them separate? Again, there were both 'pros' and 'cons' to consider. In favour of working with the agencies separately was a history of conflict that had dogged some inter-agency relationships. Indeed, the Children's

[23] The problem is, when there is abuse, young people are generally reluctant to disclose it to these two agencies, so they are nearly always returned to the abusive situation.

Society, one of the funders of our intervention, was an agency that had been a participant in this conflict, so we had to consider the possibility that we might be perceived as having a vested interest.[24] The last thing we wanted at our crucial first meeting was pseudo-dialogue, with agency representatives skating around difficult issues, knowing that they might reach a verbal agreement but actions would not follow from their words. However, in favour of bringing the agencies together was the fact that they were mostly at a very early stage in their thinking about how to offer support to homeless young people. Some had policy statements, but only the Children's Society was actually offering a service to young people on the streets, suggesting that the agency representatives might be open to setting aside previous conflicts in order to forge a new, consensual direction. In addition, time was short[25], and working with all the agencies together would mean one or two workshops rather than eight or ten. We therefore went with the idea of a multi-agency workshop, knowing that we would have to deal with the history of conflict along the way.

In making ready for this, we prepared the following set of goals which we sent through the post and asked the participants to commit to from the start:

10.13.1.1 Overall Goals for the Group

- Produce an innovative vision to guide the genuine improvement of services for young people who go missing from home or care;
- Say who should be involved in the more detailed planning of these services, and in what ways;
- Identify what can realistically be achieved;
- Identify barriers to be overcome and possible constraints to be tested; and
- Build commitment to bringing about useful change.

[24] As it happened, the fact that we were employees of a University from another city meant more to the key stakeholders in the conflict than where the money for the intervention was coming from. We were therefore viewed positively, as external facilitators able to broker change, rather than people with a vested interest in taking sides.

[25] Although we had nine months to complete the intervention, in our judgement we were trying to cram eighteen months worth of work into this time. We were explicit about this from the start in discussions with our steering group, as we knew that negotiating what we did in relation to the time available was always going to be an issue.

10.13.1.2 Goals for Individual Participants (Including the Workshop Facilitators)

- Identify changes that might usefully be made to your situation and your role in it;
- Develop productive relationships with other participants; and
- Learn new approaches for developing services and working in groups.

We also decided to build in some formal evaluation of our activities to inform both our future decision making in this intervention and our learning about systemic intervention more generally. We therefore designed a questionnaire to be filled in by participants after their involvement in a workshop. This asked questions about people's perceptions of what happened, and their commitment to change (or lack of it) based on the results.

10.13.2 Conducting the Workshops

We set out to conduct three workshops at this point: one with young people under 16 (from children's homes); one with representatives from the various agencies; and one with homeless young people over 16 who had previously been runaways. Only one homeless person over 16 turned up to the latter workshop, so we talked with her individually. However, the other two workshops were well attended: five young people aged 12–15 came to the young people's workshop (all of whom had previously lived on the streets), and twelve agency representatives came to the multi-agency workshop.

10.13.3 The Multi-Agency Workshop

The first one was the multi-agency workshop, which was attended by representatives of all the organisations listed earlier (except the Education Department). Interestingly, the Police and Children's Society representatives came early, and began to chat informally about some of the inter-agency conflicts that they had both previously been involved in. The two agencies were concerned about how the media were to be handled in this intervention. In their view, a great deal of the conflict resulted from a previous media misrepresentation of the Children's Society's point of view on a sensitive local issue, thereby causing a problem for the Police. Therefore, before the workshops had even begun, work was already in hand to set the conflict aside—and without the need for any explicit intervention from us, other than setting up a forum where contact between people was made possible. These discussions continued in the break, and then many jokes about the conflict were cracked after the workshop was over, suggesting that it was well on the way to being overcome.

The workshop started with a presentation of the key issues that had been revealed by the interviews with young people. Prior to the workshop we asked ourselves, how can we present the issues in a manner that will engage the agencies, but without breaking the confidentiality of any of the young people by revealing too much from one person's story? Also, how can we demonstrate the systemic nature of the problems? They were highly interactive, and young people often described vicious cycles that they got into. Finally we asked ourselves, how can we support the agencies in taking responsibility for their part in solving these problems?

We were aware of several techniques for presenting systemic problems. First, there is Checkland's (1981) idea of producing 'rich pictures' (visual representations of issues linked with arrows. The only drawback of this, in our experience, is that the method works best when people can produce their own rich picture: if an 'outside' analyst presents a picture to an audience, it is only meaningful if it is fairly simple. Another possibility was the use of 'signed digraphs', 'influence diagrams' or 'qualitative system dynamics models' (Coyle, 1977; Roberts *et al*, 1983; Flood and Carson, 1993; Morecroft and Sternman, 1994). These are three different terms for a basic kind of system model representing causal links between phenomena. The advantage of these modelling techniques is that they can make feedback loops (vicious or virtuous cycles) very visible—but a significant disadvantage is that they use symbolic languages which, while fairly basic, are not immediately transparent to an untrained audience. They are arguably more self-explanatory than a complex rich picture, but they lack the appeal of visual imagery. Another alternative is 'problem mapping' (Midgley *et al*, 1997), which shows the links between stakeholders' verbal problem statements—but like influence diagramming, it does not have the immediate visual appeal of pictures.[26] Eventually we decided to go for rich pictures, but endeavoured to keep them simple, using cartoons (clip-art from a computer package) linked by arrows to make the vicious cycles as immediately apparent as possible.

We used the rich pictures as a visual aid to describe the systemic problems that young people found themselves caught up in, and added quotations for emotional impact. When our presentation was finished,

[26] See Flood and Carson (1993) for a good overview of diagramming techniques (although 'problem mapping' is not included, as it was invented later).

there was a general discussion in which agreement was expressed that the issues were accurately represented and something definitely needed to be done about them. There was a strong feeling that the words of the young people were powerful and should not be ignored. We therefore moved on to the CSH/IP synergy, as planned. We asked the participants to imagine that all services for young homeless people had disappeared, and their job was to plan a new system from scratch. We presented the three constraints that were designed to stop the plans from becoming utopian (feasibility, viability and adaptability), and we also used the CSH questions to facilitate debate. However, in contrast with some of our previous interventions (Cohen and Midgley, 1994; Midgley *et al*, 1998), we did not work through the CSH questions systematically: we simply held them in mind and went with the flow of discussion.

There was some unwillingness to get into a great deal of detail about the desired properties of services, and group reflection on this brought up the issue of scepticism about multi-agency working. Several agency representatives said that there was no point getting into the detail when they were unsure about whether there was even agreement on basic values. Those from the voluntary sector were particularly sceptical, as some of the statutory agencies' roles were part of the problem described by the young people (e.g., the obligation on the Police and Social Services to return children to the places they run from). We therefore spent most of the rest of the workshop discussing values before producing some general statements about the production of new services (these were not detailed plans). Encouragingly, there did indeed seem to be agreement between the agencies: the Police and Social Services representatives made it clear that they shared the values being expressed by others, and were aware of (and wanted to do something about) their roles in the vicious cycles.

At the end of the workshop we had a list of core values and a skeletal 'ideal service' consisting of three general elements:

- *Prevention* of young people going missing;
- *Support* for young people who are missing from home or care; and
- *Information* to underpin the prevention and support services.

The agency representatives were not at all discouraged by the lack of detail. On the contrary, they were very pleased to have experienced a consensus on values, and wanted to hold a second workshop as soon as possible. We agreed to this, and in between the two workshops we produced a diagram showing how all the group's values could be logically related together (we call this technique 'values mapping'). The values

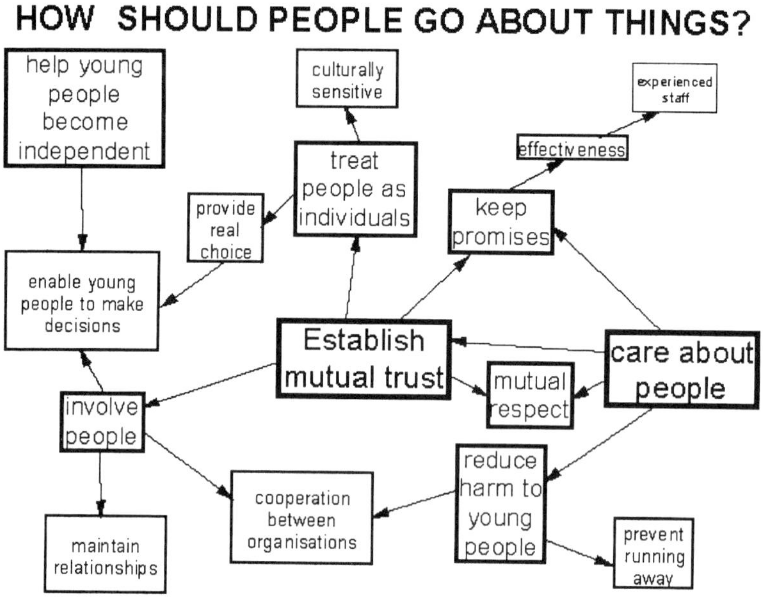

FIGURE 10.5 The values map

map is reproduced in Figure 10.5, and we sent this out by post before the second workshop was convened. All the values were taken from the flipcharts produced in the multi-agency workshop, and further value statements were derived from the transcripts of the interviews with the young people. We added these so as not to marginalise any values that might be important to young people but had not been considered by the adults. The idea was to test out through postal and e-mail correspondence, plus discussion at the next multi-agency workshop, whether the various values (and our proposed connections between them) were really acceptable to all the agency representatives.

The evaluation questionnaires filled in by the agency representatives at the end of the workshop revealed that, while everybody was indeed pleased that they were moving towards a consensus, most did not yet feel that sufficient progress had been made to enable them to commit to any action. Some barriers were mentioned. These included the lack of involvement of some key policy makers (while some people could make resource commitments themselves, others could not); the absence of the Education Authority; lack of time due to workload pressures; and a feeling that the issue was a low priority on some organisational agendas.

10.13.4 The Young People's Workshop

A few days after the multi-agency workshop, we held our workshop with young people under 16.[27] At first we were concerned that there would be almost as many adults as young people (there were five young people, three facilitators and one appropriate adult), and that the presence of so many adults would be overpowering. However, in the event, this did not seem to be even the slightest barrier to participation.

Prior to the workshop, we discussed what we wanted to do. We anticipated that, to facilitate meaningful participation, we would have to moderate our language (e.g., stop talking about 'agencies', 'stakeholders', etc.).[28] We also wanted to make sure that the participants had plenty of opportunities to question us. Furthermore, we agreed that we would use the synergy of CSH and IP methods, despite an earlier thought that we might try something more 'playful' with the young people. We made this decision on the grounds that we wanted the results of the workshop to be comparable with what the multi-agency workshops produced, and we didn't want the output from the young people to appear less 'rational'. Again, this would have replicated the marginalisation of young people—especially *these* young people, from children's homes, who could easily be stereotyped as 'disturbed adolescents'.

Another issue we were not sure of was the degree to which the young people would be able to think in the 'ideal' world (transcending the limitations of current services, but still producing plans that could be implemented). Would they plan with a clean sheet, or would they be unable to think beyond what currently exists? Ackoff (1981) describes how many adults, let alone children, have difficulty focusing on an ideal. At this point there was a difference of opinion between the facilitators about our expectations of the young people. One view was that they would be very concrete thinkers, and another was that they would find it

[27] In-between the two workshops, we were also given an unexpected opportunity to visit a Social Services children's home and meet with twelve of its staff, including the manager. However, rather than schedule a planning workshop with these staff, we decided to make this a focus group and explore their perspectives on the issues raised by the young people in the first phase of our intervention. This material added to the insights that we had already gathered from the young people and foster parents, and enabled us to enrich the narrative about the young people's problems that we wrote for our final report (Boyd *et al*, 1999a).

[28] This was important. We quickly found out that even words like 'organisation', which we used without even thinking about it, were not meaningful to a couple of the participants. However, they had no trouble challenging us when we were unclear, and we quickly adapted to each other's language.

easier than adults to think about ideals because they would not be operating with the same history of preconceptions about what is and is not possible. We discussed how we would handle the situation if indeed they could not conceptualise an ideal. We agreed that we would make comparisons between proposals for new services and their children's homes; ask them what they would change at their homes, and if the improvements they wanted would be relevant for the new services.

In the event, we need not have worried so much about this issue. The young people focused most of their attention on two issues. First, the need for some form of 'refuge' in the city which runaways could go to for food, shelter, a shower, confidential advice, etc.; and second, the need to improve children's homes so that young people would be less likely to run from them. What we found particularly striking about this workshop was the participants' ability to take an overview of the issues and still produce a great deal of detail about how these services should be run (and this detail was in marked contrast to the agency representatives' ideal design). They handled a variety of difficult and/or sensitive issues, such as producing a drugs and alcohol policy; rights to confidentiality; informing parents; the attitudes and values of care staff; managing violence; the size of children's homes; etc. In our view, they demonstrated a great deal of sense, compassion and responsibility in the production of their ideal design, and thereby produced a compelling argument for change.[29]

The evaluation questionnaires filled in by the young people revealed a much higher level of personal commitment to action than expressed by the agency representatives in the previous workshop. Most of the young people said that they would be keen to be involved in the future design of actual services.

10.13.5 The Second Multi-Agency Workshop

After the young people's workshop, we set up the second multi-agency event. The first thing that struck us about this was the fact that all the

[29] A striking example of the exercise of responsibility was when they discussed the possession and use of drugs in a refuge. In their experience, drugs and alcohol were often closely linked to violent behaviour. Therefore, while they wanted to offer support to young people with drug and alcohol problems, they also thought that some basic rules about the possession and use of drugs on the premises should be put in place to protect others. When one girl was challenged by another participant for her support for this, on the grounds that she often took drugs herself, she laughed and said 'just because I do it doesn't make it right'. There were many instances of the young people being able to separate their own individual desire to do something from what would be necessary in a refuge for the wider good.

agency representatives returned, except those from the Police and Social Services—the two agencies which had been criticised by young people for being part of the problem. The Police representative was scheduled to participate, but did not turn up on the day (later, it turned out that he had been unavoidably detained—no snub was intended). We had not been able to contact the Social Services representative, and therefore we had to go ahead without him (again it turned out that he was simply away from his office on business, and no snub was intended). There was some disquiet in the group about the fact that these two agencies were absent, especially as (at that time) no reasons were given. A view was expressed that the group was obviously not as far advanced in terms of building bridges and constructing a consensus as they had thought, and that it would not be possible to make any concrete decisions without the involvement of the Police and Social Services.

However, the participants were keen to make good use of their time together. They particularly wanted to see what the young people had proposed as their ideal, so we made a presentation of this. There was a strong feeling that the young people and professionals had similar values, and participants were also reassured to discover that the young people's desired services were similar to their own. The participants could see no reason to object to anything that the young people had asked for, and welcomed the level of detail in their ideal design—but they were concerned that Social Services in particular were not present to hear the presentation and be influenced by it.

The only difference between the young people and agency representatives that was noted by the latter was that the professionals seemed to put greater emphasis on 'prevention', which they said needed more consideration. They therefore spent most of the rest of the workshop discussing what preventative services might look like.

Finally, we brought the workshop round to the issue of how we were to use the rest of our time in this intervention, given the fact that no multi-agency agreement had yet been secured. We wanted some guidance about how to proceed. There were two reasons for asking for this guidance. First, we assumed that the agency representatives would know better than us what would overcome the obstacles we were experiencing (in terms of the absence of key agencies). Second, by giving them some control over how we used our time (by making us their instrument), we anticipated that they would be more motivated to continue with their own involvement. The response we got was that we should endeavour to work with some of the agencies individually (especially the Police, Social Services and the Education Department whom the group thought might feel threatened) to discuss possible routes for action before convening one final multi-agency workshop. The participants argued that top managers needed to be engaged, and agreement reached on principles and mechanisms for stakeholders to work together to develop and implement plans (each agency representative undertook to do this within his/her own organisation). The aim should be to develop a genuinely shared vision prior to the final multi-agency event. The participants also felt that the evidence from young people was powerful enough to persuade the other organisations to get involved once again.

Here, in offering to be the instrument of the group, we deliberately blurred the boundary between our own identity as a team, and the identity of the group as an agent of change. In terms of boundary critique (and some other theoretical ideas about boundaries discussed in Midgley, 2000), blurring these boundaries of identity had a doubly empowering effect. It gave the group a new resource (us) and it granted a new credibility to our actions, stemming from the group's consensus (which was added to the credibility stemming from the voices of the young people). We anticipated that the combined pressure from the young people, the multi-agency group and ourselves would be difficult for the non-participating organisations to ignore.

When people completed our evaluation questionnaires after the workshop, we found that the agency representatives were now much more positive, and most had taken ownership of the change process. This therefore provided a solid platform upon which to build the next phase of the intervention.

10.14 Phase Three: Action Planning

Going into the next phase, we decided not to choose or design any *general* methods: we knew that we were going to have to work separately with a variety of agencies, and that each one was likely to present different challenges. Therefore, we decided to plan each engagement individually as it became apparent what it would entail.

First, we had a meeting with our steering group. There, we discussed our proposed strategy. Our funding organisations were in agreement that we should concentrate in particular on the Police, Social Services and the Education Department. However, we also raised another issue that had surfaced through our own reflections on the boundaries of our intervention. There was a possibility of public resistance to the idea of a refuge for homeless young people (if this was indeed going to be one of the proposals to emerge from our intervention), as there could be a perception that it would attract 'undesirables' to the city.[30] To check this out,

[30] A similar objection to a refuge had been raised by a member of staff at the children's home that we visited earlier in the intervention: he said that he

we proposed that we should contact the Chamber of Commerce to discuss the matter with owners and managers of retail outlets in the city (many homeless young people get involved in shoplifting just to survive). The steering group agreed that it would be worthwhile to identify any such potential resistance so that appropriate preventative action could be taken if necessary.

We held several interviews with retailers, and were encouraged to find that everyone we spoke with had a concern for the young people, and did indeed want something to be done to help them. They thought that a refuge would be a good idea, as it would provide a route away from a hand-to-mouth existence where shoplifting became a necessity. They therefore saw it as in their interests to support such a proposal rather than resist it.

We made contact with Social Services, and they agreed that two of their officers would meet with us. This was a productive meeting for a variety of reasons. First, it emerged that their representative had not deliberately pulled out of the earlier multi-agency workshop, and they did indeed want to discuss the possibilities for change. Second, they were keen to discuss the views expressed by the young people, and were encouraging about the possibility of a positive response. In particular, they were in the process of conducting a review of the running of their children's homes, and wanted to feed the young people's ideas into this. Third, it became evident that the officers we were in touch with did not have the direct authority to implement proposals for change. We therefore worked with them to identify the appropriate route into the organisation's planning process that would have to be taken, without which change could not be achieved.

We also made contact with the Education Department who agreed to a workshop. This was a breakthrough because they had previously not been able to allocate personnel to work with the multi-agency group. We discussed in advance how we were going to facilitate this workshop, and agreed to go with the CSH/IP synergy as used previously, primarily because it would produce results that would be comparable with the material already generated. It would also bring the Education Department 'up to speed' with what we had done. This workshop

would not be in favour of such an initiative in case it *encouraged* young people to run away from their home. However, in the young people's workshop, the participants had anticipated this objection and had talked about it in quite derisive terms, saying that no young person runs away from a children's home just for a shower and a bed for the night. There is almost always a problem that, for one reason or another, cannot be addressed at the home itself.

resulted in a good number of suggestions for service developments, but was not able to identify concrete ways of moving forward on them in the short-to-medium term. The primary barrier was perceived to be resource constraints that, they said, were so severe that they had little staff time and money available for anything other than fulfilling their statutory responsibilities. This was disappointing, but we resolved to take the issue to the multi-agency group for discussion to see what support might be forthcoming.

Arguably the most significant breakthrough came in a workshop with the Police, set up by the two Police representatives who attended the first multi-agency workshop. Eight officers with different planning and managerial responsibilities connected with the issue of homeless young people participated in this workshop, most of whom had never met previously and were glad of the opportunity to work together. Again, we spent some time prior to the workshop discussing how we would facilitate it. It had been made clear by the representatives setting up the workshop that they wanted to hear about the results obtained so far. We were therefore asked to present these, and were happy to oblige: we used the various rich pictures we had developed in order to give an overview of the issues, and also introduced the idea of a refuge that had come from the young people.

We were also aware that many of the young people whom we had interviewed in the first phase of the intervention had made comments about the role of the Police, and even related specific incidents that they were unhappy about. We were keen that these voices should be heard in the workshop, and that the Police should have an opportunity to think about how the problems could be addressed. The easy option would have been simply to ignore the young people's comments and thereby avoid controversy, but we regarded the marginalisation that would have resulted as unethical. The question was, how should we go about dealing with what could be a highly sensitive issue without jeopardising the other objectives of the intervention? We considered the production of a written narrative, containing quotations from the young people, which could be circulated before the meeting. However, we were concerned that we might be perceived as constructing the issues: it would be much more powerful if our words could be excluded. We therefore prepared a handout that simply contained a list of quotations, taken from the interview transcripts and anonymised, all of which were of direct relevance to the Police. We were not selective: we simply listed every mention of the Police, thereby letting the voices of the young people speak for themselves. We gave the handout to the participants after our initial presentation, and as they began to read it, the room went very quiet. Everybody took the time to digest it thoroughly before speaking.

The first comment was striking. The man who spoke said that he recognised the validity of everything that had been said, and could even take an educated guess at the identities of some of the officers who had been involved in the incidents related in the quotations! All the participants were in agreement, and many similar comments followed. Our overview of the issues, together with the quotations, had presented a picture of an unproductive cycle of young people running away from home or care, being returned by the Police, and then running away again. This had led to a negative attitude among many officers that 'missing from homes' are a nuisance. All the participants agreed that this cycle needed to be broken.

Our plan at this point was to enter the CSH/IP synergy once again, to produce results that would be comparable with the ones from the other workshops. However, there hardly seemed any need for facilitation: the participants immediately started coming up with ideas for improving the situation. There was an atmosphere of creative excitement in the room, and one after another the ideas went down on paper. Having generated a whole set of possibilities, the participants asked themselves which ones could be pursued straight away. They identified two, both of which would need the co-operation of Social Services and the Children's Society. The main insight was that, while the Police and Social Services both have a statutory obligation to return young people to the place which they have run from, the Children's Society does not. This means that, if a Children's Society worker can be involved when the Police and/or Social Services pick up a runaway, s/he can have time alone with the young person before a decision is made about what to do with him or her. While young people are often reluctant to reveal abuse to a police officer or social worker, they are more likely to trust a youth worker from the Children's Society. Thereby, the cycle of abuse, running and return can be broken.

The participants then said that they would have to gain the support of their senior management team before going ahead, but were optimistic of success because (apart from the importance of the issue) there would be a significant resource saving if the vicious cycles could be stopped. They then suggested that the other ideas should be referred to the internal operational research unit in the Police, with a proposal to set up a pilot study (including a controlled experiment) to find out whether the ideas would indeed be worth pursuing. Again, the officers were confident about the possibility of gaining senior management support for this. Shortly after this workshop, the Police got the go-ahead to start negotiations with the Children's Society and Social Services, and made an approach straight away.

10.14.1 The Final Multi-Agency Workshop

Following the single-agency workshops, we convened a final multi-agency workshop (as previously agreed with the multi-agency group). As we were aware that this would be the final event, we contacted as many potential participants as possible. The Police, Social Services and the Education Department all came, in addition to the sponsoring organisations, the Housing Department, the Health Authority, foster carers and representatives from a variety of other voluntary organisations.

Again, we needed to design methods for this workshop that would address a variety of questions. These included, how can we make sure that young people's voices are listened to, and remain the foundation for any plans that emerge? How can we create a positive atmosphere that will make people feel that they have achieved something, and will enable them to take ownership of the process of change? Related to this, how can we make sure that the process of change is not dependent on our continued facilitation? And finally, how can we do all of this in half a day!?[31]

We decided to recap the situation as constructed in the rich pictures from the interviews with young people and others, as we were aware that some new participants would be involved. We were also aware that some successes had already been achieved, and these could also be used as the basis for a presentation. There was the agreement with Social Services to participate in making changes; a commitment from a small group of young people to future involvement in the design of specific services; the ideas from the Education Department (although the participants in that workshop were not optimistic about their realisation in the short-to-medium term); the ideas from the Police, which already had internal support and were rapidly progressing through a process of negotiation with the Children's Society and Social Services (importantly, this was now happening independently from our involvement); and the previous commitment from the multi-agency group to pursue change, which they had said was dependent for success on the participation of the Police, Social Services and the Education Department, all of whom were now on board.

In the previous set of workshops, four projects (amalgamating ideas from several sources) had been suggested:

- The design and establishment of a refuge;

[31] The time constraints of the workshops had created difficulties all along. However, we had to be realistic and accept that asking agency representatives to give up full days for the workshops, when most of them were doing this planning in addition to their normal jobs, was too much and would kill participation.

- Enhanced co-operation between the statutory and voluntary sectors to break the vicious cycle of abuse, running and return;
- The development of a long-term strategy for work in schools to support vulnerable young people; and
- The production and dissemination of information for young people about going missing.

See Boyd *et al* (1999a) for more detail. In addition to these projects, there was a general commitment to making sure that all the organisations have 'caring about people' as the principle value informing both policies and everyday practice. Also, Social Services had some related initiatives in hand (amongst other things, the review of children's homes) which were being developed independently from this intervention, but would complement the projects identified above.

The first half of the workshop was therefore presentations from us to bring everyone 'up to speed'. The question then was, how to structure the second half? We knew that there would be no time for the use of some of the problem structuring methods (e.g., Checkland, 1981) that we have used on previous occasions to move on from CSH/IP outputs. We therefore needed to be quite directive in the second half, to minimise 'drift' in the discussion and ensure that the focus was on how the group could take forward the ideas. We decided to check with the group our own perception that the four specific projects were indeed priorities, and then (if they were accepted) ask people to work in small groups on the one that they had most personal commitment to. The idea was that, if everyone focused on just one thing they were committed to, they would be more likely to take ownership of driving future action.

People did indeed agree on the four projects, and worked in small groups as we had planned. We asked the groups to elaborate the project ideas where necessary, and to make sure that they came out with two things: a clear plan to take the projects forward, and the name of a co-ordinator who was prepared to take responsibility for keeping the project on track. All the groups did this. When we came together at the end, the Children's Society representative volunteered to act as 'overall co-ordinator', and we were happy to bow out in the knowledge that there were named individuals willing to accept responsibility for making the changes happen.

10.15 Stage Four: Dissemination

While in many projects the methods of dissemination are not a subject for explicit reflection (it is generally taken for granted that a report will be produced as a record of events), this was not the case in this intervention. We were keen (as was our steering group) to ensure that the means of dissemination supported the continuation of the intervention after our departure. We therefore planned to produce 500 copies of a report (Boyd *et al*, 1999a), to be distributed free of charge to people in key agencies. However, in addition to this, we agreed to print 1,000 magazines summarising the results of the intervention (Boyd *et al*, 1999b): these were to be made available to young people as well as professionals. Finally, we produced a set of 1,000 'poster kits' (self-assembly posters) highlighting the main issues. These were for distribution to all services involved with young people, the idea being that they could be pinned on walls and notice-boards to remind professionals of the concerns of young people and the commitments that the agency representatives had made.

10.16 Outcomes after Six Months

We kept in touch with the commissioning organisations for six months after our involvement in this intervention, and were regularly informed of developments. The Police, Social Services and the Children's Society built a close working relationship in Manchester, as planned. In our view, this alone is a significant achievement: prior to our intervention, these agencies were hardly talking to one another. Active fundraising for the refuge was also underway. The information and education strategies were put on a back burner, but we expected this as they were explicitly meant to be longer-term initiatives, and there were significant barriers to be overcome in terms of raising the resources for implementation. The important thing, however, is that these ideas were still said to be 'alive', with people committed to keeping them on the agendas of their organisations. Finally, the collaboration between the statutory and voluntary sectors has spawned a whole new set of ideas which are being pursued, suggesting that the intervention is now firmly in the hands of the local participants. To us, this is the most important indicator of success, as it was always one of our primary purposes to facilitate the engagement of others in change, based on the needs expressed by young people. We were not there to direct change ourselves.

10.17 Reflections

We believe that this application of our systemic intervention methodology demonstrates the value of both boundary critique and methodological pluralism, and the synergies that can be achieved when they are brought together. Let us start our reflections by focusing on methodological pluralism.

As our intervention unfolded, we found it useful to draw upon the following methods and techniques (adapted to the contingencies of the local situation):

- Individual interviews with young people, foster carers and retailers;
- The use of photographs and saga cards to stimulate ideas;
- A focus group with care staff;
- Rich pictures;
- A synergy of the critical systems heuristics (CSH) questions with idealised planning (IP), implemented in separate stakeholder and multi-agency workshops;
- Values mapping;
- Small group, multi-agency action planning;
- The production of reports, magazines and posters; and
- Formative evaluation (feedback questionnaires filled in by participants).

As far as we are aware, no previously existing set of methods covers all of the purposes addressed by these. Indeed, the range of methods that, in principle, it is possible to draw upon is far wider than the above, including many quantitative as well as qualitative methods.[32] We therefore argue that pluralist methodologies, like our own systemic intervention approach, offer a more flexible and dynamic practice than any 'isolationist' methodology that prescribes only a narrow range of methods. This is not a new conclusion (see, for example, Jackson, 1987b; Flood, 1989; Mingers and Gill, 1997), but it is important to reiterate the value of methodological pluralism given the continuing tendency for some people to champion particular isolationist methodologies and methods as panaceas (as observed by Jackson, 1995).

However, in our view, the real value of our systemic intervention methodology becomes apparent when we reflect on the explicit use of boundary critique in our intervention. It is important to point out that we did not just use boundary critique as a sociological device, to theorise about power and marginalisation in order to produce a descriptive account of the situation of homeless young people. Rather, in every case, the use of boundary critique either *informed the creative design of methods* or

[32] See Midgley and Floyd (1990) and Cohen and Midgley (1994) for examples of systemic interventions mixing quantitative and qualitative methods.

became an integral part of the methods themselves. Here are some examples:

- Understanding the double marginalisation of homeless young people (children generally being ascribed a lower status in society, and those on the streets often being viewed as particularly profane) made us aware of the need to privilege their voices in the first phase of the intervention;
- Being aware that researchers often reduce other people to the status of 'subjects' of research by unilaterally imposing their own purposes, which is another form of marginalisation, we deliberately left space in the interviews for young people to introduce their own issues and purposes;
- Understanding that all-male or all-female research teams can introduce marginalisations based on gender, we always sought to go onto the streets in mixed-gender pairs;
- Realising that our status as paid professionals might alienate young people who have no source of income, we paid the latter for their participation;
- Being informed that professional child-care workers have a tendency to marginalise the views of foster parents led us to modify our initial decision to base the rich pictures solely on the views of the young people (but, to keep the marginalised voices central, agency representatives were still excluded at this point);
- Knowing that marginalised groups often need time and space to develop their views before communicating them to others, we decided to keep professionals and young people separate in the planning workshops so that the young people's understandings could emerge without competition from other voices;
- To facilitate both young people and agency representatives in reflecting on the boundaries of their own ideas about what ought to be designed, we employed Ulrich's (1983) CSH questions;
- To prevent the marginalisation of young people's voices in planning when it became apparent that we would not be able to hold workshops with young people on the streets, we accessed young people in children's homes;
- Being aware that part of the marginalisation of young people involves adults viewing them as less 'rational' than themselves, we chose to use the same methods (the CSH/IP synergy) with both the agency representatives and young people so that their outputs would be directly comparable;
- To ensure that the 'values map' did not marginalise the views of young people by focusing only on professional values, we deliberately added contributions from young people;

- Understanding how boundaries can be established around research teams in relation to other participants, we deliberately blurred these boundaries to make us the 'instrument' of the multi-agency group when they felt disempowered by the absence of the Police, Social Services and the Education Authority;
- So as not to marginalise the young people's comments about the Police, we presented these in an uncensored form at the Police workshop.

In our view, significant synergies can be gained by bringing together boundary critique and methodological pluralism, and it is this that makes systemic intervention of value. While others have advocated methodological pluralism, there is a danger without boundary critique that choices of methods will be based on superficial diagnoses of problem situations, failing to take account of multiple viewpoints and power relations (Midgley, 1996, 2000; Mingers, 1997; Flood, 1999; Han, 2000).

This argument has implications for Community OR more generally. While there have been a variety of previous writings on the value of methodological pluralism for Community OR (e.g., Jackson, 1987a, 1988; Gregory and Jackson, 1992a,b; Ritchie *et al*, 1994), as far as we are aware only Midgley (2000) has advocated this synergy between pluralism and boundary critique. However, a cursory overview of some Community OR projects written up in the literature (other than work from our own research group) makes it very obvious that issues of power and marginalisation are regularly encountered. Just five representative examples are:

- Supporting tenants of a mining corporation in gaining ownership of their houses, then managing their own affairs (Thunhurst and Ritchie, 1992; Thunhurst *et al*, 1992);
- Supporting a parents' campaign group seeking to keep their local school open (Mar Molinero, 1993);
- Involving Bengali speakers in assessing their need for health services (Taket and White, 1994);
- Supporting campaign groups fighting for improved public services in Venezuela (Ochoa-Arias, 1994).
- Supporting objectors in challenging the 'expert' reports provided by companies who are applying for permission to develop land against the wishes of people in local communities (Riddington, 2002).

Systemic intervention is therefore a very relevant form of methodology. Indeed, we would go further and claim that our systemic intervention approach is not only useful for *Community* OR, but for other forms of OR practice too. In an interconnected world where organisations can ill afford to ignore how they are perceived by those involved in and affected by their activities, and where environmental issues are increasingly on the agenda in both the public and private sectors, we believe that systemic intervention has potentially wide-spread relevance.

10.16 Conclusion

In this chapter we have outlined our systemic intervention methodology, focusing on how boundary critique and methodological pluralism can inform action for improvement (defined in local and temporary ways). We have also presented a detailed account of an intervention designed to support the development of services for young people (under 16) living on the streets of Manchester. All through the intervention, issues of marginalisation, power and participation (understood using boundary critique) had to be addressed by creatively mixing and adapting methods from a variety of sources. We argue that, by bringing together boundary critique and methodological pluralism in systemic intervention, significant synergies are gained, and this can be of great value for the practice of Community OR. Indeed, in an interconnected world where organisations are required to pay attention to the views of their stakeholders, and environmental issues are increasingly coming to the fore, systemic intervention may also be of use in many other application domains of OR.

References

Ackoff, R.L. (1981). *Creating the Corporate Future*. Wiley, New York.

Barter, C. (1996). *Nowhere to Hide: Giving Young Runaways a Voice*. Centrepoint, London.

Bilson, A. (1996). Bringing forth organisational realities: Guidelines for a constructivist approach to the management of change in human services. PhD Thesis, Lancaster University.

Bilson, A. (1997). Guidelines for a constructivist approach: Steps toward the adaptation of ideas from family therapy for use in organizations. *Systems Practice*, **10**, 153–177.

Booth Fowler, R. (1991). *The Dance with Community: The Contemporary Debate in American Political Thought*. University Press of Kansas, Lawrence.

Boyd, A., Brown, M., and Midgley, G. (1999a). *Home and Away: Developing Services with Young People Missing from Home or Care*. Centre for Systems Studies, Hull.

Boyd, A., Brown, M., and Midgley, G. (1999b). *Manchester Times*. Centre for Systems Studies, Hull.

Bradburn, N.M. and Sudman, S. (1980). *Improving Interview Method and Questionnaire Design*. Jossey-Bass, London.

Brown, M. (1996). A framework for assessing participation. In: *Critical Systems Thinking: Current Research and Practice*. Flood, R.L. and Romm, N.R.A. (eds.). Plenum, New York.

Brown, M. and Packham, R. (1999). Organisational learning, critical systems thinking and systemic learning. Research Memorandum #20. Centre for Systems Studies, School of Management, University of Hull, Hull.

Checkland, P. (1981). *Systems Thinking, Systems Practice*. Wiley, Chichester.
Churchman, C.W. (1970). Operations research as a profession. *Management Science*, **17**, B37–B53.
Churchman, C.W. (1979). *The Systems Approach and its Enemies*. Basic Books, New York.
Cohen, C. and Midgley, G. (1994). *The North Humberside Diversion from Custody Project for Mentally Disordered Offenders: Research Report*. Centre for Systems Studies, Hull.
Córdoba, J. and Midgley, G. (2002). Addressing organisational and societal concerns: An application of critical systems thinking to information systems planning in Colombia. In: *Critical Reflections on Information Systems: A Systemic Approach*. Idea Group, Hershey.
Córdoba, J., Midgley, G., and Torres, D. (2000). Rethinking stakeholder involvement: An application of the theories of autopoiesis and boundary critique in IS planning. In: *Human-Centred Methods in Information Systems: Current Research and Practice*. Clarke, S. and Lehaney, B. (eds.). Idea Group, Hershey, PA.
Coyle, R.G. (1977). *Management Systems Dynamics*. Wiley, New York.
Dewdney, A., Grey, C., Minnion, A., and the Residents of the Rufford Street Hostel (1994). *Down but Not Out: Young People, Photography and Images of Homelessness*. Trentham, Stoke-on-Trent.
Douglas, M. (1966). *Purity and Danger: An Analysis of the Concepts of Pollution and Taboo*. Ark, London.
Drever, E. (1995). *Using Semi-Structured Interviews in Small-Scale Research: A Teacher's Guide*. The Scottish Council for Research in Education, Edinburgh.
English, C.J. (1973). Leaving home: A typology of runaways. *Society*, **10**(5), 22–24.
Fay, B. (1975). *Social Theory and Political Practice*. George Allen and Unwin, London.
Fine, G.A. and Sandstrom, K.L. (1988). *Knowing Children: Participant Observation with Minors*. Sage, California.
Flood, R.L. (1989). Six scenarios for the future of systems 'problem solving', part 1. *Systems Practice*, **2**, 75–99.
Flood, R.L. (1999). *Rethinking the Fifth Discipline: Learning within the Unknowable*. Routledge, London.
Flood, R.L. and Carson, E.R. (1993). *Dealing with Complexity: An Introduction to the Theory and Application of Systems Science*, 2nd edn. Plenum, New York.
Flood, R.L. and Jackson, M.C. (1991a). *Creative Problem Solving: Total Systems Intervention*. Wiley, Chichester.
Flood, R.L. and Jackson, M.C. (1991b). Critical systems heuristics: Application of an emancipatory approach for police strategy toward the carrying of offensive weapons. *Systems Practice*, **4**, 283–302.
Foddy, W. (1993). *Constructing Questions for Interviews and Questionnaires*. Cambridge University Press, Cambridge.
Foucault, M. (1980). In: *Power/Knowledge: Selected Interviews and Other Writings, 1972–1977*. Gordon, C. (ed.). Harvester Press, Brighton.
Francescato, D. (1992). A multi-dimensional perspective of organizational change. *Systems Practice*, **5**, 129–146.
Freire, P. and Faundez, A. (1989). *Learning to Question: A Pedagogy of Liberation*. World Council of Churches, Geneva.
Graue, M.E. and Walsh, D.J. (1998). *Studying Children in Context: Theories, Methods, and Ethics*. Sage, California.
Gregory, A.J. and Jackson, M.C. (1992a). Evaluating organizations: A systems and contingency approach. *Systems Practice*, **5**, 37–60.
Gregory, A.J. and Jackson, M.C. (1992b). Evaluation methodologies: A system for use. *Journal of the Operational Research Society*, **43**, 19–28.

Gregory, W.J. (1992). Critical systems thinking and pluralism: A new constellation. PhD Thesis, City University, London.
Habermas, J. (1971). *Toward a Rational Society*. Beacon Press, Boston.
Han, J. (2000). A framework of value-focused systems thinking. *Systemic Practice and Action Research*, **13**, 97–109.
Heron, J. (1988). Validity in co-operative inquiry. In: *Human Inquiry in Action: Developments in New Paradigm Research*. Reason, P. (ed.). Sage, London.
Ho, C.H. (1997). A critical process for the evaluation of methodology. PhD Thesis, University of Hull.
Holstein, J.A. and Gubrium, J.F. (1995). *The Active Interview*. Sage, London.
Horkheimer, M. (1937). Traditionelle und Kritische Theorie. *Zeitschrift fur Sozialforschung*, **6**, 245–294.
Ivanov, K. (1991). Critical systems thinking and information technology. *Journal of Applied Systems Analysis*, **18**, 39–55.
Jackson, M.C. (1985). The itinerary of a critical approach: Review of Ulrich's 'Critical Heuristics of Social Planning'. *Journal of the Operational Research Society*, **36**, 878–881.
Jackson, M.C. (1987a). Community operational research: Purposes, theory and practice. *Dragon*, **2**(2), 47–73.
Jackson, M.C. (1987b). Present positions and future prospects in management science. *Omega*, **15**, 455–466.
Jackson, M.C. (1988). Some methodologies for community operational research. *Journal of the Operational Research Society*, **39**, 715–724.
Jackson, M.C. (1995). Beyond the fads: Systems thinking for managers. *Systems Research*, **12**, 25–42.
Jackson, M.C. (2000). *Systems Approaches to Management*. Kluwer/Plenum, New York.
Jackson, M.C. and Keys, P. (1984). Towards a system of systems methodologies. *Journal of the Operational Research Society*, **35**, 473–486.
Julienne, L. (1998). Homelessness and young single people from black and minority ethnic communities. *Youth & Policy*, **59**(Spring), 23–27.
Kant, I. (1787). *The Critique of Pure Reason*, 2nd edn, 1933. Smith N.K. (trans.). Macmillan, Basingstoke.
Kant, I. (1788). *Critique of Practical Reason and Other Writings in Moral Philosophy*. Beck, L. (trans. and ed.). 1949 edn. University of Chicago Press, Chicago.
Kvale, S. (1996). *InterViews: An Introduction to Qualitative Research Interviewing*. Sage, California.
Leach, E. (1976). *Culture and Communication: The Logic by which Symbols are Connected*. Cambridge University Press, Cambridge.
Liebl, F. (2002). The anatomy of complex societal problems and its implications for OR. *Journal of the Operational Research Society*, **53**, 161–184.
Mar Molinero, C. (1993). Aldermoor School: The operational researcher on the side of the community. *Journal of the Operational Research Society*, **44**, 237–245.
Maturana, H. (1988a). *Ontology of Observing: The Biological Foundations of Self Consciousness and the Physical Domain of Existence*. http://www.inteco.cl/biology/
Maturana, H. (1988b). Reality: The search for objectivity or the quest for a compelling argument. *Irish Journal of Psychology*, **9**, 25–82.
Midgley, G. (1992). The sacred and profane in critical systems thinking. *Systems Practice*, **5**, 5–16.
Midgley, G. (1994). Ecology and the poverty of humanism: A critical systems perspective. *Systems Research*, **11**, 67–76.
Midgley, G. (1996). What is this thing called critical systems thinking? In: *Critical Systems Thinking: Current Research and Practice*. Flood, R.L. and Romm, N.R.A. (eds.). Plenum, New York.

Midgley, G. (1997a). Mixing methods: Developing systemic intervention. In: *Multimethodology: The Theory and Practice of Combining Management Science Methodologies.* Mingers, J. and Gill, A. (eds.). Wiley, Chichester.
Midgley, G. (1997b). Dealing with coercion: Critical systems heuristics and beyond. *Systems Practice*, **10**, 37–57.
Midgley, G. (2000). *Systemic Intervention: Philosophy, Methodology, and Practice.* Kluwer/Plenum, New York.
Midgley, G. and Floyd, M. (1990). Vocational training in the use of new technologies for people with disabilities. *Behaviour and Information Technology*, **9**, 409–424.
Midgley, G. and Milne, A. (1995). Creating employment opportunities for people with mental health problems: A feasibility study for new initiatives. *Journal of the Operational Research Society*, **46**, 35–42.
Midgley, G. and Ochoa-Arias, A.E. (1999). Visions of community for community OR. *Omega*, **27**, 259–274.
Midgley, G., Munlo, I., and Brown, M. (1997). *Sharing Power: Integrating User Involvement and Multi-Agency Working to Improve Housing for Older People.* Policy Press, Bristol.
Midgley, G., Munlo, I., and Brown, M. (1998). The theory and practice of boundary critique: Developing housing services for older people. *Journal of the Operational Research Society*, **49**, 467–478.
Mingers, J.C. (1992). Recent developments in critical management science. *Journal of the Operational Research Society*, **43**, 1–10.
Mingers, J.C. (1997). Towards critical pluralism. In: *Multimethodology: The Theory and Practice of Combining Management Science Methodologies.* Mingers, J. and Gill, A. (eds.). Wiley, Chichester.
Mingers, J. and Brocklesby, J. (1996). Multimethodology: Towards a framework for critical pluralism. *Systemist*, **18**, 101–131.
Mingers, J. and Gill, A. (eds.) (1997). *Multimethodology: The Theory and Practice of Combining Management Science Methodologies.* Wiley, Chichester.
Morecroft, J.D.W. and Sternman, J.D. (eds.) (1994). *Modeling for Learning Organizations.* Productivity Press, Portland, Oregon.
Morgan, G. (1986). *Images of Organization.* Sage, London.
Morgan-Klein, B. (1990). *Young Runaways: A Scottish Review.* HMSO, London.
Morrissette, P. (1992). Engagement strategies with reluctant homeless young people. *Psychotherapy*, **29**, 447–451.
Ochoa-Arias, A.E. (1994). The possibilities for community OR in a third world country. *International Transactions of Operational Research*, **1**, 345–352.
Pinzón, L. and Midgley, G. (2000). Developing a systemic model for the evaluation of conflicts. *Systems Research and Behavioral Science*, **17**, 493–512.
Ploeg, van der, J. and Scholte, E. (1997). *Homeless Youth.* Sage, London.
Power, S., Whitty, G., and Youdell, D. (1995). *No Place to Learn: Homelessness and Education.* Shelter, London.
Powney, J. and Watts, M. (1987). *Interviewing in Educational Research.* Routledge and Kegan Paul, London.
Prosser, J. (ed.) (1998). *Image-Based Research: A Sourcebook for Qualitative Researchers.* Falmer Press, London.
Rich, J. (1968). *Interviewing Children and Adolescents.* Macmillan, London.
Riddington, G. (2002). Combating the experts with expertise: The role of community OR in a planning inquiry. *OR Insight*, **15**(2), 3–10.
Ritchie, C., Taket, A., and Bryant, J. (eds.) (1994). *Community Works: 26 Case Studies Showing Community Operational Research in Action.* Pavic Press, Sheffield.

Roberts, N., Andersen, D., Deal, R., Garet, M., and Shaffer, W. (1983). *Introduction to Computer Simulation: A System Dynamics Modeling Approach*. Addison-Wesley, Reading, MA.

Romm, N.R.A. (1995). Knowing as intervention: Reflections on the application of systems ideas. *Systems Practice*, **8**, 137–167.

Romm, N.R.A. (1996). Inquiry-and-intervention in systems planning: Probing methodological rationalities. *World Futures*, **47**, 25–36.

Rorty, R. (1989). *Contingency, Irony and Solidarity*. Cambridge University Press, Cambridge.

Sieber, J.E. (1992). *Planning Ethically Responsible Research: A Guide for Students and Internal Review Boards*. Sage, California.

Spash, C.L. (1997). Ethics and environmental attitudes with implications for economic valuation. *Journal of Environmental Management*, **50**, 403–416.

Stein, M., Rees, G., and Frost, N. (1994). *Running—The Risk: Young People on the Streets of Britain Today*. The Children's Society, London.

Stein, M., Rees, G., Wallis, L., Wade, J., Franks, M., Stevenson, J., Mumtaz, S., Lynes, D., Johnson, L., and Raws, P. (1999). *Still Running: Children on the Streets in the UK*. The Children's Society, London.

Taket, A. and White, L. (1994). Doing community operational research with multicultural groups. *Omega*, **22**, 579–588.

Taket, A. and White, L. (2000). *Partnership and Participation: Decision-Making in the Multi-Agency Setting*. Wiley, Chichester.

Thunhurst, C. and Ritchie, C. (1992). Housing in Dearne Valley: Doing community OR with the Thurnscoe Tenants Housing Cooperative. Part 2—An evaluation. *Journal of the Operational Research Society*, **43**, 677–690.

Thunhurst, C., Ritchie, C., Friend, J., and Booker, P. (1992). Housing in Dearne Valley: Doing community OR with the Thurnscoe Tenants Housing Cooperative. Part 1—The involvement of the Community OR Unit. *Journal of the Operational Research Society*, **43**, 81–94.

Ulrich, W. (1983). *Critical Heuristics of Social Planning: A New Approach to Practical Philosophy*. Haupt, Berne.

Ulrich, W. (1987). Critical heuristics of social systems design. *European Journal of Operational Research*, **31**, 276–283.

Ulrich, W. (1994). Can we secure future-responsive management through systems thinking and design? *Interfaces*, **24**, 26–37.

Ulrich, W. (1996). *A Primer to Critical Systems Heuristics for Action Researchers*. Centre for Systems Studies, Hull.

Ulrich, W. (2000). Reflective practice in the civil society: The contribution of critically systemic thinking. *Reflective Practice*, **1**, 247–268.

Vega-Romero, R.R. (1999). Health care and social justice evaluation: A critical and pluralist approach. PhD Thesis, University of Hull.

Weber, L.R., Miracle, A., and Skehan, T. (1994). Interviewing early adolescents: Some methodological considerations. *Human Organization*, **53**, 42–47.

Weil, S. (1998). Our concerns as researchers. Workshop with research students, held at the School of Management, University of Hull, May 15.

Wellmer, A. (1970). *The Critical Theory of Society*. Herder and Herder, New York.

Yolles, M. (2001). Viable boundary critique. *Journal of the Operational Research Society*, **52**, 35–47.

11

Playing with PANDA: The CybOrg and the Rhizome

Ann Taket and Leroy White

11.1 Introduction

We have introduced elsewhere (Taket and White, 2000), a framework that we have developed, called PANDA (participatory appraisal of needs and development of action), which we use as we plan and implement group working and group decision-making in the multi-agency setting. We developed PANDA in the process of critically reflecting on our own practice and that of others. Much of our own practice (both individually and together) has been in the field of community operational research. Underlying PANDA are a number of theoretical principles, which we label pragmatic pluralism. PANDA is based on a theoretical stance that we unashamedly label post-structuralist or postmodernist (Taket and White, 1993), and as such, makes no grand claims for being uniquely justifiable, or for guaranteeing success in application. PANDA represents one particular type of multimethodology.

In this chapter, we want to focus on just some aspects within the PANDA framework, illustrating these with two short case studies from our practice. We have chosen to focus on the issue of work in multi-organisational groups, noting in passing that most of our community OR

Accepted for publication August 2001.

Ann Taket • Faculty of Health, South Bank University, London SE1 0AA, UK.
Leroy White • Management Research Centre, University of Bristol. Wills Memorial Building Lewis Wing, Queens Road, Bristol, BS8 1RJ, UK.

Community Operational Research: OR and Systems Thinking for Community Development, edited by G. Midgley and A.E. Ochoa-Arias, Kluwer Academic/Plenum Publishers, New York, 2004

practice involves such groups. Our interest in this chapter is in extending the work of others on the metaphors for understanding the multi-organisational setting. We introduce two concepts or metaphors that we have found particularly useful, the CybOrg and the rhizome.

This chapter will start by outlining what we think is important to appreciate in working with multi-organisational groups, namely working with difference or otherness. One of the typical characteristics of multi-organisational work is the involvement of many different kinds of stake-holders, where each organisation often has multiple roles, i.e. it has a multifaceted, polyvocal nature. Each organisation is not a singular body but something that is polyvocal and fragmented. It is made up of very different people, and the situation being investigated becomes complex because different people may see the same thing differently. But not only this, the same group at different times may have a different view. Thus, the processes involved need to revolve around recognising and reacting to differences. Recognition that the groups involved will always be fractured, polyvocal, contingent and dynamic further underlies the need for such an approach. In this chapter we want to describe some aspects of how we have found it useful to think about the process of doing this kind of work.

It is common to find these issues discussed in terms of metaphor (see, for example, Morgan, 1986; Flood and Jackson, 1991). We find this unhelpful where it is used to artificially unify and homogenise what is diverse. We also find it is not helpful to think in terms of 'going into another culture'. We wish to challenge the notion that what is happening is a meeting of one culture with another, so that what is required is challenging the self-evidentness of one organisation with a homogeneous culture meeting with other (different) but unified cultures. We will challenge this in (at least) two ways. Firstly, we notice that each organisation has no fixed singular identity or culture, but rather this is always constantly created and recreated, negotiated and always fluid, dynamic and contingent. Secondly, we notice that any group contains inevitably more than a single culture (it is a collection of individuals, each complex and fractured, sharing some things but different in others); this is true of multi-organisational groups in particular. So it is always more complex than an encounter of One with Another. None of this, however, is cause for despair, since we can notice that in such a complex, contingent and dynamic situation, there is (inevitably?) scope for re-negotiation and interaction. It does however limit the possibility for (modernist) predict-and-control strategies, but this is not something that causes us disquiet.

As difference accumulates, especially in the multi-organisational context, the effect can be seen as overwhelming. Increased diversity begs for further simplification and classification in order to make it more manageable and to provide an easy route to understanding. We argue that this temptation should be resisted: classification and simplification can aid understanding, but they can exert a confining hold. The map replaces the territory. What we will argue for, is to move beyond simplification.

The view above poses a challenge which we find the traditional OR (or organisational theory) view unable to meet, because it often only deals with singular entities and immutable concepts (e.g. the decision-maker, the organisation). A number of authors have responded by trying to embrace diversity, heterogeneity, difference, multiplicity, uncertainty, chaos, anarchy and so on, but some of these attempts have resulted in what we would see as false (or incomplete) starts. For example: OR agonising over the establishment of a hard/soft divide (Taket and White, 1993); in the organisational theory literature, Morgan's (1986) use of metaphor, discussed in the next section; total systems intervention (Flood and Jackson, 1991) in its rigid use of metaphor, discussed later. Others have responded by not accepting the need to change. An example: those insisting on retaining the idealised version of the scientific method as a description of the modelling process, for example Chapman (1992). What we present below is a description of some of the ideas that underpin our approach to work in multi-organisational contexts. This involves reworking the notion of the paradigm of pluralism, and then examining different metaphors that we can use to help us act in multi-organisational contexts. The chapter concludes with two case studies that illustrate these ideas in action.

11.2 Paradigms, Metaphors and the Multi-organisational Setting

11.2.1 Paradigms

The term 'paradigm' has traditionally been used in a number of different ways to mean:

- a complete view of reality, or way of seeing;
- as relating to the social organisation of science in terms of schools of thought connected with particular kinds of scientific achievements;
- as relating to the concrete use of specific kinds of tools and texts for the process of scientific puzzle solving.

What the above have in common is that they offer a *weltanshaung*, or meta-theoretical view, and we find this problematic. Instead the term paradigm will be used here in a slightly different sense to mean a coherent and mutually supporting pattern of concepts, values, methods and actions, designed and used in a particular situation. We thus see a paradigm as locally specific, contingent and possibly ephemeral. Rorty (1991) argues that paradigms can successfully prescribe their own choices of methods and appraisal criteria in conjunction with a hermeneutic approach to other paradigms. Thus there is a pluralist approach to the choice of techniques in the sense that there is a range of techniques whose selection will be governed by the context of the issue addressed. However the potential of the pluralist paradigm is lost if the issue context is described in a singular way. Metaphors can be used in attempts to remedy this.

11.2.2 Metaphors

Metaphors function by asserting a resemblance between something to be known and something assumed as already known. There has been an increase in the use of metaphor in organisational studies, enhanced by a recognition that views of organisations emerge from discursive processes. Biology is a discourse from which early images of organisations took shape. From the Oxford English Dictionary we get a definition of organisation as 'the action of organising, or a state, in the biological sense'. Spencer later used the term to refer to 'an organised body, system or society' (Spencer, reprinted 1994). So the biological sense was used as the basis of a metaphor for a body, system or society in a general sense.

Morgan (1986) works with the idea of metaphor as an aid to understand the discursive processes from which organisations emerge. This has also been taken and used by Keys (1991) and Walsham (1991) for example. Morgan (1986) claims that organisational theory has been imprisoned by its metaphors, and, drawing on his earlier work with Burrell (Burrell and Morgan, 1979), offers a pluralist approach to escape the imprisoning characteristic of the metaphor. They argue that organisational theory could be analysed in terms of four broad world-views, each reflecting different sets of meta-theoretical assumptions (again working within the traditional notion of a paradigm). We find something uncomfortable, however, about the way these authors have worked with metaphors; namely that their pluralism is reduced to complementarism. We discuss this further elsewhere (Taket and White, 1996; Taket and White, 2000).

Morgan (1986) recognises the potential of metaphors, as well as the problems of using metaphors singly and statically to capture organisations as singular entities (e.g. as a machine). This treats metaphors as a set of neat categories. Sontag (1979, 1989) warns us of the danger of metaphors as giving us, or making up, the world. The danger is more apparent when we make the world up as a collection of singular static entities. In the multi-organisational setting, as discussed earlier, it can be seen that this is completely inappropriate where it is used to artificially unify and homogenise what is diverse. We thus find Morgan's work singular and unrich, in that his pluralism reduces to complementarism (the metaphors are seen as producing complementary pictures).

Morgan's metaphors are also used by Flood and Jackson (1991) in their total systems intervention (TSI). This is prone to the same type of problem, in that their use of a particular metaphor directs them to a singular methodology; they use an extremely limited range of metaphors; and set up a rigid hierarchy of dominant and subsidiary metaphors for a situation.

Another way of phrasing this critique is in terms of Ashby's (1958) law of requisite variety. The complementarist use of metaphor does not give us sufficient variety to deal with complex organisation, let alone the multi-organisational context. Instead what we espouse is a notion of judiciously mixing metaphors, a position we have described elsewhere as 'pragmatic pluralism' (Taket and White, 1996).

The main argument we want to make in this chapter is that the potential of metaphor as a paradigm has been lost because the use of metaphors has not yet adequately recognised the proliferation of diversity and difference. The uses of metaphors in the multi-organisational setting have yet to represent different sections of workers, different subordinated groups or different organisational forms. We think we can regain the potential of this paradigm if we 'mix our metaphors' and use them to recognise diversity and difference. Mixed metaphors, to use a figure of speech from Wittgenstein (1969), are like a spun thread or rope, which gains its strength, not from a continuous strand that runs its entire length, but from the overlapping and entwining of many separate fibres. What we would like to present below is a confusion of mixed metaphors. We now move on to introduce two new metaphors with which we have found it helpful to work: the CybOrg and the rhizome.

11.2.3 Beyond Cybernetics: Organisations as CybOrgs

In much the same way as organisms gave rise to organisations in the latter half of the nineteenth century, cybernetics as a metaphor had a hold on those interested in organisations in the latter half of the twentieth century. The cybernetic metaphor encourages us to view organisations as patterns of information, and focuses attention on the way in which states of homeostatic balance can be sustained through feedback (double loop possibly). The core insight emerging from early work was that the ability of a system to engage in self-regulating behaviour depends on processes of information exchange involving *negative feedback*. This concept is essentially one of automatic error detection and correction, so that movements beyond specific limits in one direction initiate movements in the opposite direction to maintain a desired course of action. Cybernetic principles are evident in many kinds of system.

Perhaps because the field is still young, there are many definitions of cybernetics. The usual one is that it is the science of communication and control in animal and machine. This definition was attributed to Norbert Wiener, who coined the word 'cybernetics' from the Greek word meaning steersman (Wiener, 1948). For philosopher Warren McCulloch (1965), cybernetics was an experimental epistemology concerned with the communication within an observer and between the observer and his/her environment. Stafford Beer, a management consultant, defined cybernetics as the science of effective organisation (Beer, 1966). The anthropologist Gregory Bateson (1972) noted that, whereas previous sciences dealt with matter and energy, the new science of cybernetics focuses on form and pattern; i.e. a way of looking at things and a language for expressing what one sees. Bateson, Wiener, McCulloch, Goffman and Beer, together formulated a new non-linear model of information and communication which placed 'interaction' and 'systems' at the heart of their studies (Beer 1962; Mattelart and Mattelart, 1992). This attempted to provide an antidote to the perceived danger of making up organisations based on linear mechanistic concepts of humans and communications.

It is perhaps Beer who has constructed the organisation in cybernetic terms. Writing on the impact of cybernetics on the concept of industrial organisation, Beer (1956) argued that the limit had been reached to the insights that could be gained from the use of the sort of model typified by the 'organisation chart'. 'The models are not dynamic; the structures of theory are not viable; the organisation chart is dead' (1956, p.77) Beer uses the system of the human body as his basic cybernetic analogy, and argues that organisations are also amenable to much the same description. We are concerned with, and critical of, the model offered by cybernetics, in that we see it as being prone to staticness in an unhelpful way. Double loop learning is also problematic in its self-referentiality and relativism. We see this as limited by the usual understanding of what is human, and wish to consider instead the CybOrg (Haraway, 1985, 1991).

Within cybernetics, we also find the CybOrg defined as an organism with a machine built into it with consequent modification of function, or that it is an organism which is part animal and part machine. Organisms are not seen as biological machines, but as something whose elements are the result of small loop autopoiesis. In much the same way we can define the cybernetic organisation, in which the organisation is autopoietic; i.e. it creates itself, but the parts are the result of localised processes. The organisation as a whole is not involved in the production of the parts. The organisation results from the functioning of the parts but is manifested in the external behaviour of the organisation. Here we have outlined a common boundary, that between human and machine.

Emery and Trist (1981) conceived of organisations as consisting of people who produce desirable products or services using a certain technology. The social sub-systems represent the relationships among the organisation members, and the technical sub-systems include the task to be performed as well as the equipment, tools and layout. Emery and Trist argue that the technological demands constrain the operation of the social system through tasks and work-rules, thus shaping the behaviour of the social (sub-)system, but the social sub-system has properties (socio-psychological) of its own. Thus an organisation will function 'optimally' if, and only if, the social and technical sub-systems are designed to fit the demands of each other. Thus a boundary is established between the social and the technical.

Today within organisations the focus is on communication, knowledge and decision-making, which is now at the heart of decision-support systems (Eden, 1990), computer conference systems, and electronic mail (Bryant, 1993). Communication and knowledge are seen to be increasingly important, but they are social as well as technical. Hence, with the CybOrg there is the recognition that there are not distinct boundaries between human/machine, social/technical and communications/organisation. Instead, drawing on Haraway and her version of the CybOrg, these boundaries are recognised as blurred and confused, regarded as leaky, permeable, dynamic.

So we seek an expanded flexible approach using metaphor, in particular that of the CybOrg. This metaphor of the CybOrg is rather different from the traditional definition of the cyborg, and Haraway expands it in two ways:

- for her, in that she is concerned with theorising identity in a way useful to social/political action, and her notion, very much embodied, we find much richer that the traditional cyborg;
- through a concept of leaky boundaries, three in particular: human vs. animal; animal/human (organism) vs. machine; physical vs. non-physical.

The CybOrg contains a recognition that there are no distinct boundaries, it contains a blurring or confusing of the boundaries between man/machine, communication/organisation, social/technical.

What cybernetics helps us to recognise is that organisations are complex, where inter-connections are being made piecemeal between participants at different levels and in different functions within the same organisation, and between many different levels of participants in different organisations. Haraway also argues that, underlying communication sciences and modern biologies,

> the translation of the world into a problem of coding [is] a search for the common language in which all resistance to instrumental control disappears and all heterogeneity can be submitted to disassembly, reassembly, investment and exchange. (Haraway, 1985, pp. 82–83)

Her CybOrg is conceived as one way to resist this.

Organisations do not consist solely of individuals as members, but also of communication and the discourses produced. This necessitates a view that moves on from the dichotomy between individual and collective, to dissolve the sharp boundary between them. The CybOrg as an imagery is principally concerned with the ways information/knowledge is used, and the processes of information flow within and between interconnecting systems. Using this imagery, methods for analysing the diverse processes of decision-making in groups, organisations and societies, and processes for widening access to organisations and communications systems, can be enhanced.

We recognise that Haraway's CybOrg has some historical roots in cybernetics, but is also concerned with intervening in the relationships between new technology and social relations in organisations and communities. We acknowledge that, as a metaphor, it is becoming increasingly visible across a range of texts. It operates on a number of different levels and is capable of carrying multiple and often contradictory meanings.

The CybOrg metaphor points us to what may count as nature in the techno-cyber-worlds that we now move in. It tells us of the 'implosion of the technical, textual, organic, mythic and political in the gravity wells' (Haraway, 1985) of our understanding.

Haraway's CybOrg, however, is not completely helpful to us in the context of theorising intervention and action in the multi-organisational community setting. We still find a need to loosen it up somewhat, so we wish to draw in addition on Deleuze and Guattari's (1980) notions of the rhizome.

11.2.4 Loosening the CybOrg: Introducing the Rhizome

What are we attempting to do? It is to integrate a novel form of analysis with a particular grammar, one that can be captured by the metaphor of Deleuze and Guattari's (1980) rhizome. We couple with this our development of a metaphor, mixed and differentiated, of the 'CybOrg', and find it productive if we 'mix our metaphors' and use them to recognise diversity and difference.

Let us now expand our understanding of this different metaphor, Deleuze and Guattari's rhizome:

> ... we enumerate certain approximate characteristics of the rhizome.
>
> 1 and 2. Principles of connection and heterogeneity: any point of a rhizome can be connected to anything other, and must be. ....A rhizome ceaselessly establishes connections between semiotic chains, organizations of power, and circumstances relevant to the arts, sciences, and social struggles. A semiotic chain is like a tuber agglomerating very diverse acts, not only linguistic, but also perceptive, mimetic, gestural, and cognitive: there is no language in itself, nor are there any linguistic universals, only a throng of dialect, patois, slangs, and specialized languages. There is no ideal speaker–listener, any more than there is a homogeneous linguistic community. ... There is no mother tongue, only a power takeover by a dominant language within a political multiplicity. ... It is always possible to break a language down into internal structural elements, an undertaking not fundamentally different from the search for roots. There is always something genealogical about a tree. It is not a method for the people. A method of the rhizome type, on the contrary, can analyze language only by decentering it onto other dimensions and other registers. A language is never closed upon itself, except as a function of impotence. (Deleuze and Guattari, 1980, pp.7–8)

The rhizome helps us to overcome lapsing into binaries and dualisms. Important characteristics of the rhizome, which mark significant differences for us from other systemic metaphors, are:

- its degree of connectivity (highly connected);
- the nature of connectivity (not information only);
- its dynamic nature (constantly reshaping itself);
- the objective of its use (intervention and action, not prediction and control).

One of the key characteristics of the rhizome is its multiplicative connectivity; it can be seen as anti-systemic in explicitly connecting incommensurate registers and different levels of the hierarchies set up within much of systems thinking (as in Jackson and Keys', 1984, system of systems methodologies). Within systems thinking, Maturana and Varela's (1992) autopoiesis offers something that is close to the rhizome, or put another way the rhizome may offer a useful metaphor for recursion and autopoeisis. We will explore this elsewhere.

The observation made above that the rhizome is helpful for the objective of intervention and action (not prediction and control) needs some further expansion. This derives from the difference made, in

Deleuze and Guattari's terms, between the notion of the map and the tracing.

> The rhizome is altogether different, *a map and not a tracing*. Make a map, not a tracing. The orchid does not reproduce the tracing of the wasp; it forms a map with the wasp, in a rhizome. *What distinguishes the map from the tracing is that it is entirely oriented toward an experimentation in contact with the real.* The map does not reproduce an unconscious closed in upon itself; it constructs the unconscious. It fosters connections between fields, the removal of blockages on bodies without organs, the maximum opening of bodies without organs onto a plane of consistency. It is itself a part of the rhizome. The map is open and connectable in all of its dimensions; it is detachable, reversible, susceptible to constant modification. It can be torn, reversed, adapted to any kind of mounting, reworked by an individual, group, or social formation. It can be drawn on a wall, conceived of as a work of art, constructed as a political action or as a meditation. Perhaps one of the most important characteristics of the rhizome is that it always has multiple entryways; in this sense the burrow is an animal rhizome, and sometimes maintains a clear distinction between the line of flight as passageway and storage or living strata ... A map has multiple entryways, as opposed to the tracing, which always comes 'back to the same'. The map has to do with performance, whereas the tracing always involves an alleged 'competence'. (Deleuze and Guattari, 1980, pp.12–13)

Deleuze and Guattari claim that certain disciplines, e.g. accountancy and bureaucracy, proceed by tracings. The 'model' has been central to systems thinking, with the aim of using models to trace the real world. Systems thinkers often think of their models as maps; however, often they are merely tracings, in their separation from the real. In shifting emphasis from a tracing (a model/metaphor that represents the world or part of it and how it is) to a map (a metaphor that helps us act in the world), we shift from predict and control to action and intervention.

> It is inaccurate to say that a tracing reproduces the map. ... The imitator always creates the model, and attracts it. The tracing has already translated the map into an image ... It has organised, stabilised, neutralised the multiplicities ... It has generated, structuralised ... and when it thinks it is reproducing something else it is in fact only reproducing itself. That is why the tracing is so dangerous. It injects redundancies and propagates them. (Deleuze and Guattari, 1980, p.13).

An attempt to picture a rhizome is almost a contradiction in terms of the level of connectivity involved, not all of which is equally (or simultaneously) visible. To mix metaphor further, it is rather like an iceberg, 9/10ths invisible and under the surface of the water. Figure 11.1 'represents' the rhizome drawing on yet another metaphor, that of the rabbit warren.

Our book (Taket and White, 2000) discusses the case of rhizomatic networks as one possible form of multi-organisational work. A rhizomatic

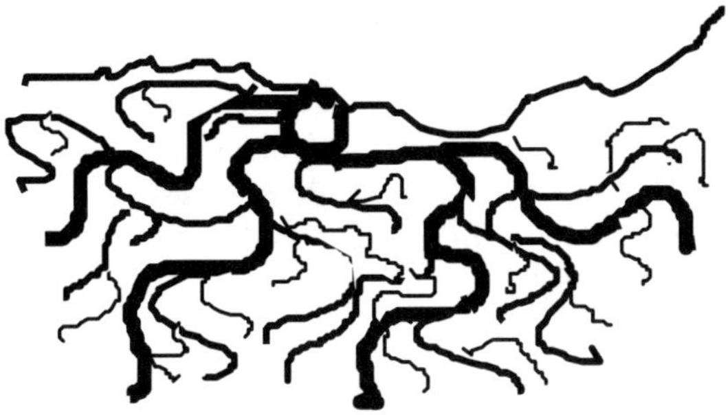

FIGURE 11.1 Rhizome

network is understood to be a network with a complex set of dynamic and contingent cross connections, many of which are unseen or unnoticed (or underground) according to the perceptions of parties involved. The example we used in our book was the case of the networked form of organisation in the protests/carnivals that took place in cities across the globe on June 18 1999, timed to coincide with the opening of the G8 summit in Cologne. A campaign to cancel Third World debt was the touchstone but the aims of the day were far wider. Organised by loose networks of small groups and individuals, with internet based communication of information, a wide variety of action took place—electronic forms of protest (a virtual sit-in), as well as actions requiring physical presence. Central co-ordination or direction was not apparent. Web sites contained links to 43 different country sites, alternative sites were made available for when the traffic on others became too great. This form of organising was heralded in some quarters as posing a unique challenge to the authorities in terms of attempting to respond. A detailed tracing of the groups involved proved impossible. The most that could be done was to map the effects of their combination.

So, what does our rhizomatic CybOrg offer? It is an ability to map the polyvocal, heterogeneous, multifaceted aspects of the multi-organisational situation. Organisations, through internal communication and discourses, shape how social structures around them are perceived. These interconnect and interact at the multi-organisational level (we note that there is a further complication in terms of the individual within the organisation and multi-organisational setting—but we will not discuss this further here). The rhizomatic CybOrg is a device we have found useful to keep us on our toes about the nature of the multi-organisational contexts we work in. It has served to keep us alert to the dynamic nature of the coalitions involved; to keep us on our toes as facilitators; to keep us alive to the opportunities offered by complex interconnectivity; etc. It is not necessarily a device that we would expect to share with our co-participants in all circumstances. It is also a device that decentres the facilitator as 'authority' or 'expert', emphasising the facilitator's involvement in the interaction and the need for recognition and reflection on this. We now consider the application of these ideas with reference to two examples drawn from our practice.

11.3 Mapping the Multi-organisational: Case Studies in Community Operational Research

11.3.1 PANDAs in Pakistan

The first case study is based on work carried out by one of the authors (Ann Taket, henceforth referred to using her initials, ART) in Pakistan. The overall subject of the work was to do with the strengthening of health services in the province, with subsidiary focuses in terms of postgraduate education in public health and health services research. The different organisations involved were as follows. The work centred on a College of Community Medicine (CCM) that was a postgraduate medical institution and health service provider, and two specific projects. The first one, the Link Project, was funded by the UK's Overseas Development Agency (ODA), as it then was, and managed by the British Council (BC). Under this project, faculty from South Bank University (SBU) were involved in a variety of curriculum and research development activities, working with the faculty and students of CCM. The second project, the Family Health Project, was also funded by the ODA and managed by the British Council (BC). The Family Health Project brings together the same organisations as the Link Project, with the addition of the provincial Ministry of Health (located in the same city) and other academic institutions, both in Pakistan and the UK. The focus of this second project is directly on health services. The diagram shown in Figure 11.2 depicts this multi-organisational context from the description thus far, and represents ART's initial view into the situation.

The situation was viewed in systemic terms by many of the participants, but there was no common view. One factor behind this was the involvement of different individuals in some of the organisations with the two projects (especially the BC, but also to a lesser extent in SBU and CCM). The tendency of many of the CCM faculty was to see the connections in terms of two distinct systems, one for each of the projects, thus they would not 'recognise' the systems' coexistence as shown in Figure 11.2. On the

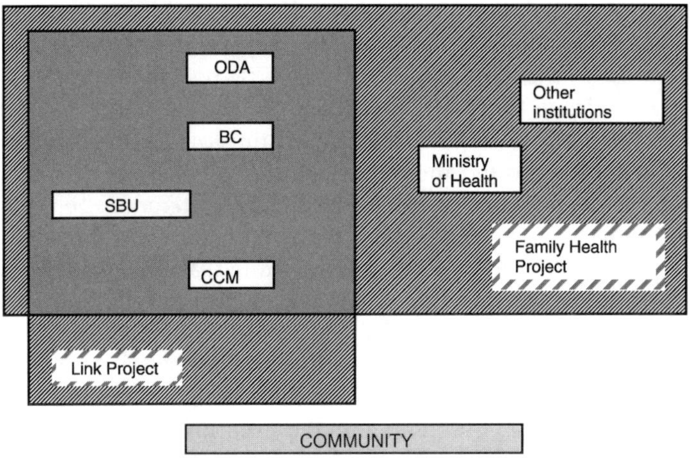

FIGURE 11.2 Case study in Pakistan: The multi-organisational context

part of those in the BC responsible for Link Project management, there was almost complete selective amnesia. The Family Health Project was viewed as not relevant to mention, or when prompted with a reminder of its existence, viewed as a completely separate entity. In contrast, those in the BC responsible for Family Health Project management saw a single hierarchical system—neurocybernetic (almost like Beer's, 1966, viable systems model)—with the Link Project as a subordinate system, feeding into, but not interacting with, the Family Health project.

Also shown on the diagram is the community, not explicitly connected into the projects, owing to a lack of formal involvement in the project structures. They were however connected in a number of diverse ways, through participation in different fora, and through provision of information and use of services. Some of the work in which ART was engaged as part of the Link Project involved identifying priorities for action research to improve the health of the local population. Here community involvement was part of the picture.

The diagram in Figure 11.2 was helpful to ART personally. However, the groups working on any of the different parts of the projects did not share the view it contained. It was only once a representation with a higher degree of cross-connectivity replaced the one described above that progress could be made. This representation was not expressed in diagrammatic form. It remained verbal only.

An important part of this was to recognise that the connectivity depended heavily on the issue under consideration. For example, the nature of cross connections for the issue of child health was quite different to those for the issue of health risks from the water supply. This meant that the nature of the hierarchies (and indeed whether they existed or not) changed from issue to issue. The concept of the CybOrg's leaky boundaries was also helpful here, in acknowledging the permeability of those boundaries shown around the projects and organisations in Figure 11.2.

In parts of the work we therefore used a different type of diagramming, recognising such a diagram as a particular slice through a more complex picture. A form of Venn diagramming was used to depict the different possible institutions that could take action in relation to a specific issue: see Figure 11.3, which presents a simplified form of the results. The size of each rectangle indicates the influence the organisation could have on action, and overlaps indicate where joint decision-making action was required. We explicitly compared here the situation for maternal and child mortality and water quality. Figure 11.3 presents a simplified version of the results, and shows how the situation regarding acting on maternal and child mortality data is relatively simple and feasible compared with that of water quality data. CCM, through its role as a service provider for some areas, can directly affect maternal and child health, while also having an effect jointly with and through other health service providers.

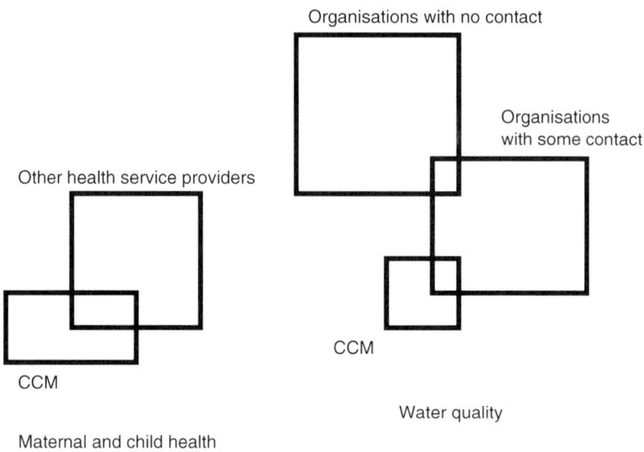

Note: in both cases the presentation has been simplified to show types of organisations, rather than the extremely messy pictures which named individual organisations

FIGURE 11.3 Discussing the possibilities of action affecting health based on the use of different data

In contrast, the effect on health that CCM could exert through taking action in relation to water quality was much smaller, the joint possibilities with other organisations was also small, and some organisations who had a considerable influence on the issue were not in contact with the Institute. The group accordingly prioritised maternal and child mortality data over that of water quality data, in terms of offering a better chance of contributing to the mission of CCM, which was to contribute to the improvement of the health status of the population.

A second shift was also necessary to facilitate progress, this time in the nature of the connectivity, distinguishing *different* connections for financing, managerial accountability, information provision, consultation and service provision. Thus the nature of the context, complexly interconnected and with dynamically evolving links, was hard to match with other systemic metaphors. Indeed progress was impeded when traditional systemic metaphors were being drawn on by some of those involved. It was also important to avoid a pessimistic view of the situation, 'just a chaotic mess', as this also impeded progress. The metaphors of the CybOrg and rhizome with shifting leaky boundaries, multiple entryways and exits, rather than a singular static structure, were helpful. Once a more rhizomatic view could be adopted, and participants felt comfortable with the complexity of the situation, discussion and prioritisation was able to proceed. This involved the participants making use of partial representations (for example Figure 11.3), in the full recognition that they were partial, and suspending the 'need' to produce a single representation of the 'whole system'.

11.3.2 PANDAs in Deptford and New Cross (UK)

Deptford and New Cross (inner city areas of London in the UK) have received more than £150 million of regeneration money over the last 12 years, yet remain recognised deprived priority areas. Most of the programmes were meant to involve the community, but in a sense failed in that local people did not participate in identifying the priorities or developing solutions.

The New Cross partnership was established in January 1999 and is a collaboration between over 10 agencies and organisations. The partnership works on the principle that regeneration and community development can be genuine learning experiences for understanding the barriers to and opportunities for community involvement. Thus, the partnership's aim is to develop and test a coherent, integrated local citizenship programme which can facilitate community development, establish structures to encourage active citizenship, and develop partnership working between the community, agencies, schools, businesses and community groups.

One of the key objectives for the partnership is to explore how the wider community could be best served by the partnership. With such a wide range of organisations there are many different views about concepts such as community, involvement and citizenship; thus it seemed impossible to find consensus on terminology, let alone a strategy. The possibilities for linkages and connections between organisations was extremely high, meaning that any view or ideas generated about community or citizenship seemed not to come from any one particular corner or group. Thus, ideas and views had the appearance of rumours and hearsay, which made the co-ordination of a collective vision for the group very difficult. The whole partnership exhibited almost rhizomatic behaviours.

The partnership meets quite regularly and has organised a number of events to link into the community and to actively recruit local people. It has also organised community fairs and has a stall in the local market. The partnership wanted to have a review and approached one of the authors (Leroy White, henceforth referred to as LW) who had worked with some of the partner agencies before. In the meeting to discuss the review, the partnership talked about its core values and the fact that they have used participatory appraisal (PA) and would like to see the review process incorporate their values and some of the PA methods. The principle was one of working with diversity. The partnership had used the 'solution tree' in a previous review, and together we thought it would be a good idea to incorporate it into the process (see below for details). A description of a process that had been used before by LW (the facilitator) was given, and this seemed to touch the felt needs of the partnership. The PANDA framework (Taket and White, 2000), and the use of Team Syntegrity (Beer, 1995), Open Space (Owen, 1992) and the Strategic Choice Approach (Friend and Hickling, 1987) were described to the partnership. The partnership members saw the process as being able to help them generate ideas on how to take things forward and to develop them as a group to deliver the options. Key for the partnership was the notion that each group, somehow, would want to maintain its own integrity while at the same time developing a collective view or consciousness. It was thought that the best way to achieve this was to assume that each representative was part of the network and that the process would allow rapid feedback of views and thoughts to engender the idea of a Cybernetic Organisation. This would also mean a de-centring of control, because some people were anxious that other groups had their own agenda to pursue, or that they wanted to have control of the outcomes.

The review took the form of an event that was held in February 2001. 30 people attended. As they arrived it was clear that they had both anxieties and hopes. They were pleased at the opportunity to be part of the process. However, they also brought their own agendas and worries. Using the solution tree concept, the participants were asked to write on individual leaves what they thought the issues were in relation to the opening question, which was *Get set for citizenship: how do we get it right this time?* They were then asked to place these views on the branches of a tree that was drawn on the side of a wall. Over 80 ideas were placed. Some were seen as options and they were placed on the branches of the tree, while the issues and problems were placed on the roots of the tree. Here the opportunity for rapid feedback of ideas and exchange of views could occur. Many ideas were borrowed and reinterpreted. New views appeared which seemed to be of a higher order. The group was constantly reflecting on what was put up on the tree.

The next part of the process was the 'problem jostle' used in Team Syntegrity (Beer, 1995). Here any participant could take one of the leaves from the tree and start a discussion. There were 10–12 places in the room where a discussion could take place, thus at any one time there were 10–12 discussions. The purpose was for the individual who had taken a leaf to be an advocate for the idea or issue placed on it, so a market place of ideas is generated. The currency of support is a signature. The participants could sign as many statements as they believed would be good topics to discuss later. For a discussion to take place, there needed to be five signatures. In all, between 20 and 25 issues were deliberated on in this way. What was interesting was that a number of connections were made between members of the group, in that people were beginning to share and understand different points about what was being discussed. But more importantly they were beginning to appreciate how the process enabled them to understand that they were part of a system. There were people who resisted the networking that was going on. However, they were encouraged by the others to engage in the process.

The ideas that were endorsed then became the issues for small group work. There were 15 topics, which were separated into 5 groups of 3, and there was a limit of 6 discussants in each group. The group allocated themselves to 2 topics each. They were then told that they were free to roam around the room and join in discussions on any topic (outside their own group), and could play at being a bumblebee (devil's advocate) or a butterfly (just moving around choosing whether to listen or not). The participants found this a fun and engaging exercise. There was a qualitatively different feel to the atmosphere of the workshop. The group appeared much more co-operative.

During the discussions an issue passing round the group was that there needed to be better links between the partner group and the other groups and community organisations in the area. It was felt that, previously, communication and information was working in only one direction. Interestingly, a number of groups of people were forming alliances and networking and were beginning to organise meetings after the event. Many people in the partnership were beginning to feel that things would get out of control if the partnership did not manage this. During the plenary session, a number of the comments from participants were about resisting the partnership managing the process, and in effect the groups were beginning to self-organise. In fact the sub-groups that formed reflected the areas identified as a priority by the partnership. These were: a lack of engagement in estate boards by residents; and a lack of involvement of black and minority ethnic communities and youths. Thus, in a rhizomatic way, the seeds were being sown for a number of forums to begin debates on priorities. The question that remains is whether the partnership can somehow allow the seeds to take root in a self-organising way.

Finally, the participants thought that the process worked really well in that it did not undermine their principles and values. Having a democratic, non-hierarchical process ensured that there were many opportunities for participation. Even though individuals had a vested interest in maintaining their own integrity, it was possible for the whole group to share ideas and develop solutions, while maintaining their own views. It was also interesting to see how individuals modified their behaviour during the event and worked towards a collective goal.

As a postscript to the case, the partnership has developed in a different way with more people being involved. The workshop was also the catalyst for different groups to form and develop. For example, on the Pepys Estate, a new forum was established and the first meeting employed many aspects of the process used for the partnership workshop. This was an example of self-organisation that was brought to the attention of the authors. We are sure there have been, or will be, others springing up.

11.4 In Conclusion and Inconclusion ....

The first general point that we would like to offer in conclusion is that the discussion of multi-organisations is particularly timely, given the current level of interest in embracing difference, diversity, decentralisation and heterogeneity. We can label this, if we wish, as a new paradigm. The potential for this paradigm is, however, lost if we do not loosen up the use of metaphor in this multi-organisational situation.

The two case studies discussed above illustrate the difficulties of trying to embrace difference, diversity, decentralisation and heterogeneity when using metaphor in a static and singular manner. In the first case study the attempt to understand the situation as two separate static systems could not deal with the variety present in the situation. In the second, the rhizomatic form of organisation, created by mixing the methods used, allowed the whole group to share ideas and develop solutions, while maintaining their own views. It also allowed individuals to modify their behaviours during the event and work towards a collective goal. Reflecting on these case studies demonstrates the value of developing different metaphors, the CybOrg and the rhizome for example, which prompt us to think flexibly about breaking down boundaries and subjecting them to critical evaluation. For instance, the CybOrg metaphor encouraged us to deconstruct the boundaries between communication and organisations. We will continue to develop these further in the future.

CybOrgs and rhizomes, to us, are also more than just metaphors, for (at least) three reasons. Firstly, their use facilitates a critical approach, deconstructing the notion of boundaries that modernists regard as sacred (for example, boundaries between nature/culture, art/science, body/machine, male/female and self/other). Secondly, they emphasise connections between technologies, actors, processes, and the importance of re-conceptualising boundaries as inherently leaky, which is to say something more than that they are dynamic. Finally, they facilitate a radical approach to working with difference while respecting diversity. This was illustrated in the second case study, where social solidarity was possible, but only in a contingent and strategic fashion, which required remaking through time as events and alliances changed. Other reasons require more exploration elsewhere based on further practice and/or reflection. Where these will lead we cannot with confidence say, and that is our joyful inconclusion.

References

Ashby, W.R. (1958). Requisite variety and its implications for the control of complex systems. *Cybernetica*, **1**, 83–99.

Bateson, G. (1972). *Steps to an Ecology of Mind*. Jason Aronson, Northvale, NJ.

Beer, S. (1956). The impact of cybernetics on the concept of industrial organization. Reprinted in *How Many Grapes went into the Wine: Stafford Beer on the Art and Science of Holistic Management*. Harnden, R. and Leonard, A. (eds.). Chichester, Wiley, 1994.

Beer, S. (1962). Towards the cybernetic factory. Reprinted in *How Many Grapes went into the Wine: Stafford Beer on the Art and Science of Holistic Management*. Harnden, R. and Leonard, A. (eds.). Chichester, Wiley, 1994.

Beer, S. (1966). *Decision and Control: The Meaning of Operational Research and Management Cybernetics*. Chichester, Wiley.

Beer, S. (1995). *Beyond Dispute*. Chichester, Wiley.

Bryant, J. (1993). Supporting management teams: AIDA, groupware and computer group decision support systems. *OR Insight*, **6**(3), 19–27.

Burrell, G. and Morgan, G. (1979). *Sociological Paradigms and Organisational Analysis*. London, Heinnemann.
Chapman, C.B. (1992). My two cents worth on how OR should develop. *Journal of the Operational Research Society*, **43**(7), 647–664.
Deleuze, G. and Guattari, F. (1980). *A Thousand Plateaus: Capitalism and Schizophrenia*. Tr. B. Massumi. London, Athlone, 1988.
Eden, C. (1990). The unfolding nature of group decision support—two dimensions of skill. In *Tackling Strategic Problems: The Role of Group Decision Support*, pp. 48–52, Eden, C. and Radford, J. (eds.). London, Sage.
Emery, F.E. and Trist, E.L. (1981). *Sociotechnical Systems. Sytsems Behaviour (ed.)*. Open Systems Group, Open University, Chapman, London.
Flood, R.L. and Jackson, M.C. (1991). *Creative Problem Solving: Total Systems Intervention*. Chichester, Wiley.
Friend, J. and Hickling, A. (1987). *Planning Under Pressure: The Strategic Choice Approach*. Pergamon, Oxford.
Haraway, D.J. (1985). A manifesto for cyborgs: Science, technology and socialist feminism in the 1980s. *Socialist Review*, **80**, 65–108.
Haraway, D.J. (1991). *Simians, Cyborgs and Women: The Reinvention of Nature*. London, Free Association Books.
Jackson, M.C. and Keys, P. (1984). Towards a system of systems methodologies. *Journal of the Operational Research Society*, **35**, 473–486.
Keys, P. (1991). Operational research in organizations: A metaphorical analysis. *Journal of the Operational Research Society*, **42**(6), 435–446.
Mattelart, A. and Mattelart, M. (1992). *Rethinking Media Theory*. University of Minnesota Press, Minneapolis.
Maturana, H.R. and Varela, F.J. (1992). *The Tree of Knowledge: The Biological Roots of Human Understanding*. Revised edition. Shambhala, Boston.
McCulloch, W. (1965). *Embodiments of Mind*. MIT Press, Cambridge, MA.
Morgan, G. (1986). *Images of Organisation*. Sage, California.
Owen, H. (1992). *Open Space Technology: A User's Guide*. Abbott, Potomac, Maryland.
Rorty, R. (1991). *Objectivity, Relativism, and Truth: Philosophical Papers, Volume 1*. Cambridge University Press, Cambridge.
Sontag, S. (1979). *Illness as Metaphor*. Allen Lane, London.
Sontag, S. (1989). *AIDS and its Metaphors*. Allen Lane, London.
Spencer, H. (1994). The man versus the state. In *Herbert Spencer: Political Writings*, Offer, J. (ed.). Cambridge University Press, Cambridge.
Taket, A.R. and White, L.A. (1993). After OR: An agenda for postmodernism and poststructuralism in OR. *Journal of the Operational Research Society*, **44**(9), 867–881.
Taket, A.R. and White, L.A. (1994). Doing community operational research with multicultural groups. *Omega: International Journal of Management Science*, **22**(6), 579–588.
Taket, A.R. and White, L.A. (1996). Pragmatic pluralism—an explication. *Systems Practice*, **9**(6), 571–586.
Taket, A.R. and White, L.A. (2000). *Partnership and Participation: Decision-making in the Multiagency Setting*. Chichester, Wiley.
Walsham, G. (1991). Organisational metaphors and information systems research. *European Journal of Information Systems*, **1**(2), 83–94.
Wiener, N. (1948). *Cybernetics: Control and Communication in the Animal and Machine*. New York, Wiley.
Wittgenstein, L. (1969). *Blue and Brown Books*. Penguin, Harmondsworth.

III

Dealing Locally with Global Issues

12

An Interpretive Systemic Exploration of Community Action in Venezuela

Alejandro E. Ochoa-Arias

12.1 Introduction

Over the last 12 years, I have been involved in Community OR activities with a variety of community organisations (COs) in Venezuela (e.g., Ochoa-Arias, 1994, 1996, 1998, 2000). As part of this work I have been studying community action using the inquiring framework of Interpretive Systemology: a systems approach that focuses on disclosing different possible interpretations of phenomena, exploring the grounds for these interpretations as an aspect of the inquiry. For more details of the philosophy and methodology of Interpretive Systemology see, for example, Fuenmayor (1991a–c); López-Garay (1991); Fuenmayor *et al* (1991); and Ochoa-Arias (1996).

My studies of community action have been devoted to critically understanding how the phenomenon of COs can be understood in relation to current discourses in political philosophy. This political research is an *integral part* of my Community OR practice. The questions I seek to address come out of dialogues around the problems experienced by people in COs, and the answers developed through the research are mostly

Final version accepted for publication March 2002.

Alejandro E. Ochoa-Arias • Centro de Investigaciones en Sistemologia Interpretativa, Universidad de Los Andes, Mérida, Venezuela.

Community Operational Research: OR and Systems Thinking for Community Development, edited by G. Midgley and A.E. Ochoa-Arias, Kluwer Academic/Plenum Publishers, New York, 2004

co-evolved in partnership with those people. The only exception to this is when I encounter a really significant issue in my practice that I cannot deal with or account for using my current understandings, and this leads me to conduct political inquiries separately for a while (for an example see later in this chapter). However, the idea is always for the research to reshape my Community OR practice in the longer term, and the practice of the COs I am working with.

This chapter focuses on the substance of my political inquiries in the context of Venezuelan society. Venezuela is, of course, a country characterised by significant inequalities; political confrontations and social unrest; and deeply distrustful relationships between the government, COs, multi-national companies and some overseas powers, especially over the ownership and use of the country's oil wealth. There is also a gulf between political discourses and people's actions in organisations, suggesting to me a state of 'institutional schizophrenia' (more on this later).

However, the political analysis unfolded in this chapter takes us beyond Venezuela alone. I believe it can help us to identify an issue that is becoming fundamental to understanding the current crisis faced by western societies in general: the legitimation crisis. I would say that this crisis is deepest in those societies, like the developing ones, that are at the margins of the political discourses I am talking about, which is why Venezuela provides such an interesting arena for research.

Below, I use a case study of the design of a participative Neighbourhood Association in Venezuela as a vehicle for critically unveiling the role of COs in a would-be modern and democratic society. This work started out as a typical Community OR project, engaging with COs in the city where I live, yet it evolved into the political inquiry that is the focus of this chapter.

12.2 *Community Action in Venezuela*

Before plunging into the process of presenting the case study, it is important to briefly discuss the emergence and development of COs in Venezuela in order to provide some context. The presence of organisations devoted to assisting disadvantaged sectors of Venezuelan society can be traced back to the nineteenth century (Miliani, 2000). However, I want to focus on more recent times when the establishment of COs came to be identified as a response by civil society to issues that were either neglected or poorly dealt with by the State. My starting definition of COs is therefore that they 'are created by sectors of civil society for the purpose of providing a minimum set of conditions that make possible the social integration of the marginal or excluded sectors of society' (Ochoa-Arias, 2000). This definition is useful as it allows us to distinguish COs from governmental initiatives devoted to social welfare. It also opens the possibility of considering COs as relevant actors in the processes of both designing and challenging social policies proposed by the government, political parties and others.

However, this connection with civil society requires further elucidation, as the concept of 'civil society' itself is disputed. Understandings of the concept range from, at one extreme, alliances between private interests defending themselves against intervention by the State to, at the other extreme, forms of organisation devoted to defending and promoting civic virtues, such as patriotism and the defence of traditions. The relevance of this dispute will become clear when I present the case study.

An important aspect of my own definition is the inevitably public nature of COs; the definition refers to the alleviation of marginalisation and social exclusion. Therefore, it implies a political commitment to challenge or surpass those conditions that brought about the social exclusion in the first place. How far this process of political engagement goes depends on the nature of the issues that each CO is addressing. However, this political dimension has often been diminished by rhetoric that describes COs as needing to preserve the current social order and promote taken-for-granted positive values: solidarity, charity and co-operation. As Midgley and Ochoa-Arias (1999) point out, the ways in which notions of community (and hence the politics of COs) are constructed are highly relevant for Community OR practice. If COs are legitimately seen as preserving the current social order, this would suggest a different role for the Community OR practitioner than if they are seen as agents working for a more inclusive society.

In my view, in those societies like Venezuela where social, economic and political problems have reached crisis point, there is a need to orchestrate the emergence of a new minimum social order that addresses the conditions that have given rise to the crisis. Accordingly, COs should become active in the process of building up political mechanisms to articulate the search for a common ground which the society can defend as essential for its own existence. This was the thinking behind the case study I will present.

The case study went through two stages or (in the language of Interpretive Systemology) 'layers of unveiling'. The first stage was devoted to understanding and participating in the process of organising one particular neighbourhood in Mérida (a medium sized city in Venezuela). I worked with a local neighbourhood association (NA), a kind of Venezuelan CO (explained below), and one of our central tasks was to critique the meaning of 'participation' in relation to some key political discourses. The second stage was an interpretive-historical inquiry that helped us build a better understanding of the role of civil society in the institutionalisation of democracy in Venezuela, and this threw new light on the policies of the NA in relation to issues of participation and governance. I am currently working with the NA on a third 'layer of unveiling', but this is not yet sufficiently advanced to write up for publication.

12.3 Neighbourhood Associations as Community Actors

The setting up of neighbourhood associations (NAs) in Venezuela began during the 1960s as an urban movement concerned with the development of residential areas in major Venezuelan cities (Zambrano, 1986). However, in the last 20 years, they have become more concerned with problems surrounding the provision of public utilities in defined geographical areas, and the decision-making procedures of local and regional governments (Santana, 1983; Zambrano, 1986; Lander, 1987; Ovalles, 1987).

Using my earlier definition of a CO, NAs can be seen as COs because they are rooted in the actions of social groups from civil society concerned with the quality of life in their local areas, and they explicitly address issues of social exclusion. It is important to mention that, although earlier NAs were devoted to ensuring that their residential areas were kept safe and quiet (an essentially conservative agenda), there has been a change in their role. This change relates to the emergence of the *barrios*: a Spanish word referring to non-regulated urban settlements, often without 'basic' amenities like sewage and electricity that most people in the developed world take for granted. The emergence and growth of the barrios as the most common type of urban settlement in Venezuela compelled the NAs to reorient their efforts. Now they are principally concerned with securing access to basic services for the people in the barrios. In this regard, they play the role of attending to the needs of marginal sectors of society (defined as such according to their geographical and demographic conditions).

NAs are the most common form of CO in Venezuela: at the last count I am aware of, in 1993, there were nearly 12,000 of them. Their rapid spread since the 1960s led, in the 1990s, to the formal recognition of the NA as a means of participation through which communities are able to influence the design and execution of policies devoted to enhancing the living conditions of the population.

The first stage of my research was defined in terms of a Community OR project helping a specific NA. It was set up and run by a mixture of working and middle class residents of one neighbourhood in Mérida, and they called me in to help enhance its performance. In a nutshell, the problems defined by the NA were:

(1) A lack of influence or power, as requests made to local and regional authorities to address issues affecting the community

(health, housing, security and water supply) were not receiving a satisfactory response.

(2) A lack of participation by the neighbours of those involved in the NA, especially when the NA was dealing with the regional authorities and needed to demonstrate solid support for its arguments. In 1991, the NA only had 7 fully active members (5 belonging to political parties) drawn from an estimated population of 3,000 adults.

It was evident from the beginning that the poor level of participation was a crucial problem for this NA. It is common practice among public officials in Venezuela to show indifference to any claim from NAs and other pressure groups that is not going to affect the public image of the people or political party in charge. NAs therefore need to be able to exert significant public pressure in order to be heard, and a NA with only 7 active members does not have the credibility (in the eyes of the officials) to do this.

Because participation was a key issue, it was decided to take this as the starting point for the project. However, rather than simply launch into a set of activities designed to increase participation, we thought it would be worthwhile beginning with a critical inquiry into the modes of participation exercised by this NA. After all, given that the existing level of participation was so low, simply trying to get more people to do what these 7 were already doing might be problematic. Perhaps there was something wrong with the existing mode of participation that was causing, or contributing to, the problem.

The inquiry began by developing two contrasting, general accounts of the meaning of participation, which could be used to evaluate the specific claims about participation being made by the members of the NA. This is the usual starting point in an inquiry conducted using Interpretive Systemology: it is necessary to explore and clarify the different possible discourses that can make sense (in different ways) of a given problematic phenomenon. More than two discourses can be explored, but a minimum of two is necessary (Midgley and Ochoa-Arias, 2001). The activity of constructing the two accounts of participation departed from an analysis of the legislation that, at the time, constituted NAs as legitimate actors in Venezuelan society and stipulated democratic processes of governance.[1] The first account was explicit in the wording of the legislation, and focused on participation through representative democracy. A second account was based on viewing participation as a process that enhances a culture of dialogue and debate.

1 This legislation was made redundant by the introduction of a new constitution in 1999. However, it is expected that a new law regulating the participation of citizens will be introduced in the near future.

12.4 The First Layer of Unveiling: The Notions of Participation in the Discourses of Liberal and Deliberative Democracy

In the process of unveiling some taken-for-granted assumptions regarding the issue of participation, a series of workshops were carried out in the community to explore with members of the NA the strengths and weaknesses of their own practice of participation. The two accounts of participation were developed in partnership with the members of the NA, drawing in and interpreting research I conducted outside the context of the workshops. I therefore acted as both the NA's researcher and a facilitator of the inquiry process.

The starting point for constructing the two accounts of participation was the legislation regulating the NA. According to the law, the NA should be run democratically. The Board of Directors should be elected by the wider membership, and the Board should use democratic decision-making procedures (people should be allowed to vote on issues of contention). Formally, one of the basic tasks ascribed to the NA in each specific neighbourhood is to be a watchdog for the local public services (water and electricity supply, education, healthcare and security). In order to carry out these tasks there are Co-ordination Committees devoted to specific issues. Their role is explicitly formulated in the law as: 'to collaborate with public entities in taking the necessary steps to secure the satisfaction of public needs in a specific locality' (LORM, 1978).

The two accounts of participation (or 'contexts of meaning' in the language of Interpretive Systemology) were developed once we were clear on the requirements of the legislation. The accounts were labelled 'liberal democracy' and 'deliberative democracy'. Both contexts of meaning were structured using the same set of sub-headings to aid comparison: 'The Notion of Self-Realisation in Society'; 'The Exercise of Citizenship'; and 'Relationship with the State'. Importantly, the production of these accounts was not divorced from any consideration of their consequences for action. As each account was elaborated, people in the workshops considered its implications for the role of the NA and defined the tasks it would be necessary to undertake if they were to accept that account as legitimate. Below, brief summaries of the accounts are presented, along with people's thoughts about the roles and tasks implied by them (see Ochoa-Arias, 2000, for a fuller presentation).

12.4.1 Liberal Democracy

1. *The Notion of Self-Realisation.* The individual is viewed as a more or less autonomous, rational agent whose voluntary involvement in society

is motivated by the use of the latter as an instrument for its own purposes (Nozick, 1974). This means that social networks are conceived as instrumental webs for the fulfilment of individual projects.

2. *The Exercise of Citizenship*. The notion of individual rights is important because it emphasises the law as the means through which the State can regulate interaction among individuals (people may use the law to claim their rights). Property rights constitute the core for the material realisation of a society in which commodity exchange constitutes the main mechanism of social interaction. To be an 'owner' means to be capable of bargaining with others. Even more, the very *being* of an individual is associated with having rights: to *be* an individual human being means to *have* human rights. These rights are interpreted principally as a way of constraining the intervention of the State in the private lives of autonomous individuals. In addition, the assumption that economic relationships are governed by independent 'laws' which develop in strategic interactions beyond the control of the State, means that rights can only be endorsed at the level of the individual (Dworkin, 1977).

3. *Relationship with the State*. The State becomes functionally defined and takes on the role of facilitating the development of the economic sphere as a way of securing the self-realisation of individuals who pursue their own projects. The State safeguards the conditions necessary for the realisation of these individual projects (Nozick, 1974). In this regard, the main issues to be addressed by the State are the risks and uncertainties that permeate the economic milieu. The specific policies of the State are defined through strategic interactions among different interest groups, and one of its tasks is to regulate the potential and actual side effects of this interaction (such as the emergence of monopolistic control by a single economic interest).

4. *Implications for the Role of a NA*. In a liberal democracy, NAs should be conceived as organizations enabling the pursuit of individual projects concerned with enhancing material conditions. However, the pursuit of individual projects should be seen as a common goal. This common goal arises because there is the potential for individual projects to conflict with one another. Strategic calculations of the costs and benefits of collective action in the pursuit of the common goal are therefore necessary, together with evaluations of the uncertainties implied by the fact that each person is dependent on the involvement of others in realising his/her own project. Facilitating these calculations and evaluations could be the remit of NAs.

5. *Basic Tasks for the Mérida NA given this Context of Meaning*. The participants defined the following:

(a) The NA should act according to the norms and procedures defined by the State in order to be acknowledged as a representative of sectional interests. Such norms exist to reduce the uncertainty among different interest groups.

(b) The NA should conduct cost-benefit analyses of proposed mobilisation efforts. Mobilisations are acts of collective action orientated to the fulfilment of a specific need via claims made to the local and regional authorities. Mobilisations attempt to make the State aware that some needs should be considered as public needs when their fulfilment supports individuals in pursuing their projects.
(c) The NA should do its part to ensure that the information flow between itself and the various State bodies is regular and accurate, thereby enabling both parties to properly understand the requirements of the community.
(d) The NA should continually update information channels between itself and the community it represents in order to preserve the support of community members. Because instrumental social relationships can be uncertain, the NA should act to reduce perceived uncertainty in its own interactions with members of the local community.

12.4.2 Deliberative Democracy

1. *The Notion of Self-Realisation*. The development of human beings is considered to be an outcome of a social process in which the individual identifies him/herself with a socially constituted set of norms and values. This set of norms and values defines the framework in which people pursue their well being. The social network is conceived as having an ethical dimension. This ethic asks people to collectively 'ensure the inviolability of socialised individuals by requiring equal treatment and respect for the dignity of each, and protect the intersubjective relations of reciprocal recognition by requiring solidarity among individuals as members of a community in which they are socialised' (Habermas, 1990, pp. 243–244).

2. *The Exercise of Citizenship*. The process of communicative debate is the nucleus for the practice of citizenship. 'The only way to find out what, if anything, is common to us all, what should be the domain of legal regulation, what forms of political decision-making are legitimate, what should be left to the autonomous subject's personal judgement, and what must be compromised … [is through practical debate]' (Cohen and Arato, 1992, p. 355). A conscious composition of a collective identity is required which provides the 'minimum criterion, with respect to content, of the legitimacy of norms in the negative sense as *that which cannot be violated*' (Cohen and Arato, 1992, p. 368). This collective identity is composed of the discovery or re-appropriation of traditions, memories, values and social practices that are sources of solidarity. Solidarity is needed in order to

sustain the rational core of a political collective identity, which is subject to the procedures of public communication (Cohen and Arato, 1992). It is assumed that individualisation is achievable through the communicative processes of socialisation in the context of a speech community and in an inter-subjectively shared 'lifeworld' (set of meanings). Individuals simultaneously acquire unique personal and group identities, but can only do so through their membership of a collective.

3. *Relationship with the State.* The main mechanism of decision-making in collective affairs is democracy. Democracy is understood as *all political orders that satisfy a procedural type of legitimacy, in the sense of the procedures validated by discourse ethics.* This implies a 'self-controlled learning process' (Habermas, 1975) that is capable of promoting institutional change. The process of learning through which social interaction develops is based on the presence of institutions devoted to a discursive validation and justification that ensures the democratic participation of all concerned citizens, in order to define laws that can be accepted as legitimate by members of society more generally. It includes the organisations of civil society devoted to defending and promoting the fulfillment of their interests from a standpoint of tolerance and practical debate. In particular, it should provide grounds for the legitimation of a key component in the development of a democratic order: the definition of rights. The issue of rights becomes paramount in this context because it implies the exercise of control over the State by largely autonomous individuals through processes of (i) legislation and (ii) debate about the binding character of laws. The State in this context can be defined as *the set of institutions devoted to guarantee the existence of a public sphere in which a search for the formation of the public will is conducted under conditions of fairness, equality and justice. These conditions constitute the primary rights for every individual, and any subsequent rights people want to define should flow from these.* The duties of the State unfold in two dimensions. The first one is the guarantee and preservation of a public sphere structured through an aggregation process from local to national institutions. To make these institutions work effectively, the State should reinforce the means of political involvement through civic education allowing the enrichment of a democratic culture in society. The second dimension is the guarantee of the minimum set of material conditions necessary to ensure that any conditions of inequality and marginalisation emerging from the operations of the market will not distort the public sphere. This recognises that the operations of an unconstrained free market can, over time, result in some people gaining monopolistic financial power while others end up with so little that their ability to satisfy their basic needs is compromised. Both extreme wealth and poverty can distort the public sphere: the rich can buy undue influence (for instance

through advertising) and the poor will normally (and understandably) choose to prioritise meeting their basic needs over participation in public debates. Therefore, the State has a responsibility to establish the material conditions to prevent or counteract these effects.

4. *Implications for the Role of a NA*. From the above, it becomes clear that NAs should be orientated to support the building of a collective identity from a self-reflective and critical standpoint. They should be viewed as institutions of civil society. Civil society can be conceived as the space in which the exercise of democratic legitimating procedures occurs. This means that the State and the economic sphere are not considered as autonomous systems ruled by 'laws' that operate wholly independently of human intentions. They should be made accountable in debate through public legitimation processes, and the NAs can help to facilitate the operation of these.

5. *Basic Tasks for the Mérida NA given this Context of Meaning*. The main task for the NA in this framework is to become engaged in social learning processes and identity formation through the promotion of the public sphere and a practice of discursive ethics within a wide range of institutions. More specifically, the NA should:

(a) Engage in the redefinition of cultural norms, individual and collective identities, and the form and content of discourses (*the politics of identity*).
(b) Search for and empower new political actors to obtain benefits for those whom they represent (*the politics of inclusion*).
(c) Work to modify the universe of political discourse to incorporate new identities, needs and norms (*the politics of influence)*.
(d) Support the democratisation of political and economic institutions in order to assure the longevity of successes accomplished in the previous three processes (*the politics of reformation*) (Cohen and Arato, 1992).

These tasks imply the development of an organisation that can exert pressure for inclusion in the economy and public administration (a strategic/instrumental mode of action is indispensable at this level) in order to ensure that money and administrative power are not the only steering media in society.

12.5 *Choosing Between the Contexts of Meaning: The Emergence of an 'Institutional Schizophrenia'*

The development of these two contexts of meaning stimulated debate and discussion within the NA around two main themes. The first one was the relationship of the NA with the State and other sectors of society. The second was the management of the NA's internal affairs. I will briefly present the outcomes of these debates.

Concerning the NA's relationships with the State and other sectors of society, it became evident that the 'deliberative democracy' model fitted nicely with the desire of the members of the NA to develop a fairer and more egalitarian process of participation when conducting public affairs in their neighbourhood. The emphasis on the definition and defence of rights was recognised as a way of dealing with the lack of responsiveness from local authorities: claims should be phrased in terms of people's rights. When comparing the current practice of the NA with those tasks identified as necessary for it in a deliberative democracy, the workshop participants expressed a strong commitment to change their orientation. However, people complained about the external constraints they anticipated, given that they were a small organisation and the prevailing political discourse at the time was liberal democracy, which carries the assumption that the evolution of economic relationships is outside of people's control. Further discussion resulted in a clear commitment amongst the members of the NA to constitute a network of organisations (including other NAs), on the grounds that a co-ordinated network would have greater strategic influence than a single, small NA.

The 'liberal democracy' model was refused because it essentially made the NA an instrument of individuals, with the only common goal being to secure the means for individual freedom. The members of the NA were concerned about the well being of their community as a whole, rather than as a simple aggregation of individuals. Here there was a recognition that a community has properties that emerge from its operation as a system, and not through the actions of its individual members in isolation (interestingly, this mirrors Ackoff's, 1970, finding that the Mantua community leaders in Philadelphia, USA, had a good systemic awareness). They were also particularly keen to increase the influence of their neighbourhood in the conduct of public affairs, for both local and more general benefit. The 'deliberative democracy' model seemed to them better able to offer this than liberal democracy which limits collective influence to securing the conditions for individuals to pursue their own projects.

Regarding the management of their internal affairs, however, the debate in the NA became problematic, contradictory and conflictual. Although they claimed that participation was a key issue for the survival and further development of their NA, some members only saw the politics of inclusion, influence and reformation as relevant to their relationships with *outside* bodies. These members refused to accept that their own organisation also had to embrace these politics when dealing with its internal affairs. Only the politics of identity was unanimously accepted for the NA itself, on the grounds that it would help the members with team building. Even so, further discussion revealed problems here too: when they were asked to define the identity of their neighbourhood, it became apparent that this identity was based on the exclusion of some local people with different economic backgrounds—and, more importantly, the affiliation of these people to different political parties. Even worse, when they were asked if they wanted to invite other organisations in the neighbourhood to participate with the NA (a group devoted to cultural promotion and a CO devoted to community healthcare), they said 'no'. It appeared that the group's expressed desire to build a network of COs to gain greater influence was fine in theory, but in practice other concerns started to intrude. Debating these concerns revealed that they were worried about the NA falling into the 'wrong hands'. Although this reaction had some foundation in people's experiences of local politics (political parties can sometimes vie for control of NAs), in my view it was also the case that some deep-seated prejudices were involved, and these were leading to active political discrimination and the refusal of debate.

In the end, they decided that they would manage their internal affairs by seeking to orchestrate the whole neighbourhood to pursue some goals defined by the Board of Directors without further discussion. This was considered to be the best option for the success of the NA. The neighbourhood was viewed as an instrument of mobilisations determined by the Board, and local people were not permitted to debate the pertinence and appropriateness of such mobilisations.

How was this possible? Where did the will for debate and practical discussion go when it came to determining the NA's own policies and practices? Most importantly, how could the members of the NA accept what seemed to me to be a blatant contradiction between their political aspirations for the wider society and their own modus operandi? This is what I describe as 'institutional schizophrenia'.

At this point, I felt obliged to raise the issue for further discussion rather than simply let it drop. The NA agreed to convene a further workshop. This workshop revealed to us all that the problem was a little more complex than I had at first thought. It appeared that our earlier discussions about the two contexts of meaning (liberal and deliberative democracy) had not surfaced some problematic assumptions about the nature of participation that were causing the apparent contradiction to manifest itself. Essentially, the members of the NA assumed that (i) people in the neighbourhood had common problems, and (ii) they had an emotional commitment to addressing these that would guarantee their support.

Given these two assumptions, it is not surprising that the NA had failed to notice that they were institutionalising a contradiction between their espoused political theory and their own practice. They thought that deliberative democracy was only relevant to the wider society because, at the local level, there were no significant disagreements to deliberate.

Surfacing these assumptions allowed the group to reflect on them, and (after some debate) they realised that they were indeed problematic. A new design for a participatory organisation was therefore developed (more details will be provided shortly). However, the institutional schizophrenia that I had witnessed troubled me greatly, so I decided it would be worth conducting some research (not involving the NA that I had already worked with) to find out if it was a wider phenomenon. If it was not unique, it might signal the need for deeper political inquiries that could be of longer-term value to Community OR practice in Venezuela (and possibly also in other countries too).

A survey of other NAs in Mérida (see Ochoa-Arias, 1996, for details) showed that the same institutional schizophrenia could be identified in a range of different social and economic settings. Additionally, the survey revealed that NAs were rapidly developing a new relationship with the State: they were beginning to see themselves as an integral part of the governmental structures in Venezuelan society. Their members saw the NAs as part of the State apparatus, not as independent from it. If citizens in a local community wanted to gain access to public officials, the presence of a NA representative had become compulsory. The mechanisms of influence and complaint were rapidly being narrowed to only forms of negotiation governed by legislation, and the practice of 'political clientelism' (making the provision of resources and services conditional on electoral support for a particular political party) was common. It was evident that further political inquiry, focused initially on the contradictions surrounding the issue of participation, would indeed be worthwhile.

12.6 *The Second Layer of Unveiling: The Historical Contradictions in Venezuelan Civil Society*

My survey demonstrated that participation in NAs was usually constrained to the right to vote. People reported that political debate was a rare and sporadic practice in most public forums in Venezuelan society, not just the NAs. However, at the same time, a discourse asking for more egalitarian forms of participation, and the engagement of non-governmental organisations (NGOs) and COs in running public affairs, was prevalent even among those sectors that had opposed the involvement of NGOs and COs in the past. The institutional schizophrenia, characterised by arguments for greater participation but little evidence of its practice (beyond the exercise of the right to vote) in any forum in Venezuelan society (including the NAs) was even more wide spread than I had anticipated. The contradiction in the discourses within the NA I worked with before might have been explained by the erroneous assumptions about local participation made by its members. However, I wondered if there might also be deeper reasons for the lack of political debate that was being observed, both in the NAs and in other public forums that espoused the rhetoric of a participative civil society. Was this rhetoric even authentic?

My political inquiries took a new turn. I had begun facilitating a design exercise with the NA I had worked with previously to develop a new, more participative approach to its structure and policies. In parallel with this, I started researching the historical development of political participation in Venezuelan civil society, focusing on the notions of democracy and rights. Again I used the inquiry framework of Interpretive Systemology, unveiling two different understandings of the history of Venezuelan democratic institutions that might shed further light on the problematic phenomenon of participation.

First let me explain what happened with the participative design work. Although the process of design proved to be successful as a device for engaging different parties in the neighbourhood, and for expanding their horizons of expectations and concerns, it failed when it came to the moment when the NA had to implement the proposed changes. The main reason for this failure was a cultural and political incompatibility between the organisation that the participants designed and the structure and function assigned to the NA according to the law. Once people realised there was an incompatibility, they turned away from debating forms of participation beyond those enshrined in the legal definition of democracy for NAs. The realisation that there was a conflict between what the law prescribed and the participation people actually wanted made my new political inquiries into the history of democratic institutions in Venezuela even more relevant, not just to me but to the NA.

A full presentation of the two contrasting historical interpretations of the evolution of democratic practices and their influence and presence in Venezuelan civil society can be found in Ochoa-Arias (2000). Only a summary is provided below.

The first historical account was developed with reference to the 'official' discourse about pluralism and political participation in Venezuelan democracy. It was claimed that this pluralism and participation replaced the previous dictatorship in the mid-20th Century, and was then enhanced when the new Venezuelan political elite defined a set of rights and enshrined these in law. In this account, the development of Venezuelan democracy is also conceived as stemming from the emergence over time of different actors in civil society who built relationships with the State. These actors included trade unions, professional associations and (more recently) voluntary and community organisations. Their emergence and constructive engagement with the State was considered to indicate the existence of a healthy democratic regime.

According to this historical account, Venezuelan society deteriorated in the second half of the 20th Century because of an inappropriate management of public resources. Therefore, the solution devised in the early 1980s was to make the State more efficient. The State was to withdraw from some of its social duties. However, the impoverishment suffered by large sectors of society was said to be the result of economic failure. Consequently, the ensuing social and political unrest was also interpreted as stemming from the failure of the economic system. Indeed, the dominance of the discourse of economic failure continued until 1989 when there were violent riots in Caracas. It then became clear that there was a need for a political change so as to articulate the interests of large sectors of the population whose voices had been marginalised. However the change came too late, and the 1998 elections saw the defeat of all the political parties that had held office during the last forty years, and a new, left-wing government came to power.

The new regime promotes a discourse that is closely related to the second account of the development of Venezuelan democratic institutions unveiled through the historical research I am describing. This second account is based on the assumption that the previous 'democratic' regime in Venezuela had long been illegitimate. It is argued that democracy became a façade for the enrichment of an elite in Venezuelan society based on an illegal and immoral take-over of public goods (in particular, revenues from the oil industries that were owned by the State found their way into the pockets of the elite). This historical account claims that the action of defining citizens' rights (in particular rights to political participation) was carried out without considering (i) the peculiarities of a society emerging from a dictatorship and (ii) the lack of intermediate institutions between individuals and the State. Indeed, it is argued that the 'official' history of Venezuelan democracy is disingenuous as it falsely claims that it allows the majority of the population to articulate their political views. However, the events of 1989 revealed the absence of mechanisms of social articulation for the majority of people in an allegedly democratic society. Did these mechanisms ever exist beyond a formal statement? And if these mechanisms were ever there, how did they come to disappear?

The disclosure of an alternative historical account of the illegitimacy of the Venezuelan 'democratic' regime provided the Mérida NA with a meaningful framework to comprehend the process through which NAs and other civil society organisations were becoming absorbed into the Venezuelan political system. Indeed, from the perspective of the second account, the idea that NAs could ever legitimately maintain their independence from the State was always going to be problematic. This independence would be too much of a threat to the political status quo, and the absence of other mediating institutions between the individual and the State meant that the co-option of NAs to play this role was actually necessary from the State's point of view. In this regard, the alternative historical account not only disclosed some unfair social and political relationships in Venezuelan society, but it also posed a new challenge to the members of the NA: if they accepted it, they would have to rethink their own roles in relation to it. This is precisely what has happened for the new people in charge of the Venezuelan government: they have had to rethink what government is actually for.

The challenge for the government and for the NAs goes far beyond adding participatory democracy to the already existing democratic institutions. The current process of transformation in Venezuela involves the redefinition of political and social institutions in relation to a new National Constitution. This Constitution promotes participatory democracy as the nucleus of the legitimation and action of the Venezuelan State. However, as with any transformation, bringing about the desired change is not a linear process, and is not free from turbulence. This implies the need to identify opportunities and obstacles to developing a new discourse about participation.

It is becoming clear that the history of understandings and actions of COs and NGOs, especially during the time of the previous regime, will influence and condition the 'new' discourse about participation in Venezuela. In this regard, the relevance of Interpretive Systemology was twofold: the process of using the interpretive historical framework and the content of the analysis were both helpful. First, let us consider the process. The use of Interpretive Systemology provided a historical critique that linked the specific conditions experienced by the NA to understandings of trends related to the evolution of Venezuelan civil society in general. Importantly, it did this without making any claims to absolute truth: both histories were presented as interpretations, thereby allowing their relevance to be assessed against the experiences of the members of the NA. This in itself aids democratic deliberation because interpretations are more obviously open to critical inquiry than truths. Linking the general interpretations of history to the specific experiences of people in the NA also had the added advantage of making the historical analysis directly relevant to the consideration of future actions in relation to the development of civil society as a space for political debate.

The other useful aspect of the interpretive historical approach came through consideration of the content of the analysis. The alternative discourse unveiled Venezuelan civil society as being both the deceived and the deceiver at the same time. By becoming mediators between the individual and the State, and slipping into political clientelism, civil society organisations were helping to silence the voices of socially excluded sectors of society, making it difficult for them to criticise the status quo. This realisation not only allowed the members of the NA to reconsider the role of their organisation, but also set them on a path of further political analysis and debate. In particular, problems surrounding the notion of participation were examined in more detail. Some of the ideas that came out of this examination are presented below. Others are still being debated, and are therefore not included in this chapter.

In our inquiry, the core concept of legitimation in a democratic society came under scrutiny. It became evident to the members of the NA that representative democracy in a country like Venezuela with weak political institutions can actually be an obstacle to the development of a strong civil society capable of pursuing policies oriented to democratising the whole political system. However, the alternative of participatory/deliberative democracy implies a set of conditions that do not yet exist in Venezuelan society. In particular, deliberative democracy requires the 'right to have rights' as a foundation for political debate, yet the Venezuelan State has long been reluctant to, and/or has been incapable of, guaranteeing human rights for every citizen. Even now, when there is a new discourse asking for the Venezuelan State to provide basic social welfare as a right, it is being held back by the forms of participation in civil society nurtured in the past. Moves to debate welfare rights in a deliberative democratic mode are being frustrated by the lack of an effective legal/institutional framework to facilitate this (although this is being addressed), and by a dominant understanding of participation that has been left in place by the previous political administration. The latter point requires some clarification, because our political inquiry revealed that it is a crucial problem.

Notwithstanding the currently dominant left-wing political discourse in Venezuela, there is still a strong alternative discourse about participation that continues to affect civil society. According to this discourse, participation refers to something other than people coming together in a search for the common good, and it takes place in spaces beyond the public arena. This notion of participation (often promoted by governments, international bodies and important sectors of organised civil society world-wide) refers to mechanisms through which sectors of society become responsible for the provision of services previously regarded as public (healthcare, education and housing amongst others). If participation means a local community designing and providing services for itself, this might explain why self-support and self-sufficiency have become such important measures of success among NGOs and COs all around the world. In Venezuela, the trend of giving responsibility for the running of public services directly to local COs is clearly evident. Those who champion this form of participation claim that, when each community is running its own affairs, transparency and a sense of belonging are beneficial by-products.

It is important to point out that this mode of participation was strongly supported by neo-liberal governments during the reform of the Venezuelan State mentioned in the first historical account presented earlier. The new discourse on participation was introduced in opposition to the previously dominant Welfare State discourse, and it was used (along with arguments for economic efficiency) to legitimate the withdrawal of the State from the provision of many public services.

But, what exactly is this notion of participation legitimating? I believe the answer is not simple or straightforward. On the one hand, it seems that participation is an instrument used to strengthen communities and neighbourhoods (especially disadvantaged ones), and it is seen as an essential ingredient in the move to ensure communities have the necessary capabilities to solve their own problems. In this sense, it is the development of strong and largely self-sufficient local communities that is being legitimated by this discourse of participation, along with the idea of improving provision in the areas of health, education, housing, access to food and other essential goods and services. This being the case, it would appear that, underlying the discourse of participation, is the legitimation of moves to remove or alleviate the conditions giving rise to disadvantage. It seems, then, to be a beneficial discourse.

On the other hand, participation (understood as the devolution of responsibility for service provision to local communities) could be seen as a mechanism to diminish the burden on those institutions that were originally established to assist disadvantaged sectors of society—in particular, State organisations. If the State no longer has responsibility for remedying disadvantage, and this falls to local communities providing services that are not regarded as 'public' (in the State-provided sense), then whether or not disadvantaged sectors are able to meet their needs is left to market forces. The risk here is that disadvantaged sectors of society already have inferior buying power in the market, so their disadvantage is most likely to intensify. Given that the neo-liberal State now gets its main legitimacy from its contribution to economic performance in the service of the individual, rather than for what it does to remedy disadvantage (see the 'first layer of unveiling', reported earlier), it cannot intervene in the worsening situation. To do so would provoke a 'legitimation crisis' (Habermas, 1988), or perhaps intensify an existing one, where the role of the State is brought into question through the clash of discourses.

From the above, it would seem that the notions of community empowerment, civil society and especially participation are seriously contested because we are in the process of a transition towards a new rationality. The 'we' who are in transition include many developing *and* developed countries. This new rationality is essentially about dealing with disadvantage in ways that do not threaten the expansion and growth of the market. While there appear to be positive aspects to the new discourse in terms of enhancing capacity in local communities, the exposure of disadvantaged sectors of society to unfettered market forces brings with it considerable risks and, for many people, undermines any serious consideration of social justice at the societal level.

The problem for Venezuelan society is that the alternative understanding of participation—building a political debate concerned with social justice that is capable of integrating disadvantaged sectors into the process of dialogue—flies in the face of the neo-liberal view of participation, which is already institutionalised. The dominance of this neo-liberal view is sustained by the discourse of self-sufficiency that obliterates any pursuit of a common good beyond the boundaries of each local community or interest group. Attempts to build participatory democracy therefore founder, despite the change of government.

It would appear that the attempt to rescue the notion of public debate, not only to confront narrow sectional interests but also to build up common and mutual understanding, is in peril at the present time. For many people, participation understood as self-empowerment and self-affirmation (rather than as deliberative democracy that has the potential to transcend narrow boundaries) has become a taken-for-granted good. This raises a vital question: how strongly embedded is the neo-liberal view of participation? It was this question that motivated the third 'layer of unveiling' that is still being researched, and is therefore not reported here.

12.7 *Lessons for the Practice of Community OR*

From this exploration, I believe two important lessons can be derived for Community OR practice. First, it is evident that unveiling historical interpretations about how a given community became concerned with solving a particular problem can play a key role in increasing the community's political awareness and its capability to understand the space in which its actions, goals and hopes are defined and pursued. Second, viewing the work of a CO in terms of historical and political interpretations of community action can help in the identification of strengths and weaknesses of the CO, and also reveal new possibilities for future action.

In summary, I believe it is essential for community development practitioners to ask themselves about the sources of legitimation for their own actions, and the actions of others, which affect the well being of their communities. To those who call this kind of thinking 'navel gazing' I say, consider how the meaning of 'participation' has been understood through the research reported in this chapter. Had the NA simply attempted to increase participation in its own affairs without further reflection, they would no doubt have continued down the road of promoting a kind of participation that is only concerned with local community self-management and which contradicts the aspiration for a more genuine participatory democracy. I worry that too many community development projects promote participation as an unassailable good without considering the possibility that there may be different kinds of participation that work in the interests of different political philosophies.

One way to question taken for granted ideas is to surface historical interpretations of how the identity of the community was built up in the past. Another way is to look at ideas in the light of different political discourses. The historical and political interpretations and discourses can then be used for the assessment of current policies and practices, and they can also help people question issues of legitimacy and define new ways forward. The methodology that enabled me to question the meaning of participation when working with the Mérida NA was Interpretive Systemology (e.g., Fuenmayor *et al*, 1991; Fuenmayor, 1991a–c; López-Garay, 1991; Ochoa-Arias, 1996). I therefore want to end by recommending this to Community OR practitioners as a useful framework of inquiry that can be employed (either alone or in conjunction with other approaches) to produce more penetrating political understandings than might otherwise be generated.

12.8 Conclusion

In this chapter, I have described my use of Interpretive Systemology with a Venezuelan CO to surface different interpretations of the history and politics of participation—the longer term goal being to support the CO in rethinking its political role.

I suggest that some of the contradictions and problems of 'participation' and 'democracy' are most apparent in societies developing at the periphery of modern political discourses, simply because of the extent of the inequalities in those countries and/or the relative immaturity of their political institutions. However, this is not to say that the results of inquiries with COs in countries like Venezuela lack wider potential relevance. On the contrary, given the current globalisation of political discourses, it would be surprising to me if similar issues concerning participation in civil society were not pertinent in the developed world too, even if they are more deeply hidden. Indeed, I believe that, unless historical and political critiques are conducted with COs wherever they are to be found, there is a danger that they can become players in the unfolding of 'invisible' political rationalities, advancing policy agendas that (if made explicit) they would prefer to challenge.

I consider this kind of political and historical critique to be necessary as a first step that is useful for problematising our own roles and identities as community organisers. This can help reveal what it might be (in terms of political rationalities and power relations) that legitimates our activities. It is a first step, not an end in itself, because the greater political awareness it gives rise to should ultimately inform the development of new paths for action to bring about new political relationships.

In this sense, I believe that Interpretive Systemology can make a useful contribution by mapping out a methodological path for re-constituting our lives in terms of new histories that are explicitly of our own making. In the process of reconstitution, visions of 'progress', 'democracy', 'rights', 'civil society' and even 'equality' come to be seen as interpretations and thereby lose their supposedly universal and unassailable character. They are revealed as the specific inventions of local and temporary political discourses that we can accept, reject or re-work as appropriate. This is not critique in the sense of simply rejecting the political status quo. Ultimately, it reveals positive alternatives for understanding, action and social change.

Acknowledgement

The first draft of this chapter was written during a sabbatical visit to the Centre for Systems Studies, Business School, University of Hull, UK (September 2000 to June 2001), and I would like to thank the researchers in that institution for their hospitality and feedback on my ideas.

References

Ackoff, R.L. (1970). A black ghetto's research on a university. *Operations Research*, **18**, 761–771.

Cohen, J. and Arato, A. (1992). *Civil Society and Political Theory*. MIT Press, Cambridge, MA.

Dworkin, R. (1977). *Taking Rights Seriously*. Harvard University Press, Cambridge, MA.

Fuenmayor, R.L. (1991a). The roots of reductionism: A counter-ontoepistemology for a systems approach. *Systems Practice*, **4**, 419–448.

Fuenmayor, R.L. (1991b). The self-referential structure of an everyday living situation: A phenomenological ontology for interpretive systemology. *Systems Practice*, **4**, 449–472.

Fuenmayor, R.L. (1991c). Truth and openness: An epistemology for interpretive systemology. *Systems Practice*, **4**, 473–490.

Fuenmayor, R.L., Bonucci, M., and López-Garay, H. (1991). An interpretive-systemic study of the University of Los Andes. *Systems Practice*, **4**, 507–525.

Habermas, J. (1975). *Legitimation Crisis*. Polity Press, London.

Habermas, J. (1988). *Legitimation Crisis*. Polity Press, Cambridge.

Lander, E. (1987). La confederación de trabajadores de Venezuela ante la crisis: El mito del movimiento obrero en Venezuela. In: *Crisis y Movimientos Sociales en Venezuela*. Gomez Calcaño, L. and Lander, E. (eds.). Tropykos, Caracas.

López-Garay, H. (1991). An interpretive-systemic study of the Regional Planning Corporation of Los Andes in Venezuela. *Systems Practice*, **4**, 491–505.

LORM (1978). *Ley Orgánica de Regimen Municipal*. Congreso de la República de Venezuela, Caracas.

Midgley, G. and Ochoa-Arias, A.E. (1999). Visions of community for community OR. *Omega*, **27**, 259–274.

Midgley, G. and Ochoa-Arias, A.E. (2001). Unfolding a theory of systemic intervention. *Systemic Practice and Action Research*, **14**, 615–649.

Miliani, M. (2000). *Una Interpretación histórica de la Protección de la Infancia Abandonada en Venezuela*. Tesis de Maestria Postgrado en Sistemología Interpretativa. Universidad de Los Andes, Mérida, Venezuela.

Nozick, R. (1974). *Anarchy, State and Utopia*. Basic Books, New York.

Ochoa-Arias, A.E. (1994). The possibilities of community OR in a third world country. *International Transactions in Operational Research*, **1**, 345–352.

Ochoa-Arias, A.E. (1996). An interpretive-systemic study of community organizations in Venezuela. PhD Thesis, University of Hull.

Ochoa-Arias, A.E. (1998). An interpretive-systemic framework for the study of community organizations. *Systemic Practice and Action Research*, **11**, 543–562.

Ochoa-Arias, A.E. (2000). Community organizations in Venezuela: Toward the disintegration of the modern state and the emergence of a 'community organized' society? *Systemic Practice and Action Research*, **13**, 165–187.

Ovalles, O. (1987). Movimientos de cuadro de vida en la Venezuela urbana actual: Posibles actitudes para enfrentar la crisis. In: *Crisis y Movimientos Sociales en Venezuela*. Gomez Calcaño, L. and Lander, E. (eds.). Tropykos, Caracas.

Santana, E. (1983). *El Poder de los Vecinos*. Ediciones Ecotopia, Caracas.

Zambrano, A. (1986). *Asociaciones de Vecinos y el Proceso de Democratización*. CENDES, Venezuela.

13

Community and Environmental OR: Towards a New Agenda

Gerald Midgley and Martin Reynolds

13.1 Introduction

In 1999, the UK-based Operational Research Society granted charitable funding to a systemic intervention project, based in the Centre for Systems Studies at the University of Hull, designed to create an agenda for the future role of operational research (OR) in environmental planning and management. Our final report on the project was published in 2001. This chapter summarises our findings, then focuses in detail on one aspect of these: the outputs from a mini-conference held with OR practitioners who participated in developing the agenda.

Of course, the primary focus of these practitioners was how to develop OR to make it more relevant to environmental planning and management rather than community development. However, it was strikingly obvious to us that their prescriptions for change would, if implemented, make OR for environmental management (henceforth called 'Environmental OR' for short) a very similar practice to Community OR. Arguably, this should not come as much of a surprise: Community and Environmental

Final version accepted for publication March 2003.

Gerald Midgley • Institute of Environmental Science and Research (ESR) Ltd., Christchurch Science Centre, 27 Creyke Road, PO Box 29-181, Christchurch, New Zealand. Centre for Systems Studies, Business School, University of Hull, Hull, HU6 7RX, UK.
Martin Reynolds • Centre for Complexity and Change, Open University, Walton Hall, Milton Keynes, MK7 6AA, UK.

Community Operational Research: OR and Systems Thinking for Community Development, edited by G. Midgley and A.E. Ochoa-Arias, Kluwer Academic/Plenum Publishers, New York, 2004

OR are both concerned with issues of complexity, stakeholder involvement and the political effects of change. Indeed, writers on systems thinking have been arguing for some time that many environmental issues interact with social ones, and these interactions need to be taken into account if we are not to experience major, unanticipated side-effects of intervention (e.g., Midgley, 1992).

Following our presentation of the new agenda for Environmental OR developed by the participants in our project, we argue that there is a real need for an alliance between practitioners of Environmental and Community OR to support implementation of the agenda. There are three reasons for pursuing this alliance. First, to forge a critical mass of activists who can build a stronger movement for sustainable community development. Second, to give due recognition to the importance of environmental issues for local communities, and ultimately the whole human species. Third, to enable Environmental OR practitioners to learn more about methodologies and methods for participative community development (the province of Community OR). We suggest that an alliance could considerably strengthen both communities of practice.

13.2 *The Context of our Project*

Although OR techniques have been widely used in environmental planning and management, the label 'OR' appears to have a low profile in the discourse about appropriate methodologies and methods (Bloemhof-Ruwaard *et al*, 1995; Daniel *et al*, 1997). Our own literature search reveals that, for every paper on environmental planning and management that is explicit about using OR methods, there are at least five making claims to methodological innovation that are using the same or similar methods without any reference to OR at all (Midgley and Reynolds, 2001a).

Nevertheless, in our view, the methodological focus of OR is of great value, primarily because it is so broad: it embraces a technical focus (e.g., on the assimilation and control of relevant information); it provides ideas about participation and communication between stakeholders; *and* it can invite reflection on values. Therefore, it is our belief that OR can help make environmental management a broad-based, dynamic, applied practice of central relevance to government, industry and the voluntary sector. This is why we decided to initiate the research reported here, and it is also why the Operational Research Society provided financial support for it. Essentially, a need was identified for an agenda for the use of OR in environmental planning and management that makes the actual and potential contribution of OR more visible, and which sets out the developments needed in OR if its potential is to be realised.

13.3 *The Purposes of the Project*

Our systemic intervention project was designed to support OR practitioners in creating just such an agenda for developing their discipline. The project had three primary objectives:

- To make existing good OR practice in environmental planning and management more visible;
- To explore the further potential of using OR techniques for environmental planning; and
- To ask how OR would have to be further developed if it is to make an increased and sustained contribution to environmental management.

Points two and three represent an agenda for development and change. Two action-orientated, subsidiary aims also informed the study:

- To engender commitment from OR practitioners to the agenda through a process by which they could participate in its generation; and
- To produce a development plan for improving the institutional infrastructure to enhance the ability of interested OR practitioners to undertake the work set out in the agenda.

13.4 *Methodology*

Critical Systems Thinking (CST), as represented in the work of Midgley (1996, 2000), provided the guiding methodological framework for our systemic intervention. Midgley lists the key CST principles as:

1. *Improvement*—defined temporarily and locally, but in a widely informed manner, taking issues of power (which may affect the definition) into account;
2. *Boundary critique*—regularly questioning and exploring value and boundary judgements, both with respect to the methodological approach adopted and the substantive subject matter being investigated; and
3. *Methodological pluralism*—learning from other methodologies and drawing in methods from those methodologies.

Participants in this study not only defined improvement in terms of environmental protection, but also the more proactive improvement of approaches to environmental and associated social development. Boundary critique proved crucial, as what counts as an environmental issue was a thorny and recurring question addressed in locally meaningful ways throughout the project. Also, participants generated many insights into the ways in which OR methods can either marginalise or empower stakeholders in environmental management projects. Finally, the practice of methodological pluralism enabled us to ensure that our methods remained flexible and responsive to the great variety of situations we faced. The methods we used were drawn from Qualitative Applied Social Science (e.g., Silverman, 2000), Interactive Planning (Ackoff, 1974, 1981), Soft Systems Methodology (Checkland, 1981; Checkland and Scholes, 1990) and Critical Systems Heuristics (Ulrich, 1983).

We have chosen not to provide any further details of the enactment of the CST principles in this chapter, as the primary focus is on the outcomes of the project. For more information about CST, see Flood and Romm (1996), Jackson (2000) and Midgley (2000).

13.5 Research Process

Our project ran through four phases, as follows:

13.5.1 Phase 1: Groundwork and Stakeholder Analysis

A multi-agency steering group was established, and a review of the literature on OR and environmental planning was undertaken. A stakeholder analysis was then initiated, drawing upon both the literature and the views of our steering group. Four stakeholder groups were identified: *professional experts* (operational researchers) associated with environmental planning; and users of professional expertise including agencies of *government, business,* and the *voluntary sector* (each operating at local, national, and international levels of planning).

13.5.2 Phase 2: Interviews

Two cycles of semi-structured interviews were undertaken: one with stakeholders identified in phase 1, and a second with significant others suggested by those interviewed in the first cycle. 50 respondents agreed to be interviewed in 46 interview sessions. A sectoral breakdown of agencies reveals 11 government, 13 business, 13 voluntary sector, and

11 academic. Phase 2 culminated in the production of an interim report offering feedback to respondents and providing a stimulus to launch phase 3.

13.5.3 *Phase 3: Workshops and Mini-conference*

Two one-day workshops (in London and Sheffield) took place. Interested parties were invited to explore how better OR support could be provided, based on the outputs of Phase 2. The workshops were designed to establish ideal 'mission statements' associated with possible future agendas, and to explore the parameters in which such statements might be realised. We used a synergy of Idealised Design (from Ackoff's, 1981, methodology of Interactive Planning) and Critical Systems Heuristics (Ulrich, 1983) to structure the two regional workshops (see Midgley and Reynolds, 2001a, for full details). These provided source material for a two-day mini-conference in Hull (employing adaptations of some of Checkland's, 1981, methods from Soft Systems Methodology) at which OR practitioners and academics developed a fuller agenda for future collaboration.

13.5.4 *Phase 4: Reporting*

Three working documents were produced during the course of the study: an interim report, a workshops report and a mini-conference report. These were designed to elicit feedback from interviewees and other participants to support the learning process. Presentations (eliciting feedback) were also made to the Manchester Chamber of Commerce and Industry Environmental Committee; a meeting of the Operational Research Society Environmental Study Group; OR42 (the Operational Research Society Annual Conference); and the Industrial Ecology Conference in Berkeley, USA. The main written output is our final report (Midgley and Reynolds, 2001a), and a series of articles for both practitioner and academic audiences are also being produced (so far, Midgley and Reynolds, 2001b, 2003; Reynolds and Midgley, 2002; and this chapter).

13.6 *Findings from Phase Two*

Our literature review revealed that the traditions of OR and environmental planning share some common concerns. First, both have wide boundaries in terms of clientele, the range of methodological approaches used, and attention to multiple (and often conflicting) values. Second, both traditions have an interest in fostering purposeful interdisciplinarity.

Third, both OR and environmental planning are concerned with the *implementation*, as well as the *design*, of planning strategies.

Three generic issues were found to recur in both the environmental management literature and the interview data generated in our study:

1. *Complexity and uncertainty* (regarding the unpredictability of natural and social phenomena);
2. *Multiple and often conflicting values* (of those involved in environmental planning); and
3. *Political effects* (on those not involved in planning processes, including non-human nature).

An examination of how these generic issues are perceived in the public, business and voluntary sectors revealed clear patterns, summarised in the following four points:

First, the three issue categories (complexity and uncertainty; multiple, conflicting values; and political effects) are all of concern in all three sectors.

Second, for each sector there is considerable conflict between interpretations of how each issue category should be addressed. For example, in dealing with issues of complexity and uncertainty, some businesses are seeking to adopt and promote a 'learning culture', taking heed of wider economic, social and environmental affairs in long-term planning. However, others still reduce the idea of 'sustainability' to short-term economic prosperity.

Third, issues of complexity and uncertainty dominate the public sector, with attention primarily focused on developing appropriate 'indicators'. Competing values are the main concern of business organisations, with attention being paid to minimising risks by improving stakeholder interaction. Political issues dominate the voluntary sector, with concerns about representing marginalised interests and widening the net of meaningful participation in planning processes. These might be termed the *primary* issue categories associated with each sector.

Fourth, for each sector the two *secondary* issue categories tend to cluster around the primary issue category. For example, in the public sector, conflicting values and issues of social exclusion tend to be dealt with *in relation* to the formation of indicators to deal with complex and uncertain realities.

The issues discussed above are generic and therefore arguably quite abstract. Substantive issues like transport, green belt policy, pollution, energy, waste, genetically modified organisms—and even wider concerns relating to sustainable development, global warming, world trade, population growth, the elimination of poverty, etc.—can be more specifically examined using the same parameters. That is, any environmental issue being addressed could potentially involve each of the three user groups (from the public, business and voluntary sectors), as well as some form of 'expert' function (whether operational research or another form of expertise). Likewise, any substantive problematic situation might be analysed in terms of all three generic issue types: complexity and uncertainty; competing values; and political effects.

In our interim report, produced after the interviews with stakeholders but before the workshops with practitioners (designed to develop the new agenda for Environmental OR), we reached the following conclusion. In the increasingly complex, interdisciplinary and politicised world of environmental planning, if we want to enhance expert support using OR, it will be vital to do more than just deal with the technical difficulties associated with modelling the natural world. This is not to say that the technical issues are trivial or unimportant (far from it), but it will also be necessary to address the more messy social worlds of values and ethics in which both OR support and environmental issues are embedded. We also identified a major challenge for OR practitioners: to develop methodologies and methods that are capable of dealing with *all three* of the generic themes identified in our research—complexity and uncertainty, multiple values and political effects.

13.7 Developing the Agenda for Operational Research

Through the workshops and mini-conference, three distinct (though strongly interrelated) agendas took shape:

1. *Develop OR* (with a focus on methodological issues);
2. *Promote Interaction* (with a focus on issues of interdisciplinarity, intersectoral co-operation, etc.); and
3. *Promote Public Participation* (with a focus on issues of accountability and social inclusion).

Each agenda was subject to a process of analysis and 'conceptual modelling' (Checkland, 1981; Checkland and Scholes, 1990) at the mini-conference. Participants asked themselves, what is the transformation being sought? Who are the intended beneficiaries? Who or what might be made a victim (and should something be done about this)? Who should act to implement the agenda? What worldview underlies the agenda? Who should those implementing the agenda be accountable to? And what environmental constraints will have to be taken as given? The answers to these questions led the group to define key activities needed to realise the stated purposes of the agendas, and necessary links between these activities (expressed in the 'conceptual models' to be reproduced over the coming pages).

13.7.1 *Agenda 1: Develop OR*

For agenda 1 (*Develop OR*), the activities centred on establishing an on-going research project to relate methods with problem situations relevant to environmental management. The need for extensive testing of OR methods in case studies was stressed, as was the need to communicate the results of these tests to enhance the OR knowledge base for environmental management. Importantly, however, the idea of relating methods to problem contexts was not conceived as the production of a mechanical rule book for OR applications. Rather, it was seen as involving the reconceptualisation of OR as a *reflective practice*. Amongst other things, this will involve questioning purposes (rather than taking them as given); focusing on the big picture; multi-sectoral thinking; including multiple agents in defining problems; drawing upon and mixing multiple methods; and embracing environmental issues alongside social ones (rather than taking either environmental or social issues as prime). All the key activities identified as necessary to develop OR are represented diagrammatically in Figure 13.1.

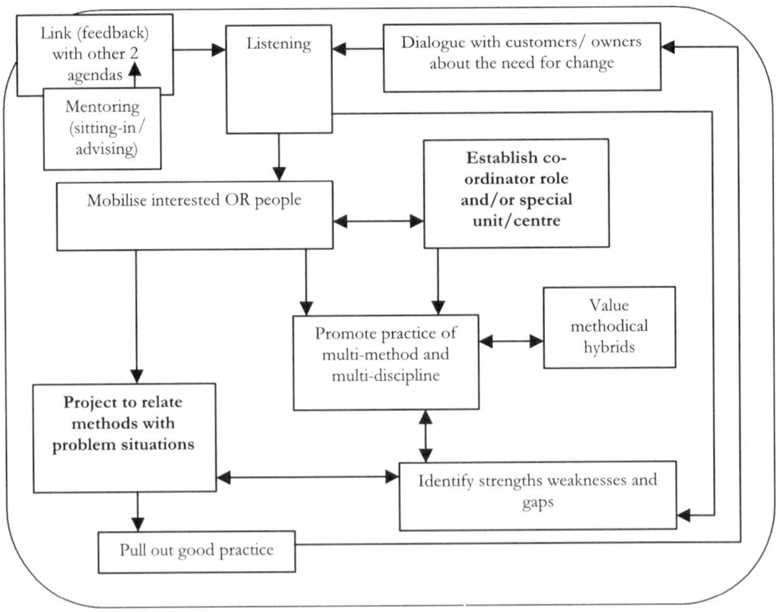

FIGURE 13.1 Conceptual model of 'Develop OR'

13.7.2 *Agenda 2: Promote Interaction*

In agenda 2 (*Promote Interaction*), the activities centred on developing 'skills', 'knowledge' and 'communication channels'. Whilst interaction was mainly focused on important issues of interdisciplinarity, the agenda was also concerned with promoting relationships between public, private and voluntary sector organisations. People said that OR needed to move from being a primarily 'backroom', problem-solving form of expertise to being a more pro-active discipline where both problem structuring and *raising awareness* of issues amongst stakeholders are key activities. Also, it will require OR practitioners to be more outward looking and facilitative than is currently the norm. See Figure 13.2 for a diagrammatic representation of these activities and their necessary interactions.

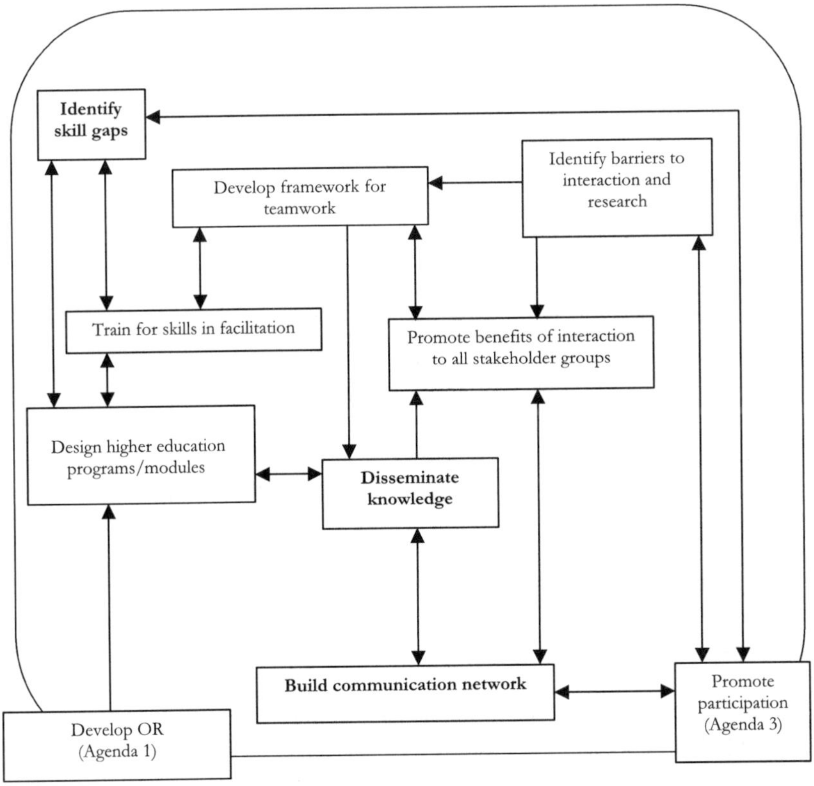

FIGURE 13.2 Conceptual model of 'Promote Interaction'

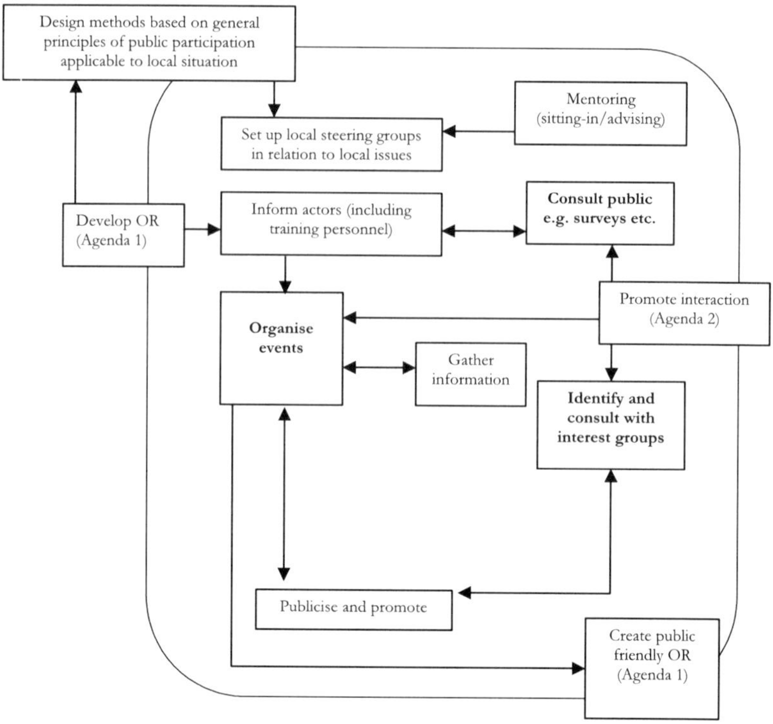

FIGURE 13.3 Conceptual model of 'Promote Public Participation'

13.7.3 Agenda 3: Promote Public Participation

Agenda 3 (*Promote Public Participation*) recognised the difficulties of having a catch-all public participation remit: it is not realistic to try to engage 'the public' in improving OR in general. Rather, the emphasis needs to be on local participation in projects, taking care to differentiate between general public expressions of concern and special interest group involvements. The key activities that will need to be undertaken to promote public participation can be seen in Figure 13.3.

13.7.4 Mapping the 'Whole System'

The three agendas can usefully be regarded as nested human activity systems: agenda 3 nesting in agenda 2, which in turn nests in agenda 1. Therefore, ensuring local public participation in projects is one aspect of keeping OR interactive and outward looking, and should have an impact

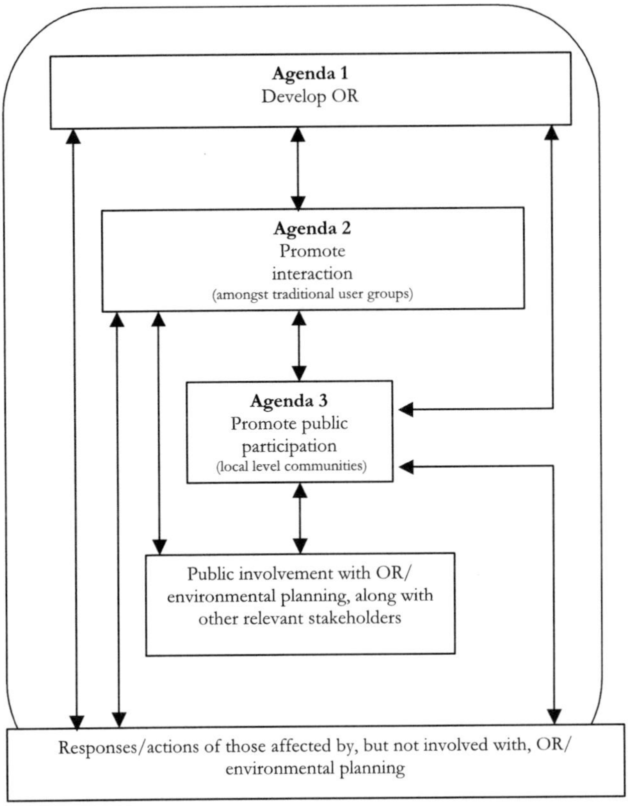

FIGURE 13.4 Relationships between the three agendas

on how interdisciplinary and intersectoral communications are conducted. Similarly, both of these agendas have important implications for developing the methodology of OR. The key interactions between the three agendas are mapped in Figure 13.4. While the participants in the mini-conference developed Figures 13.1–13.3, we produced Figure 13.4 ourselves immediately following the event (we did, however, mail it to the participants for feedback and secured unanimous agreement on it).

13.8 *Recommendations*

The agendas are arguably the primary outputs of this research, together with the bond that formed amongst the participating OR practitioners. However, the group was intent on keeping its feet on the ground, and realised that enthusiasm in a mini-conference will not automatically translate into actual change unless some preparatory actions are taken. These were identified as necessary primarily because the group realised that there is currently an insufficient critical mass of activists in the UK operational research community with an interest in environmental management to make all the plans a reality. However, concrete steps for changing this situation, with (in our view) a high probability of success, were identified.

These steps were translated in our final report into a series of recommendations to both the Operational Research Society and OR practitioners more generally (see Midgley and Reynolds, 2001a, for full details). Here, however, we will only concentrate on one recommendation: that a critical mass of activists to implement the agenda for change should be built up through an alliance between different groups of OR practitioners with common interests. These groups include those working in environmental management, agriculture, international development and Community OR.

13.9 *Reasons for an Alliance*

Essentially, there are three grounds for an alliance: the need for a critical mass of activists; the importance of bringing environmental issues into Community OR; and the need for Environmental OR practitioners to learn more about the structured facilitation of community development (the province of Community OR). Each of these is discussed in more detail below.

13.9.1 A Critical Mass of Activists

Bringing together a critical mass of activists is a particular issue for the UK operational research community because of its size (and we suspect it will be an issue in many other countries too). The OR Society is the main professional body for operational researchers in the UK, and it has a total of around 3,000 members, including some living overseas. The majority of these people have, at best, only a passing interest in Community and/or Environmental OR: most have solely business or academic interests. While the OR Society runs a number of study groups (specialising in development, health, agriculture, complex systems, community, environment, etc.), each of these has a relatively small number of activists. It has been a perennial complaint that none of these study groups are really thriving, despite the fact that each one has a reasonably substantial mailing list. Most members of the study groups seem content to be passive members. The participants in our mini-conference identified the lack of a critical mass of activists in any one specialism as a major contributing factor to the under-achievement of the study groups. At one point in our project, the idea was floated of amalgamating several study groups to overcome this problem. However, we explored this possibility with various study group representatives and encountered enthusiasm amongst some, and resistance amongst others. There was a fear that more popular concerns would dominate any amalgamated group, and minority interests (which are nevertheless vitally important) could get lost. In particular, there was a worry that environmental issues could be subsumed within a social agenda. For this reason, it makes sense to talk of *alliances* between groups with parallel interests rather than their amalgamation.

Of course, alliances need not be restricted to members of the OR Society (or similar societies in other countries). Since finishing our project, we have been involved in several activities designed to build capacity in 'systems thinking for social and environmental responsibility' (note that the 'OR' label has not been used in this phrase). Working with two colleagues (Wendy Gregory and Ian Roderick), we have run three streams at international conferences; edited a special issue of a journal (*Systems Research and Behavioral Science*, volume 20, part 2); and have participated in the development of an interactive web site. Each of these initiatives has had roots in the OR community, with a significant number of Community OR practitioners participating, but has extended outwards into the systems and other research communities. The point is that, if we are serious about pursuing an ambitious agenda for developing OR, both Environmental and Community OR practitioners need to look beyond the boundaries of their own relatively small research communities and work with others in pursuit of their substantial common interests. Even when their interests diverge, this need not pose a threat: divergence can be a valuable source of learning when groups are willing to respect each other's autonomy.

13.9.2 The Importance of Environmental Issues

The second reason for building an alliance between Community and Environmental OR practitioners is the increasing importance of environmental issues at local, regional, national and international levels. It is possible to regard this as a purely strategic issue, by pointing out that significant sums of money are available to undertake projects promoting 'Agenda 21' (one of the principle outputs from the 1992 Rio Earth

Summit). Agenda 21 consists of a large set of declarations including, for example:

- Adoption of the 'precautionary principle' (the principle that technological innovations should not be introduced unless or until there is evidence of their safety);
- A commitment to the eradication of poverty and reduction of disparities in living standards;
- A commitment to reducing and eliminating unsustainable patterns of production and consumption; and
- A statement supporting citizen participation in environmental management.

The implementation of these and the other Agenda 21 policies is being funded in the UK by the European Union, national government and local authorities (Department of the Environment, Regions and Transport, 1999)—and resources are also being made available for this in other parts of the world. Arguably, for Community OR practitioners, one of the key principles that makes their skills relevant to Agenda 21 is that citizen participation should be part of environmental management. There is now wide spread recognition that environmental protection policies require commitment from local communities if they are to be implemented effectively, and the best way to gain that commitment is through direct participation. Also, when addressing environmental issues at the local and regional levels, the unique features of particular ecosystems have to be accounted for, and it is usually local communities who know most about these.

However, phrasing this as a purely strategic issue—getting access to resources by taking on the environmental agenda—really misses the point. In the vast majority of cases, people get involved in Community OR because they want their practice to be of non-trivial benefit to local people—or even the whole of humanity. Their underlying motivations may be political, religious or humanitarian (Wong and Mingers, 1994; Midgley and Ochoa-Arias, 1999), but in a general sense they all have the desire to do something socially meaningful. There is now a wide spread consensus that the global economy is on an unsustainable trajectory. Many people claim that the ever-increasing consumption of non-renewable energy, and the production of environmental side effects of economic development, will eventually push against the ecological limits of the Earth (e.g., Meadows *et al*, 1972, 1992). Humanity therefore faces a choice: do more of the same with the global economy, and experience the limits to growth as a crisis or even an environmental and economic catastrophe, or work for greater sustainability ahead of time, thereby averting the problem

(Meadows *et al*, 1992). Notwithstanding some dissent concerning the idea that there are immutable 'limits to growth' (e.g., Dryzek, 1997), it is still widely accepted that sustainability is a major political issue. In addition, ecological problems are made all the more difficult to address by the fact that the global financial institutions appear to be resisting all but minor, market-driven changes to their ways of working (for an interesting explanation of why capitalist institutions are so resistant, see Luhmann, 1986). Indeed, an argument is beginning to emerge that some major international bodies set up to enforce environmental conservation measures are actually acting as protectors of free market forces, regardless of the ecological consequences (Young, 2002). The primary motivation for Community OR practitioners to build an alliance with the Environmental OR community is therefore that environmental issues *matter*. They are complex and multi-dimensional, with political ramifications, and they impact on both local communities (hence the imperative for local participation in policy development) and potentially the whole human species.

It is very interesting to note that one of the original motivations for Jonathan Rosenhead coining the term 'Community OR', and setting out to build a Community OR movement, was to make OR relevant to his own Marxist politics (partially expressed in Rosenhead, 1986, 1987). The idea was that Community OR would contribute to building a working class consciousness in local communities, and might ultimately help empower people to move for radical, bottom–up change. However, most of those who joined the Community OR movement in the late 1980s and early 1990s either disagreed with Rosenhead's political agenda, or were largely unconcerned about it. Community OR very quickly became a broad church where many different motivations for working for socially meaningful change, interpreted in an equally broad number of ways, were accepted. Potentially, the environmental agenda offers another 'grand' political project for people to work on (particularly as envisaged by Meadows *et al*, 1992). This is not to say that all Community OR practitioners should be committed to this, or even all those interested in the interface between community and environmental issues, but we argue that it will be a significant motivator for some people to build an alliance between Community and Environmental OR. For others, environmental issues may have local relevance disconnected from any 'grand' political narrative, but the alliance will still be useful. Our own position (following Midgley and Ochoa-Arias, 1999) is that all Community OR practitioners should be clear that their projects *do* have political implications, whether they want them to or not. There is therefore an onus on them to reflect on and justify their political assumptions (particularly the vision of community they are promoting) as part of their practice. We are happy to see a diversity of political visions, but not the abdication of responsibility for these. Community and Environmental OR can both be enriched by a debate that makes explicit the implications of various political positions for OR practice.

13.9.3 The Need for Community OR in Environmental Management

The third reason why Community and Environmental OR practitioners should, in our view, build an alliance is the wide spread recognition (mentioned earlier) that citizen participation should be integral to environmental management (this has been codified into policy in the form of Agenda 21). We have seen that dealing with environmental issues makes sense for Community OR practitioners who are concerned with facilitating meaningful community development. However, it is also the case that Environmental OR needs what Community OR practitioners can bring to the alliance in the form of methodologies and methods for facilitating debate.

It is still the case that a clear majority of Environmental OR projects focus on the use of quantitative methods. An example is cost-benefit analysis, incorporating costs to the 'external' environment (e.g., Rycroft *et al*, 1988; Pearce *et al*, 1990; Butler and Nelson, 1994; Lindsey *et al*, 1995; Mirasgedis and Diakoulaki, 1997; Parkinson, 2000). Another is the production of sustainability indicators for performance measurement (e.g., Qingzhen *et al*, 1991; Zhao *et al*, 1991; Ellis *et al*, 1996; Spengler *et al*, 1997). These methods *can* be implemented in a participative manner, for instance by involving stakeholders in the definition of indicators, but there is not necessarily any methodological onus on the practitioner to do so—and this practice also begs the question of what should happen when stakeholders hold conflicting values. We are not arguing *against* the employment of quantitative methods: on the contrary they can be very useful, especially when there is wide spread agreement on the need for intervention, but the intervention is difficult to pursue in the face of great complexity (Midgley and Reynolds, 2001a). However, several decades of research on environmental management have demonstrated very clearly that defining ecological problems in purely scientific terms, and providing technical solutions to them while ignoring management issues and local community values, can lead to major problems with implementation (e.g., Bosch *et al*, 2003). In other words, purely technical solutions are only credible to all stakeholders when these stakeholders are in agreement that the issues have been defined correctly. Securing such agreements is often problematic—and indeed, sometimes people have such divergent values that they don't even agree on what the most important issue is, let alone whether it has been defined correctly. Therefore, the experience that Community OR practitioners can offer of facilitating dialogue on complex social issues, primarily using problem structuring methods (Jackson, 1988), should be invaluable for Environmental OR. Some examples of this kind of dialogue do already exist in the Environmental OR literature (e.g., Njiforti *et al*, 1991; Fischer, 1995; Brown and Jacobs, 1996; Sudhir *et al*, 1996; Frederickson and Frederickson, 1997; McClintock *et al*, 1997). However, in our view, community facilitation is not yet a sufficiently wide spread practice, and there are many useful Community OR methods yet to be tried on environmental issues. Given this scenario, an alliance between Environmental and Community OR practitioners could be very fruitful indeed.

13.10 Conclusion

In this chapter, we have described a project in which practitioners developed an agenda for enhancing the role of OR in environmental planning and management. There were three parts to the agenda: developing OR (with a focus on methodological issues); promoting interaction (with a focus on issues of interdisciplinarity, intersectoral co-operation, etc.); and promoting public participation (with a focus on issues of accountability and social inclusion).

One of the recommendations coming out of this project was that Environmental and Community OR practitioners (amongst others) could usefully form an alliance to implement the agenda. We surfaced three reasons why this alliance would be beneficial: the need for a critical mass of activists; the importance of bringing environmental issues into Community OR; and the need for Environmental OR practitioners to learn more about the structured facilitation of community development.

We hope that this chapter will spark some debate about political and practical priorities in both Community and Environmental OR, and eventually give rise to co-operation across the boundaries of these hitherto mostly separate enterprises.

References

Ackoff, R. (1974). *Redesigning the Future: A Systems Approach to Societal Problems*. Wiley, Chichester.

Ackoff, R. (1981). *Creating the Corporate Future*. Wiley, Chichester.

Bloemhof-Ruwaard, J.M., VanBeek, P., Hordijk, L., and Van Wassenhove, L.N. (1995). Interactions between operational research and environmental management. *European Journal of Operational Research*, **85**, 229–243.

Bosch, O.J.H., Ross, A.H., and Beeton, R.J.S. (2003). Integrating science and management through collaborative learning and better information management. *Systems Research and Behavioral Science*, **20**, 107–118.

Brown, D. and Jacobs, P. (1996). Adapting environmental impact assessment to sustain the community development process. *Habitat International*, **20**, 493–507.

Butler, R.W. and Nelson, J.G. (1994). Evaluating environmental planning and management: The case of the Shetland Islands. *Geoforum*, **25**, 57–72.

Checkland, P. (1981). *Systems Thinking, Systems Practice*. Wiley, Chichester.

Checkland, P. and Scholes, J. (1990). *Soft Systems Methodology in Action*. Wiley, Chichester.

Daniel, S.E., Diakoulaki, D.C., and Pappis, C.P. (1997). Operations research and environment planning. *European Journal of Operational Research*, **102**, 248–263.

Department for Environment, Transport and the Regions. (1999). *A Better Quality of Life: A Strategy for Sustainable Development in the UK*. DETR, London.

Dryzek, J.S. (1997). *The Politics of the Earth: Environmental Discussions*. Oxford University Press, Oxford.

Ellis, H., Ringold, P.L., and Holden G.R. Jr. (1996). Emission reductions and ecological response: Management models for acid rain control. *Socio-Economic Planning Sciences*, **30**, 15–36.

Fischer, F. (1995). Hazardous waste policy, community movements and the politics of nimby: Participatory risk assessment in the USA and Canada. In: *Greening Environmental Policy: The Politics of a Sustainable Future*. Fischer, F. and Black, M. (eds.). Paul Chapman, London.

Flood, R.L. and Romm, N.R.A. (eds.) (1996). *Critical Systems Thinking: Current Research and Practice*. Plenum, New York.

Frederickson, J. and Frederickson, N. (1997). The integration of systemic and scientific thinking in the development of an innovative process for environmental management. In: *Systems for Sustainability: People, Organizations, and Environments*. Stowell, F.A., Ison, R.L., Armson, R., Holloway, J., Jackson, S., and McRobb, S. (eds.). Plenum, New York.

Jackson, M.C. (1988). Some methodologies for community operational research. *Journal of the Operational Research Society*, **39**, 715–724.

Jackson, M.C. (2000). *Systems Approaches to Management*. Kluwer/Plenum, New York.

Lindsey, G., Paterson, R.G., and Luger, M.I. (1995). Using contingent valuation in environmental planning. *Journal of the American Planning Association*, **61**, 252–262.

Luhmann, N. (1986). *Ecological Communication*. 1989 trans. University of Chicago Press, Chicago.

McClintock, D., Kersten, S., and Ison, R. (1997). Emergent practice in agricultural research: Using metaphors and dialogue to create a space for understanding. In: *Systems for Sustainability: People, Organizations, and Environments*. Stowell, F.A., Ison, R.L., Armson, R., Holloway, J., Jackson, S., and McRobb, S. (eds.). Plenum, New York.

Meadows, D.H., Meadows, D.L., Randers, J., and Behrens, W.W. (1972). *The Limits to Growth*. Earth Island, London.

Meadows, D.H., Meadows, D.L., and Randers, J. (1992). *Beyond the Limits to Growth*. Chelsea Green, Post Mills.

Midgley, G. (1992). Pluralism and the legitimation of systems science. *Systems Practice*, **5**, 147–172.

Midgley, G. (1996). What is this thing called critical systems thinking? In: *Critical Systems Thinking: Current Research and Practice*. Flood, R.L. and Romm, N.R.A. (eds.). Plenum, New York.

Midgley, G. (2000). *Systemic Intervention: Philosophy, Methodology, and Practice*. Kluwer/Plenum, New York.

Midgley, G. and Ochoa-Arias, A.E. (1999). Visions of community for community OR. *Omega*, **27**, 259–274.

Midgley, G. and Reynolds, M. (2001a). *Operational Research and Environmental Management: A New Agenda*. Operational Research Society, Birmingham.

Midgley, G. and Reynolds, M. (2001b). Greening OR. *OR Newsletter* #370, 15–18.

Midgley, G. and Reynolds, M. (2004). Systems/operational research and sustainable development: Towards a new agenda. *Sustainable Development*, **12**, 56–64.

Mirasgedis, S. and Diakoulaki, D. (1997). Multicriteria analysis verses externalities assessment for the comparative evaluation of electricity generation systems. *European Journal of Operational Research*, **102**, 364–379.

Njiforti, H., Schrader, T.H., and Toornstra, F.H. (1991). LEARN: A methodological challenge for rapid environmental assessment. *Landscape and Urban Planning*, **20**, 173–181.

Parkinson, S. (2000). The clean development mechanism—Opportunity or loophole? *Scientists for Global Responsibility Newsletter*, **20**, 7–9.

Pearce, D., Barbier, E., and Markandya, A. (eds.). (1990). *Sustainable Development: Economics and Environment in the Third World*. Earthscan, London.

Qingzhen, Z., Changyu, W., and Zhimin, Z. (1991). The application of operations research in the optimization of agricultural production. *Operations Research*, **39**, 194–205.

Reynolds, M. and Midgley, G. (2002). Systems/OR and environmental management: A new agenda. In: *Proceedings of the Eighth International Sustainable Development Research Conference*, held in Manchester, UK, 8–9 April.

Rosenhead, J. (1986). Custom and practice. *Journal of the Operational Research Society*, **37**, 335–343.

Rosenhead, J. (1987). From management science to workers' science. In: *New Directions in Management Science*. Jackson, M.C. and Keys, P. (eds.). Gower, Aldershot.

Rycroft, R.W., Regens, J.L., and Dietz, T. (1988). Incorporating risk management and benefit-cost analysis in environmental management. *Risk Analysis*, **8**, 415–422.

Silverman, D. (2000). *Doing Qualitative Research: A Practical Handbook*. Sage, London.

Spengler, T., Richert, H., Penkuhn, T., and Renz, O. (1997). Environmental integrated production and recycling management. *European Journal of Operational Research*, **97**, 308–326.

Sudhir, V., Muraleedharan, V.R., and Srinivasan, G. (1996). Integrated solid waste management in urban India: A critical operational research framework. *Socio-Economic Planning Sciences*, **30**, 163–181.

Ulrich, W. (1983). *Critical Heuristics of Social Planning: A New Approach to Practical Philosophy*. Haupt, Berne.

Wong, N. and Mingers, J. (1994). The nature of community OR. *Journal of the Operational Research Society*, **45**, 245–254.

Young, Z. (2002). *A New Green Order? The World Bank and the Politics of the Global Environment Facility*. Pluto Press, London.

Zhao, Q., Wang, C., Zhang, Z., Zhang, Y., and Wang, C. (1991). The application of operations research in the optimization of agricultural production. *Operations Research*, **39**, 194–205.

14

Adaptive Methodology for Ecosystem Sustainability and Health (AMESH): An Introduction

David Waltner-Toews, James Kay, Tamsyn P. Murray and Cynthia Neudoerffer[1]

14.1 Introduction: Coping with Complexity

Researchers and development practitioners have, since at least 1987, struggled to operationalise the concept of sustainable development with a view to improving human health and nutritional security without compromising the integrity of underlying ecological structures and processes.

[1] All four authors can be contacted through the web site of the Network for Ecosystem Sustainability and Health (www.nesh.ca).

Accepted for publication August 2001.

David Waltner-Toews • Department of Population Medicine, University of Guelph, Guelph, Ontario, N1G 2W1, Canada.
James Kay • Environment and Resource Studies, University of Waterloo, Waterloo, Ontario, N2L 3G1, Canada.
Tamsyn P. Murray • Centre for Systemic Development, University of Western Sydney, 3 Congham Road, West Pymble, NSW 2073, Australia.
Cynthia Neudoerffer • Faculty of Environmental Design and Rural Development, University of Guelph, Guelph, Ontario, N1G 2W1, Canada.
Community Operational Research: OR and Systems Thinking for Community Development, edited by G. Midgley and A.E. Ochoa-Arias, Kluwer Academic/Plenum Publishers, New York, 2004

Many of these efforts have resulted in lists of quantitative variables, expensive to measure and difficult to integrate, and qualitative models, often restricted to heuristic devices and context-specific applications. To go beyond the current doldrums we will need to:

- Elaborate a fundamental systemic theory for complex socio-ecological realities;
- Define a theoretical framework for identifying, in any given situation, the practical implications of that theoretical understanding for sustainable practices; and
- Based on 1 and 2, elaborate methodological processes that enable decision-makers to evaluate implications, make 'best possible' decisions, monitor the impacts of those decisions, and adapt to changes in context.

To complicate matters, scholars and practitioners are faced with the prospect of dealing with all three of the above simultaneously, in a context of great urgency and uncertainty, and working with research and governance institutions that are ill-equipped for the job.

Fortunately, most researchers and development workers see this as a challenge to rise to rather than as a fatal impediment. This chapter describes an Adaptive Methodology for Ecosystem Sustainability and Health (AMESH), developed in the context of several international development projects, which we believe incorporates much of our best current understanding in each of these domains. AMESH is rooted in theories of complex adaptive systems; the implications of this for practice are elaborated in a set of 'guiding principles'; and the methodological processes are described in terms of sets of activities, themselves depicted in terms of 'guiding questions'. Given the large body of literature already available on complex systems theories, this chapter focuses on their implications for practice (items 2 and 3 above). We incorporate examples from several projects in Peru, Kenya and Nepal to illustrate particular points.

14.2 Theoretical Basis

von Bertalanffy's (1968) General System Theory (GST) was one of the first schools of thought that provided alternative models and modes of inquiry to the reductionist methods of disciplinary science. GST emphasises connectedness, context and feedback. Research questions identify and explain interactions, relationships and patterns. The essential properties of the parts of a system can only be understood from the organisation of the whole, as they arise from the configuration of ordered relationships that is specific to that particular system (von Bertalanffy, 1968). Understanding comes from looking at how the parts operate together rather than from teasing them apart.

With the advent of the notion of complexity, a new understanding of systems, and in particular ecosystems, is emerging to augment GST (Holling, 1986; Kay and Schneider, 1994). Complexity is an umbrella concept that incorporates the new insights of hierarchy theory, catastrophe theory, self-organisation theory (including non-equilibrium thermodynamics), and chaos theory. It takes in the associated issues of uncertainty, surprise, and emergence across scales.

Several insights from the research into complexity and complex systems are particularly relevant to efforts to define and implement sustainable development (Casti, 1986; Nicolis and Prigogine, 1989; Kay *et al*, 1999). These include the notion that a truly complex system cannot be adequately 'captured' or represented from any single perspective. The sheer diversity of elements and interactions requires a plurality of perspectives. Indeed, when we speak of 'complex systems' in this context, we are referring to systemic ways of organising our thinking about the complex socio-ecological reality within which we live. We cannot verify in any objective way that the reality is a system; however, we can usefully describe it using a variety of systemic models, narratives and pictures. We shall refer to the collective sense of reality that emerges from these as a complex system. Not only can complex systems be described from a variety of perspectives, but new properties emerge into view as one considers these systems at different scales within a nested hierarchy. Also, complex systems, with their positive and negative feedback loops, self-organise into relatively resilient patterns (attractors), which are the result of particular, context-specific histories. Nevertheless, historical trajectories may change, and any given ecosystem may organise itself in more than one way; changes from one organisational state to another, when they do occur, may be sudden, with the particular outcome uncertain. These new insights into the nature of complexity necessitate a different approach to research into ecosystem sustainability and health.

A variety of scholarly efforts have been made to link basic insights drawn from the study of complexity to issues raised by people trying to define and implement sustainable development. The resulting theories and methodologies include: Complex Systems Theory (Schneider and Kay, 1994); Ecosystem-based Management (Allen and Hoekstra, 1992); Post-Normal Science (Funtowizc and Ravetz, 1994); Complex Adaptive Systems and Resilience (Gunderson *et al*, 1995); Social and Collaborative Learning (Roling and Wagemakers, 1998); Soft Systems Methodology

(Checkland, 1981; Checkland and Scholes, 1990); and the Sustainable Livelihoods Approach (Chambers and Conway, 1992; Singh and Wanmali, 1998). Kay *et al* (1999) summarise a theoretical approach, based on a Self-Organising Holarchic Open (SOHO) Systems model, which has informed much of the development of AMESH. This model provides a conceptual basis and framework for an ecosystem description of a situation and, in particular, for the development of narrative descriptions of the potential future pathways of ecosystem organisation which are relevant to the issues at hand.

AMESH has been developed in the context of research projects in the Ucayali District of the Peruvian Amazon (Rowley *et al*, 1997; Murray, 2000; Murray *et al*, 2001); the Kiambu District in the Kenyan highlands (Gitau, 2000); and Wards 19 and 20 in old urban Kathmandu (Waltner-Toews, 1996). The Peruvian work started with a focus on complex systems theories and shifted, in a second phase, to addressing health, nutrition and natural resource management in Amazonian villages. The Kenyan project was designed to work with agriculturally based villages in the highlands to help them determine viable pathways to achieve healthy agroecosystems. The Nepalese project, building on earlier community health initiatives in two urban communities, is working with those communities to promote ecosystem health. All three projects have received basic funding from the International Development Research Centre in Ottawa, under their Ecosystem Approaches to Health Program Initiative. Since all three projects were only recently completed, much of the information about them has yet to appear in the peer reviewed literature. Ongoing reports from the projects are being posted on the Network for Ecosystem Sustainability and Health (NESH) web site (www.nesh.ca) as they become available.

In developing AMESH, concepts, assumptions and methods were drawn from all of the above approaches and integrated into a coherent approach, and then modified and adapted in the context of the specific projects. AMESH itself is a complex set of activities, and there is no 'correct' way to present it. Fundamental to both the theory and practice, which demand context specificity, no single case study would be expected to reflect all the principles we wish to illustrate, nor can there be any straightforward technical (cook-book) systemic description. In order to clarify the approach, we have divided our discussion into a section which elaborates guiding principles, each based on a particular relevant insight from complex systems theory; a series of activities which are the essential elements of the approach; and questions to guide those activities. Examples, to illustrate these, are drawn from our three case studies. We have set these activities out in Figure 14.1 in such a way as to emphasise the key elements of AMESH, and to clarify its on-going, adaptive nature.

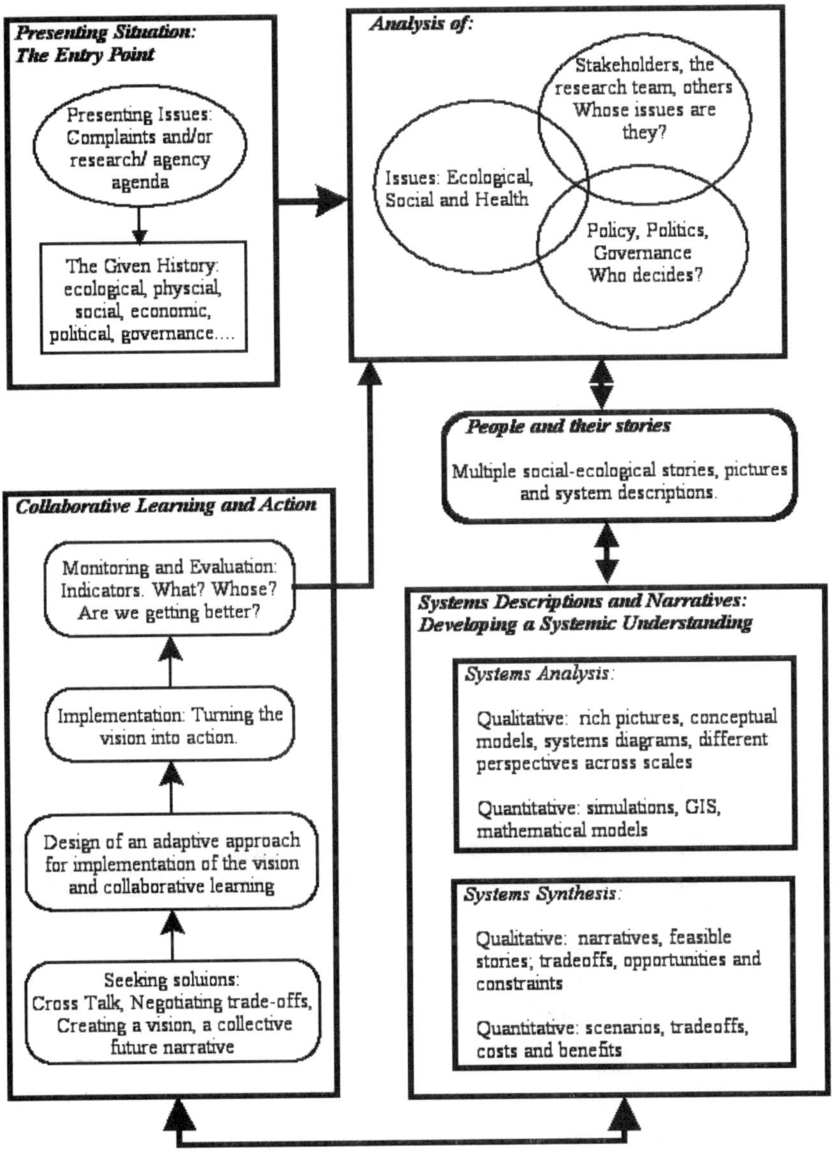

FIGURE 14.1 Adaptive methodology for ecosystem sustainability and health (AMESH): The research process

14.3 *Guiding Principles*

14.3.1 *Methodological Pluralism and Locally Grounded Multiple Perspectives*

> A complex system is one which allows us to discern many subsystems (a subsystem is the description of the system determined by a particular choice of mapping only a certain set of its qualities or properties) depending entirely on how we choose to interact with the system (Rosen, 1977).

Complex socio-ecological systems can be understood only from multiple, non-equivalent perspectives, with no single perception adequate to provide a comprehensive or adequate view of reality (Puccia and Levins, 1985). Indeed, the degree of complexity has been equated with the number of non-equivalent descriptions of a system (Casti, 1986). These non-equivalent descriptions may furthermore result in non-equivalent evaluations, and hence imply different strategies for future development.

The need to incorporate multiple perspectives has important methodological implications. The plurality of different legitimate perspectives and the inability of one particular view to capture the whole, necessitates a variety of forms of inquiry, plus inclusion of, and dialogue with, persons representing different interests and different world views (Waltner-Toews and Wall, 1997). Multiple modes of investigation and multiple sources of evidence are necessary to understand the system (Holling, 1995).

However, this assertion, often repeated in recent environmental, health and sustainable development writings, leaves unanswered the question as to how various perspectives are selected and legitimized. Greater multi- and inter-disciplinarity in the research into and management of complex socio-ecological systems is necessary, but not sufficient. Left to themselves, researchers will inquire into those aspects of the system which they, for their own personal, disciplinary or political reasons, deem to be important. This is not a very sure foundation for learning our way into a sustainable future. Because these systems develop out of particular historical conditions, with no single preferred future state, and the goals selected for management are necessarily value-laden, the values, concerns and knowledge of local stakeholders and actors must be central to any inquiry.

AMESH has incorporated these characteristics of complex systems, not just by creating spaces for qualitative and quantitative modes of inquiry, but by making involvement of local stakeholders through participatory methods central to the process of inquiry and management (Pretty, 1995, 1998; Ramirez, 1999).

The importance of this was made apparent to some of us during the first phases of the research in Ucayali District in Peru. Conventional, multi-disciplinary analyses of land use in the Ucayali region described an ecosystem dominated by small-scale cattle ranchers struggling to maintain productivity in the face of declining soil fertility, and focused on increasing commercial production in an area with major market infrastructure, technology and capital constraints. Figure 14.2 is a typical, technical, land-use systems model that could be used to describe agricultural activities in the area. Although this research provided many important insights into farm level production, land use change and ecological constraints, it failed to link land use processes with other resource sectors and actors at various scales in the region.

As the researchers using AMESH began working in the area, we discovered that cattle ranchers were only about one percent of the total farm population (only 1 in 5 farmers with pastures possess cattle). However, they owned significantly larger farms than the average; were located almost exclusively along the main road to Lima in one sub-region of Ucayali; and had differential access to government subsidies and tax relief. Cattle ranchers did not represent the diversity of livelihood systems that most often involved (in addition to farming) fishing, hunting, gathering and logging activities. These farmers have different priorities and technological demands than cattle ranchers and therefore their needs were not met by the research agenda. Looking within the household, the stakeholder analysis highlighted the important role of women in production activities. They are the critical link between resource use, nutrition and health. Native communities represent another stakeholder group that although comprising over 10% of the regional population, were not included in the research agenda.

As the description of the system was expanded to include stakeholder groups other than the cattle ranchers, three main findings emerged that described very different forces shaping the development and dynamics of the region. First, timber extraction is the main economic activity in Ucayali. It employs the largest percentage of people, generates the greatest amount of revenue and represents Ucayali's main export product. Second, annual crops provide the majority of the staple foods consumed and household income, and fish is the key source of protein and other micronutrients. On the average farm, cattle do not play an important role in either food production or income generation. Lastly, coca, the invisible crop, is cultivated over an area larger than that of the two main legal crops (plantains and bananas) combined.

This example demonstrates the need to include a diversity of local perspectives in order to capture the complexity of the situation.

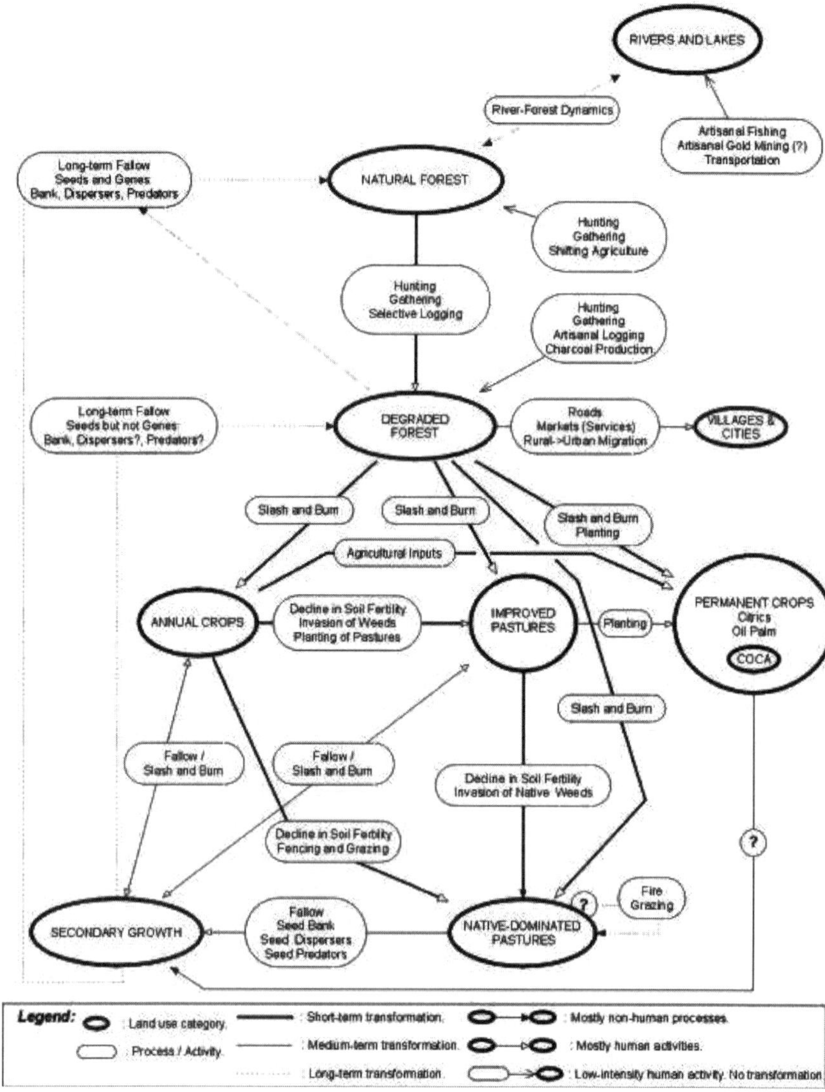

FIGURE 14.2 Land-use transformations in the Ucayali region

14.3.2 Hierarchical and Cross-Scale Interactions

Although local participation is central to the process, this primacy of the local is not without constraints. Complex socio-ecological systems can be described as nested hierarchies, in which each layer in the system has a dual nature, both as a whole (organism, population, community) and a part (Allen and Hoesktra, 1992). Each component of such a nested hierarchical system is called a holon, the overall nesting being referred to as a holarchy (Koestler, 1978). The different layers (household, community, region) evolve within a variety of ecological and socio-economic contexts and constraints (Allen and Star, 1982). In general, the smaller holons (genes, individuals) provide creative opportunities, whereas the larger holons serve as constraints (individuals to genes, communities to individuals).

In general, hierarchy is about both *scale*, the level in the hierarchy at which the system is located (i.e. farm, village or region), and *type*, the perspective of the observer (i.e. agricultural, forestry, urban) (Allen and Hoekstra, 1993). Scale considers ecological and socio-cultural holarchies as well as more conventional political hierarchies.

As one passes from one hierarchical level to another there are systemic properties that are 'emergent', since they only come into view at a particular level (Broad, 1923; Holland, 1995). Examining only the parts of the system would not capture such properties. For example, the potential of a home computer connected to the internet is an emergent property of the different parts (computers all over the world) when put together in a specific way to make a structured and functional whole (the WWW). Conversely, knowledge of all the individual computers, world wide, and the mechanisms for connecting them in a network (the internet), would not allow one to predict the emergence of the World Wide Web (WWW). The knowledge of individual risk factors for health (poor diet; smoking; 'lifestyle') is not necessarily useful for describing population health risk factors (disparities in income; power relationships; poverty) or global determinants of health (climate change; ozone depletion) (McMichael, 1999).

There are several methodological implications of the multi-scaled, hierarchical and emergent nature of ecosystems. Understanding of ecosystems requires more than one level of analysis, or implementation method. The system must be investigated at different scales and from the perspective of different types.[1] This allows for the discovery of emergent properties that are present at only one level and describes the interactions across multiple scales that frequently represent the defining contextual relationships of the system.

In our project in Peru, expanding the frame of questioning to include different scales (e.g. sub-region; region; inter-region) and type (e.g. fisheries;

[1] It should be noted that the necessity of hierarchy theory for different types of perspectives substantiates the first guiding principle.

forestry; coca), revealed a complex array of cross-scale interactions that affect the sustainability of the region as a whole. As shown in Figure 14.3 and Map 14.1, Ucayali can be divided into several sub-regions: (i) the Pucallpa sub-region, the heavily populated area along the Pucallpa–Lima road and the Aguaytia river; (ii) the natural forest and main timber extraction areas towards the Tamaya and Upper Ucayali rivers; (iii) the extensive swamps and lakes of the Ucayali floodplain; and (iv) the isolated uplands and hills of relatively undisturbed forest in the Purus river valley. Understanding Ucayali in terms of these sub-regions explains both farm and regional sustainability. At the farm level, because farmers migrate to different

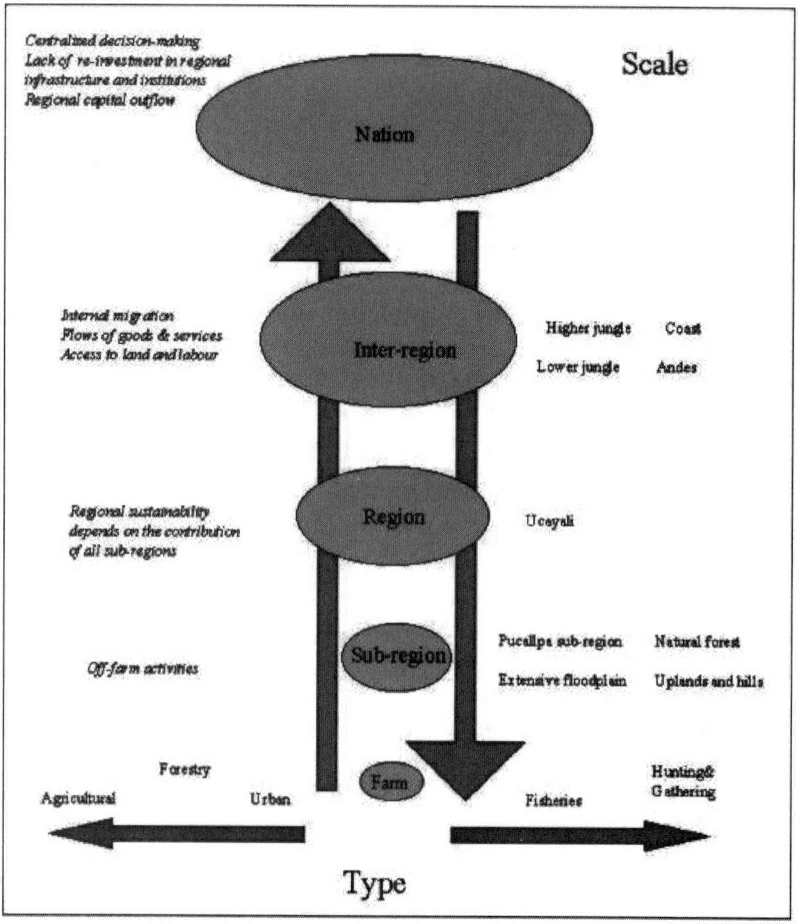

FIGURE 14.3 The hierarchical and multi-scaled ecosystems of Ucayali

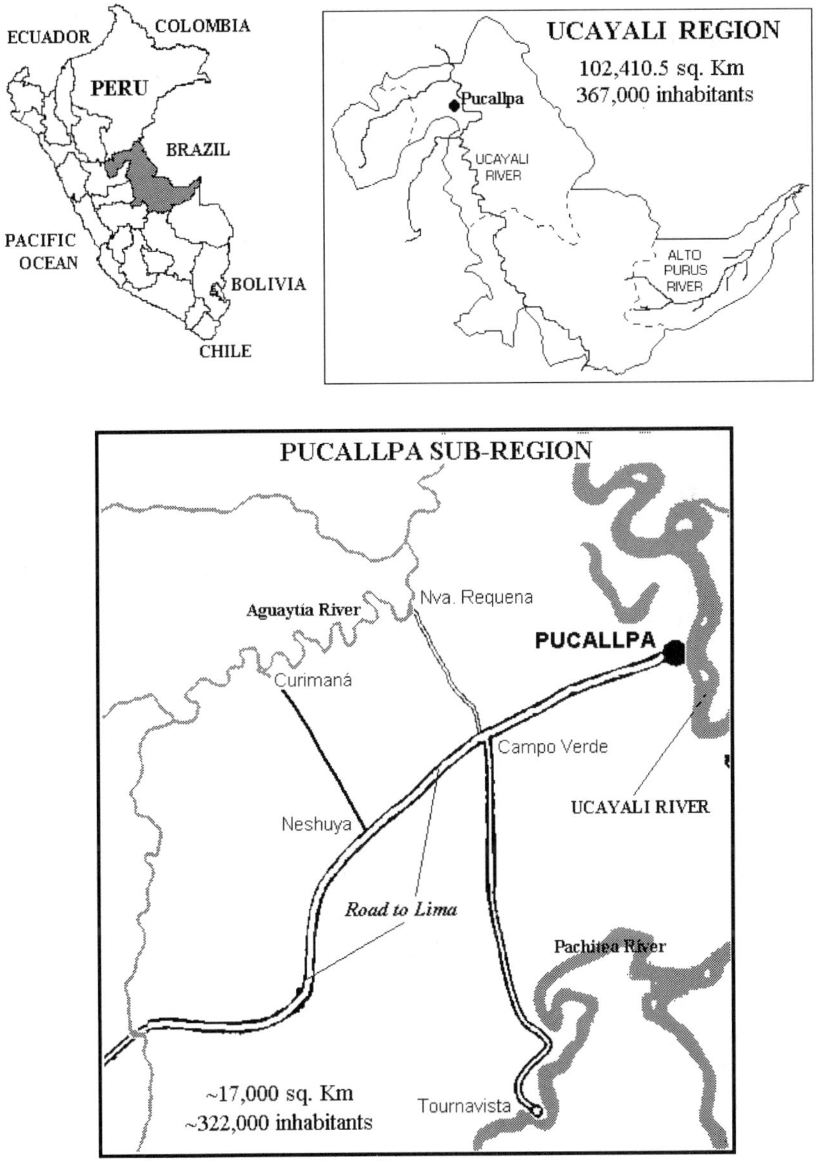

MAP 14.1

sub-regions depending on the season, they are able to exploit the differences in agricultural, timber, hydrological and fish harvest cycles to supplement and diversify their income. Farm level sustainability is essentially 'subsidised' by the extraction of forest and aquatic resources from other regions and ecosystems. At the regional level, the more densely populated areas along the road and in the city are sustained by the vast timber and fish resources dispersed across the great area of Ucayali and the coca growing regions in the southern uplands.

At the inter-regional level, the flow of goods and people link Ucayali with other departments in the upper jungle, Andean slopes and the Amazon lowlands. This inter-regional network increases access to commercial opportunities, land and labour. Seasonal and annual migration to these different regions and ecosystems diversifies resource use activities and income sources and reduces risk for the family. This migration of colonists affects the socio-cultural environment of Ucayali by increasing the transient nature of communities and altering the overall level of cohesion of social networks among community members.

Highly centralised economic and political structures at the national level restrict local capacity to manage and regulate resource use in Ucayali. As most commercial interests are owned and managed from Lima, in particular by the timber industrialists, re-investment in local infrastructure and the accumulation of regional capital (social and economic) are not a priority. Any efforts to address regional sustainability must take into account this national-regional relationship and confront the imbalances of power that affect regional development.

14.3.3 Self-Organisation

One of the defining characteristics of complex systems is self-organisation. This self-organisation occurs because of feedback loops among variables within a particular set of constraints. For instance, the use of protein supplements to increase milk and cattle production, coupled with economic and environmental incentives, led to the recycling of ruminant protein in the British (and other) agri-food systems, which created the necessary conditions for the emergence of the Bovine Spongiform Encephalopathy (Mad Cow Disease) epidemic. A traditional analysis of the problem focuses only on the recycling of the ruminant protein, rather than on the internal logic of the system within which this activity naturally emerged. Similarly, large-scale food-borne disease epidemics, such as salmonellosis, are embedded in self-organising systems. The salmonellosis system is characterised by interactions between demand for fast food; a desire for fresh produce and low consumer prices; consolidation and mass production; ecological simplification within the food system; economies of scale; and pressure for lower wages for agri-food industry workers (Waltner-Toews and Lang, 2000). For both Mad Cow Disease and salmonellosis, the agri-food system self-organised around a narrow set of economic efficiency goals without consideration for the social and ecological feedbacks which this also entailed.

In the project in Kenya, researchers worked with villagers to create influence diagrams tracing feedback loops between important issues (both positively and negatively viewed outcomes). Based on these diagrams, new understandings were created both within the agricultural communities, and among the researchers. In one village, for instance, the goal of improving the local school facilities was achieved by improving water distribution, which would reduce water-hauling time for women and children, increase agricultural productivity, and increase income available for investment in school facilities. In another village, their understanding of feedbacks led them to focus on road improvement as a way to increase income and allow them to obtain the materials and labour to expand classroom space. In Nepal, community members conceptually linked recycling of organic waste (through composting) to gardening and tree planting. These activities, in turn, were expected to decrease direct run-off and flooding, and increase ground water recharge. This, in turn, would provide more water for improving hygiene and reducing disease. In all these cases, it became clear that working toward specific desirable outcomes, such as improved education or hygiene, required an understanding of the kinds of systemic feedback patterns within which these activities were embedded.

14.3.4 Unpredictability and Uncertainty

Not only do complex systems self-organise through feedback loops, but their openness predisposes them to dramatic re-organisations at critical points of instability (Nicolis and Prigogine, 1977). These instabilities and the resulting jumps, or abrupt changes in the system, are caused by self-amplified internal fluctuations mediated especially through positive feedback loops. These give rise to the spontaneous emergence of new structures and forms of behaviour. Amplification is thus a source of new organisation and complexity in the system. At the points at which these new structures emerge, the system may branch off into one of a number of quite different organisational states (often referred to in the complexity literature as 'attractors'). The existence of multiple possible stable states necessarily implies indeterminacy, as which path is taken depends on the system's history, and various external conditions that can never be completely predicted (Nicolis and Prigogine, 1989). Hence, the unpredictable nature of complex systems. For example, shallow lakes have been described as having two alternative organisational states. One state is dominated by a benthic (bottom vegetation) association and the other by a pelagic (eutrophic) association (Scheffer, 1998). In a given summer, which association dominates in some lakes has been shown to be a consequence of the nature and timing of the spring melt—an unpredictable event.

Thus, in principle, in many situations it will not be possible to make quantitative models that accurately predict the future. As the transition to different states is unpredictable, the more conventional approach to decision making, that of using cause and effect explanations to build forecasting models which will anticipate the consequences of decisions, is not sufficient to deal with the complexity associated with socio-ecological systems.

The implications of this for research and management are profound. The role of the scientist in decision making shifts from inferring what will happen—that is, making predictions which are the basis of decisions—to providing decisions makers and the community with an appreciation, through narrative descriptions, of how the future self-organisation of the socio-ecological systems might unfold. Decision-making and management processes must be adaptable and flexible, able to deal with surprises and promote a capacity to adapt to changing environmental conditions (Holling, 1995; Kay *et al*, 1999). In all of our projects, this has reinforced the need to keep the process open-ended, with a monitoring program clearly part of the process; that is, there is no finite end-point to the research or management, only continual re-evaluation and adjustment. It has also reinforced the need to involve all the legitimate stakeholders, so that the tradeoffs, risks and benefits can be better understood and more equitably distributed.

14.4 Methodology

From a practical viewpoint, there are two important aspects we have had to consider as we implemented AMESH. In the first place, the boundaries between research and development, and analysis and action, had to be constructed (through negotiation) as part of the process, rather than being taken as given. Secondly, the roles and responsibilities of the actors had to be defined at each step. In general, the steps related to describing the socio-ecological system (historical reconstruction; stakeholder and governance analysis; analysing and synthesising systems descriptions) tend to be the primary responsibility of the researchers. However, seeking solutions; implementing them; and monitoring and adjustment tend to be the primary responsibilities of the community members. Nevertheless, our experience has been that scientists, villagers, planners, bureaucrats and development practitioners work together throughout the process with a lot of give and take.

From the outset, necessary institutional arrangements and mechanisms are required to guide and control the entire process; ensure clear 'rules' of negotiation; ways to involve stakeholders at the appropriate times; means to address conflict; and sufficient organisation to maintain the continued involvement of the stakeholders until the process is complete. Collaborative platforms, such as specific forums, workshops, or regular group meetings, are needed to facilitate dialogue among relevant stakeholders (e.g. producers, scientists and governmental agents) and provide the necessary environment in which they can genuinely participate in the research and development activities.

The research process is comprised of a variety of activities, each characterised by a set of guiding research questions. Although they are depicted graphically as sequential, they often, in practice, occur simultaneously or 'short cycle' a few times before moving to other activities (Figure 14.1). We may group these activities into five major areas. These are: an entry point and initial description; an analysis of the relationships among stakeholders, issues and governance; the multiple narratives and systems descriptions which emerge from the previous analysis; a full systems analysis and synthesis; and the translation of this understanding into ongoing collaborative learning and action.

14.4.1 The Presenting Situation

We intrude on various situations—communities, ecosystems—because of some research or development agenda, presenting complaints or perceived problems. There are essentially two parts to this element of the process. First, we need to articulate what the agenda is (what problems are we trying to help resolve? What research or development issues are we addressing?). This step is important so that unspoken, assumed or hidden agendas, with their associated power roles, do not 'hi-jack' the process.

Once the presenting issues and agendas are on the table, we need to bring together the given history of the community we are working with.

14.4.1.1 The Given History

The goal of this step is to develop a narrative of the key changes, overall trends and patterns, as perceived by current and/or past stakeholders and researchers. This provides the baseline understanding of how the system has evolved to its current state. It is not exhaustive; rather it establishes a rich context in which more specific research questions can be located.[2] Relevant questions include:

- What has been the overall historical development of the system; e.g. the establishment of a frontier town, expansion of the frontier, economic booms, etc.
- What have been the key ecological, economic, demographic, political, governance and social developments in the system? What kinds of rich inter-connections have occurred?
- At what level have the changes taken place (e.g. farm, village, region or nation)?
- What or who have been the agents of change?
- Which are the critical fast, medium and slow variables within this overall development?
- Is there historical evidence suggesting sudden shifts or changes in the structure or behaviour of the system?
- What has stabilised the system? What has tended to destabilise the system? Have there been surprising changes? Perverse stability?
- What are the positive and negative feedback loops?

Projects may be undertaken at the behest of a research agency. In this case another important historical issue is the history of the research agency in the community. The past experience of the community with the research agency will impact upon all future work the research agency hopes to do. It is also important to understand past projects that other groups have attempted, especially in a similar problem area. Past failures will colour the community's perception of the utility of working in a given area, while past successes might narrow the range of alternatives a community is willing to consider.

- What past projects have the research agency, or other groups, undertaken with/in the community?
- Who was involved in these projects?
- What were the outcomes of these projects?

Most of the work of constructing a history is based on synthesising secondary sources of data such as government reports, maps, and previous

[2] It is important to note that any history reflects the interests of the researchers as well as the adequacy of the records available. They are open to re-orientation as the research unfolds, as involvement of new stakeholder groups may cast different patterns of light and shadow on the available evidence.

research done in the area. In all the geographical areas in which we have worked, research projects based on more narrowly defined (or at least different) questions preceded our projects.

In Nepal, a three-year project to understand and prevent transmission of the parasitic zoonosis echinococcosis (Hydatid Disease in people) concluded that more systemic, participatory and integrative approaches were required to resolve even narrowly defined medical problems. This initial project was followed by a consultancy to gather and synthesise background information needed for the new project, and it involved both key-informant interviews and secondary data sources. This was followed by a major workshop, with representatives from the communities, government agencies and several non-governmental organisations. The second, expanded community-action phase of the research was designed at this workshop. In Kenya, previous research with smallholder dairy farmers in the area had established a high level of trust between the Kenyan researchers and the villagers, as well as providing baseline information. The project in the Peruvian Amazon was based on a wide variety of research projects in the area. Using complex systems theories, a synthesis of these projects was assembled and served as the starting point for understanding what had occurred. In both Kenya and Peru, a series of participatory activities and workshops were held in each village to supplement secondary data. In all cases, we found that the research team needed to include people fluent in local languages (Kikuyu, Newari and Amerindian languages) as well as the more global languages (Swahili, Nepali, English and Spanish).

Historical data were not gathered with a complex systems framework in mind, and indeed only in Peru were they organised in such a way that they could be interpreted in complex systems terms before the more participatory aspects of the project commenced. This is particularly true for the description of feedback loops and attractors. In all cases, holarchies and hierarchies were relatively easy to define, and informed both the range of stakeholders (individuals, families, neighbourhoods, etc.), and the decision-making spheres of influence in terms of the ecosystem. In both Nepal and Kenya, a complex systems interpretation of both historical and current information emerged from the multiple narratives of the stakeholders, rather than from the prior data collection. The advantage of this is that the community members can be part of the process; the disadvantage is that the historical interpretation has less opportunity to influence the design of the research itself.

The goal of this historical work is to gain a deeper understanding of the dynamics, constraints and opportunities that have characterised the system, and perhaps offer some insights into future options.

14.4.2 *Analysis of Stakeholders, Issues and Governance, and their Relationships*

Once a situation has been identified as being problematic (for instance, 'the health in our village is poor') and an effort has been made to understand how the situation came into being, the next stage involves three concurrent and mutually informing activities. These are:

- identifying who is involved in the situation now, and hence will influence resolving the situation;
- exploring what the problems and related underlying issues are understood to be; and
- investigating how decisions get made and implemented.

These activities lay the groundwork for developing an understanding of the situation and its resolution.

14.4.2.1 Stakeholder Analysis

Any complex system can be viewed through multiple windows of observation (Engel, 1997). Each 'window' may be seen as representing the perspectives of a variety of stakeholder groups. Sometimes, a stakeholder group may be characterised as having a single perspective, but this is uncommon and problematic, in that it conflates conflicts between issues with conflicts between groups. Since the stakeholders define the problems and issues, and hence the solutions, identifying who the stakeholders are (as well as their issues and powers in the system) is a critical step in the process. The goal of this step is to determine who could, or should, take part in the further steps of the research and management. Stakeholders perceive their reality through different methods and means and describe reality according to different objectives (Ramirez, 1999). The stakeholder analysis draws attention to different 'versions' of reality and provides a systemic means to expose, incorporate and reconcile them within the research agenda. AMESH is by definition interventionist and therefore must address the concept of *agency* or capacity to intervene. The stakeholder analysis establishes who has the legitimacy to intervene, under what conditions and with what purpose (Ramirez, 1999). Guiding questions in this section include:

- Who are the principle stakeholders?
- What is the process for identifying stakeholders?
- Under what and whose selection criteria will they be considered stakeholders?

- What are the critical discriminating features that allow us to differentiate the groups of stakeholders?
- What other groups, perhaps not directly related to the issue at hand, exist in the community?
- Which groups have been excluded?
- What are the factors that stakeholders perceive as being within and outside of their control (exogenous and endogenous)?
- What are the power dynamics or relationships among and between stakeholder groups?
- What are the barriers and constraints to communication and collaboration between the different stakeholder groups?
- What are the main resource coalitions?
- What goals/interests do these coalitions share?
- Which stakeholders are excluded from these coalitions and what are the consequences of this?
- What are the conflicts of interest? Which organisations are in conflict and why?
- What are the important gender relations among and between stakeholder groups? How do gender differences define people's rights, responsibilities and opportunities to participate?
- What are the assets and resources (skills, financial, materials, connections, etc.) of each stakeholder group?

The research team is a stakeholder in its own right and needs to be examined as such, with similar guiding questions:

- What is the role of the researcher as stakeholder, and what power does the researcher wield to convene the others?
- How was the research team selected?
- Are they local or non-local?
- What is their level of experience with the research issue?
- What training have they received?
- What skills training does the team need?
- Does the research team have the necessary skills to conduct the AMESH analysis?
- What is the research agency's mandate?
- How might this mandate bias the AMESH process?
- What is the power relationship between the research team and the local stakeholders?
- What gives the research team legitimacy to work in the community?
- What power structures does the research represent to the community?

Local leaders are used as entry points into the community, and as the means for compiling a list of potential key informants for the research team. This was the case in Nepal: the Ward Committee recommended potential candidates for the four Community Researcher positions, so the bias of the Ward Committee or local leaders, in terms of who they knew in the community, entered the research process. Realistically, using the local leaders to recommend key informants may be the only way to successfully gain entry into the community. Thus it becomes very important for the research team to understand the political dynamics of the community early on. This way, they can aim to contact as broad a cross-section of the community as possible and ensure that they are not favouring one group over another, or only listening to one political perspective.

Because both these sets of questions (about stakeholders and the research team) have important ramifications for the entire direction of the process, they tend to be politically loaded, and need to be asked in a variety of ways. They are naturally asked as one develops the historical picture of the area, as well as in the participatory research (interviews, structured workshops, etc.). In Kenya, a relatively straightforward picture emerged; there were no formal governance structures at the village level, for instance. Since the villages we were working with were quite small (between hundreds and a few thousand people), and local village organisations were few and obvious (churches, community-based organisations), we were able to work with stakeholders to develop new, inclusive, Agro-ecosystem Health Committees.

In all cases, we needed to revisit these questions about stakeholders as the research progressed. In Peru, we found that a 'stakeholder representative' group created by an earlier research project reflected intellectual and business interests based outside the Amazon (in Lima). In Phase II of the project, our team leader had to begin again, from the ground up, creating a coalition of local government officials, individuals, and representatives of non-government women's, Amerindian, and other organisations.

An important aspect of stakeholder analysis has been to identify the levels of holarchy/hierarchy in which the participants are active. Thus, in working at the neighbourhood level in Kathmandu, we needed to identify national players who were making decisions that strongly affected the communities; ward level organisations (governmental and non-governmental); representatives at the neighbourhood level (toles and marabuhils); households; and individuals. Given the fairly rigid social stratification, we also needed to identify, for instance, hierarchies of street cleaners, shopkeepers, and butchers.

14.4.2.2 Ecological and Social Issue Analysis

Once key stakeholder groups have been identified, several participatory techniques can be used to determine the issues (problems and opportunities) most affecting them and their vision of the future. The relevance of an issue stems directly from its relationship with a stakeholder group. As in the historical reconstruction, the scale of each issue must be made explicit. This will help identify the endogenous and exogenous variables of each issue and the cross-scale interactions among issues. The goal at this point is to identify the key elements in the systems descriptions that local stakeholders consider important, and hence represent the motivating force behind making changes or protecting the status quo. These issues are 'mapped on' to systems descriptions arrived at through more conventional kinds of ecological and sociological inquiry. Here, relevant questions include:

- What are the critical issues?
- How do they relate to one another?
- At what spatial and temporal scale do they occur?
- For which stakeholder groups are these issues critical?
- How does the resolution of these issues come together in a vision for the future?

The issues of importance to the stakeholders emerge as a normal part of most participatory action research processes: that is what they are designed to do. What most of these processes do not elicit, however, is a clear sense of how local people see the connections between, say, education, water quality, childhood disease, soil erosion, roads and agricultural production. We found that in Kenya, villagers were able to move well beyond lists of issues, constraints and action plans to systemic loop or influence diagrams linking both positive and negative outcomes (see the next stages). In Nepal, community members and researchers saw these interactions as part of a kind of meta-narrative; only later in the project did systems diagrams emerge, and conflicts, interactions and complementarities were clarified. As described earlier, as our systemic understanding of the Peruvian situation was enriched, the issues of interest broadened from cattle and pastures to include fish and river systems.

It is important to separate the perceptions the research team has of these issues and the issues as they are identified and articulated by the community. A part of this analysis needs to be a comparison of the two sets of issues with an attempt to reconcile differences in perceptions regarding the magnitude and importance of specific issues. However, all research teams are mandated to work on a pre-defined set of issues. The most important ecological and social issues for the community will not necessarily coincide with the mandate of the research team. The research team needs to be honest with the community regarding their own capabilities and limitations, so they do not set false expectations.

14.4.2.3 Policy and Governance Analysis

The identification of stakeholders is not only closely related to identifying important issues, but also to describing the relevant policies and governance structures that constrain or facilitate the local capacity to cope with such issues. The goal at this point is to arrive at an understanding of the mechanisms by which solutions are currently sought and the feasibility of introducing changes. We also need to look more broadly at the policy context, which may need to be altered to allow local communities to achieve sustainable goals. Here, key questions include:

- What are the key policies affecting the capacity of local stakeholders to sustainably manage their natural resources?
- At what level are these policies formulated and implemented?
- How do informal and formal governance structures affect local capacity to adapt to and deal with stresses?
- At what level are these governance structures formulated and implemented?

The researchers need to be aware of their own roles in the politics, both formal and informal, of the situation. A research team often needs to gain the sanction of local leaders before starting its work, especially in a very hierarchical society. However, that very process can instantly toss the research team into the political landscape of the community. The community might quickly identify the research team with a particular political group in the community, thereafter influencing how the community perceives the team.

14.4.3 People and their Stories

What emerges from the previous stage of teasing out stakeholders, their issues and their policy contexts, are a series of narratives, pictures and systems descriptions of both what the situation is, and a vision of what the situation ought to change to. These do not emerge without some deliberate efforts, of course, on the part of the researchers. Many tools from a variety of soft systems methodologies and Participatory Action Research (PAR) approaches (mapping; time lines; Venn diagrams; influence diagrams; rich pictures; story telling; Freireian consciousness-raising; visioning exercises;

and the like) may come into play here. See the PAR website at http://www.rcpla.org/, as well as Pretty *et al* (1995), for further details. These stories, in their various forms, serve as the foundation for a more formal process of developing a systematic understanding of the situation.

14.4.4 Systems Descriptions and Narratives: Developing a Systemic Understanding

This part of the process has two major components: analysing the system, both qualitatively and quantitatively, and synthesising the system—that is, bringing together our best current understanding of how the various narratives interact with each other to create the emergent whole which we recognise as 'the system'. The purpose of this exercise is to generate an understanding of the situation that suggests some possible directions forward.

14.4.4.1 Systems Analysis

By combining the results of the stakeholder analysis and the issue analysis with the story lines within which they are embedded, researchers can begin to distinguish foreground from background, and the stakeholders as a group can decide what is to be ignored and what is to be focused upon. Stakeholder groups, including poor villagers, are quite capable of creating sophisticated loop diagrams of their systems. Researchers can help them go a step further, however, to detect influential points in the system and some possible consequences of intervening at those points. In both Kenya and Nepal, Thomas Gitau and Cynthia Neudoerffer respectively, took complex systems diagrams, drawn from conversations and workshops with local villagers, and identified where possible feedback loops amplify or dampen relevant social or ecological processes.

Stakeholder participation is essential to understand what the important issues are, and how decisions regarding them are made. Once the issues as raised by various stakeholder groups are identified, however, they need to be linked to systemic understandings from various perspectives. During this stage of developing system descriptions there tends to be a transition from active stakeholder involvement to an activity done in the more traditional mode of academic, 'non-engaged' research.

In Kenya, the complex loop diagrams were first assessed qualitatively by the field research team and villagers, and then used as a basis for implementing some actions. In the meantime, members of the research team subjected the diagrams to more formal analyses, using methods described by Puccia and Levins (1985).

In another project in Chennai, India, the Cooum River Environmental System was studied in order to identify means of improving the health of the river and its environs, including human settlements (Bunch, 2000). After the process of issue identification, generating rich pictures and influence diagrams and developing system descriptions, the researcher returned to his office and generated a Geographical Information System (GIS) and environmental modelling system based on the community input. These tools allowed him to perform a systems analysis of the situation. One surprising result was that the analysis demonstrated that one particular intervention, which was commonly felt to be badly needed, would make the situation worse. Specifically it was felt to be important to put a sewage collection system into slum areas. Currently these slums act as multiple diffuse sources of pollution of the river. However it was demonstrated that putting in place a sewage system in the slums would overload the existing treatment plants resulting in *all* the sewage entering the treatment plants flowing to the river untreated. The effect of adding a sewage system into the slums would have been to increase the total pollutants entering the river and also create one very large point source of pollution, which would overwhelm the assimilative capacity of the river. This pointed out the need to dramatically increase the sewage treatment capacity before adding sewage systems to the slums. The tools developed by the researcher to examine these interactions were subsequently taken over by the community to allow them to design their own interventions and to integrate their monitoring efforts.

Systems analysis is about building a systems description—essentially a conceptual model of the situation which describes what the key elements of the situation are and how they are interconnected and interrelated. This description will be multi-layered, both in spatial and temporal scale and type of description. The systems description is used to identify the important processes that shape the situation. The behaviour and dynamics of these processes are then explored and narratives about the system's behaviour are constructed.

The questioning in systems analysis can begin in a qualitative and relatively simple manner, and this alone often provides very powerful insights and suggestions for action. When information and/or theory allow, the analysis can proceed to a quantitative stage involving statistical models, simulation and spatial analysis using a GIS.

Usually the researchers begin with an analysis of the system from each perspective, before integrating them. For each of these analyses it is important to answer the following questions:

- What is the core purpose or essence of the system from each perspective?

- To whom are they of interest and why?
- How do we define and delimit the system?
- What are the spatial and temporal boundaries and scales of observation?
- What are the key ecological and social processes that define the system?
- What are the different subsystems of interest?
- What are the relevant subsystem behaviours?
- What are the relevant subsystem variables?
- What is the nested hierarchy in which the system is situated?
- What are the defining contextual relationships between the system and its subsystems and the larger system in which it is embedded?

In some cases, where very good historical data are available, one may even begin to frame the questions in explicitly complex systems terms and to build narratives about how the system behaves. For instance:

- What are the discrete organisational states (often referred to as 'attractors') which may be accessible to the system?
- How does the system behave in the neighbourhood of each discrete organisational state, potentially in terms of a quantitative simulation model?
- What are the positive and negative feedbacks and autocatalytic loops and associated gradients that organise the system around a discrete organisational state?
- What might enable and disable these loops, and hence might promote or discourage the system from being in the neighbourhood of a discrete organisational state?
- What might be likely to precipitate flips between discrete organisational states?

After these systems analyses, from different perspectives, have been undertaken, it is necessary to synthesis them into a multi-layered, coherent understanding of the situation.

14.4.4.2 Synthesizing the Systems Descriptions

In systems synthesis, the different system analyses are rejoined and assessed in terms of the health and sustainability of the whole system. Again, this may be done in terms of searching out the linkages among narratives, and the way in which the larger narrative of the community emerges. It may also be done using a variety of sophisticated models. This synthesis enables researchers to identify what are thought to be key points of intersection between various sub-models and various worldviews. This 'reconstructed' model of the whole or future narrative forms the basis for making hypotheses about probable outcomes as the result of particular interventions. These models and narratives are fundamental to the development of policies that are capable of accounting for multiple perspectives and multiple goals. They will enable decision-makers to define their range of feasible management options and assess trade-offs. The following questions may be relevant:

- How do the different system models articulate with each other?
- How do the parts of the system that can be modeled (e.g. hard systems, historical components) articulate with those that cannot (e.g. soft systems, future-expectation models)?
- What are the possible future states of organisation of the system?
- What is our understanding of conditions under which these states might occur?
- What are the tradeoffs that the different states represent?
- What is the range of feasible and desirable management actions?
- What are appropriate schemes for ensuring the ability to adapt to different situations?
- What is the appropriate level of confidence that the narrative deserves (this is our degree of uncertainty)?

The description, analysis and synthesis of the system can be done as part of a systemic re-interpretation of history, and hence can (and should) be done from the 'outside' as well as the 'inside'. However, the remainder of the process, in which research slips into an interventionist mode, can *only* be done with the full participation of local stakeholders and actors.

14.4.5 Collaborative Learning and Action

With a rich systems description and understanding in hand, we can begin to work with ecosystem stakeholders to find ways to negotiate trade-offs, generate visions, implement changes and monitor those changes in such a way that we learn from them. In short, we can help the community put in place an adaptive, collaborative learning process for ecosystem sustainability and health.

14.4.5.1 Cross Talk and Seeking Solutions

Using the systems understanding developed in the previous stage, the next more delicate step involves reconciling the different perspectives of reality that bring together the different groups of stakeholders. This is a social process that involves conflict resolution and compromise, and is very much dependent on the decision-making capacities and processes of the different stakeholders. In the end, the process should generate a common vision for the future. Here, relevant questions include:

- What are the areas of conflict and complementarity among stakeholders?
- What are the feasible interventions that can take into account the multiple perspective goals and multiple perspective impacts?
- How can the different perspectives be synthesised?
- How are the different perspectives and descriptions of reality reconciled?
- What are the necessary trade-offs?
- What are the desirable management options?
- What is the combined vision of the stakeholders?
- What are commonalities and differences within that vision?
- How can the stakeholders design a more sustainable future?

The visions and aspirations that the stakeholders hold for the future may be elicited by asking questions such as:

- How can the stakeholders design a more sustainable future?
- What story do you want your grandchildren to tell about their home? How would you like your community to look in twenty years? What changes would you like to see?
- What needs to happen to realise these goals?
- What barriers and constraints are stopping you from achieving these goals today?
- How might these be overcome?

In Kathmandu, systems diagrams of actors, activities, concerns, impacts and needs identified differing perspectives as to who was 'at fault' for the accumulation of solid waste in the street. The low class street sweepers tended to see the problem as one of irresponsible actions on the part of residents, while wealthier residents saw the issue as one of inefficient garbage collection. These differing perceptions were laid out explicitly at a workshop attended by members of all castes and social classes (from political leaders to professionals, street sweepers and butchers). We asked them the question: 'what do you want your grand-children to say about the community you live in?' They then broke into small groups to work on this, and devise pictures of where they wanted to go, and possible strategies for getting there. The amount of time allotted did not allow for a full future development strategy; rather, it served to re-focus their thoughts on many of the common ideals and plans they had articulated throughout the years of the project. It thus served as a heuristic and motivational tool. Later, members of the Kadgi caste, which includes the butchers (and who are in the contradictory position of being both low caste and the source of the economic wealth for the communities) held a special ceremony to thank the researchers for having created a sense of collective community which transcended caste.

14.4.5.2 Design of an Adaptive Approach for Implementation of the Vision and Collaborative Learning

It is one thing to negotiate a set of trade-offs and possible futures that the entire community is willing to strive towards, or at least to live with. It is quite another to come up with explicit strategies and organisational plans for achieving that future. In many situations, especially in industrialised countries, detailed engineering plans, complete with recommendations on land-use zoning and institutional/budgetary arrangements, need by law to be made explicit. This was certainly true for the development of the master plan for the Huron Natural Area in Kitchener, Ontario, Canada, which was the result of a long series of studies and meetings involving local politicians, land-owners, developers and researchers (http://www.fes.uwaterloo.ca/u/jjkay/HNA/).

In other situations, a direct and explicit approach may be too threatening to local political power structures, and an indirect method of working with emergent, virtual governance structures may be more appropriate. Also, appropriate institutional structures may not exist, requiring new ones to be invented, and these may have to exist outside the traditional infrastructure. In Kathmandu, one of the strategies for dealing with solid waste was to have community clubs—usually made up of activist young people—facilitate recycling and garbage collection activities. This kind of emergent, 'virtual' governance enabled them to avoid direct confrontation with authority, and find ways for political leaders, shop keepers, butchers and community residents to work together more effectively. They were therefore successful in building riverside parks and recycling programs. Such strategies may involve working with community service clubs, as in Kathmandu, or semi-official non-government or quasi-governmental groups such as 'round tables', including business and public health or community activist leaders.

Some of the questions that might be asked at this point are:

- What are the management directives, goals and objectives?
- What are the desirable and feasible changes in the systems, in order to reach the identified goals?

- What institutional arrangements are necessary to support these changes?
- How can they be implemented?

14.4.5.3 Implementation

Once a plan—official and explicit or unofficial but understood—has been created, it must still be implemented. At this point in the process, it becomes clear just how successful (or not) one has been in recruiting local actors and stakeholders, and the extent to which they have come to 'own' the successes and failures of systemic change. In the Kenya project, we considered it a success when one of the villages attempted an intervention: confronting the railway company which built the embankment that was seen as the cause of the road erosion. This confrontational approach failed, after which the villagers concluded that they, not the researchers, needed to find a new strategy. They then approached the research team for advice and facilitation to identify engineers in the railway company who could help them, which the researchers did. This second cooperative strategy is still underway. There is much in the Operational Research and Management Science literature about change, agents of change, organisational development, and innovation. Some of it will be helpful at this stage in the process. Some key questions that need to be asked when considering implementation include:

- What are the steps, and in which order, that need to be part of the implementation process?
- Who will take responsibility for ensuring that each of those steps is implemented?
- How can they be implemented?
- How can people be motivated to adopt the suggested changes?
- How will implementation be sustained, both in terms of finances, and in terms of personnel and logistics?

14.4.5.4 Monitoring and Evaluation

In an adaptive process, monitoring and evaluation provide the feedback to close the loop on the process so that it truly becomes ongoing and adaptive. Appropriate indicators to monitor progress and performance of the newly developed plans and strategies are developed by local stakeholders and researchers (for a review of the literature on this topic, see Boyle, 1998). These are incorporated into an ongoing monitoring and evaluation process in which the current situation and pertinent issues, described at the beginning of the research process, are re-assessed and consequently the AMESH cycle may begin again. In Kenya, the multi-disciplinary research team and the villagers each developed separate lists of indicators representing what were deemed to be important attributes of agro-ecosystem health. We found that villagers, not surprisingly, were more likely to measure and respond to their own indicators rather than those devised by researchers. Although scientifically validated, many researcher-identified indicators proved difficult, costly and time-consuming to measure. A more feasible design would seem to be that villagers identify what they see as being important, and how to measure it (e.g. soil quality, assessed by running it from hand to hand); the researchers then determine what this indicator means in more quantitative scientific terms, as well as scanning for variables that are being missed. In Kathmandu, a mix of scientific monitoring of water quality by one of our Nepalese research partners, with community monitoring of visible outcomes such as street cleanliness, has emerged.

A whole series of questions can and should be asked about indicators, regarding both practical and scientific issues (see Boyle *et al*, 2001, for a detailed discussion). However, we might highlight a few of the questions here. For instance:

- What are the relevant indicators that can measure performance in terms of the identified goals?
- Who 'owns' the indicators—that is, who will use them for making decisions?
- Who will measure them?
- How will decisions be made on when to take action?
- Who will take action?

In closing the loop of this approach, researchers will find that they have become part of the community and its history. Even when they leave, it is important to be clear and cognizant about what they have changed by their interventions, and the sustainability of the processes they leave behind. At its best, AMESH will have altered the community's perception of itself, to the point where they see themselves more eco-systemically, and have the confidence and organisational structures to make decisions and take actions despite the uncertainty of the outcomes. While the creation of riverside parks in Kathmandu and water management projects in Kenya are gratifying, it is this aspect of AMESH—the emergence of eco-systemic vision and self-reliance—that we believe are its most important legacy.

14.5 Conclusion

AMESH is a methodology that is theoretically grounded in our most up-to-date understandings of complex systems, and is, at the same time, practical and participatory. It is currently being tested in urban, rural and frontier contexts, and appears to be adaptable to a wide variety of settings. Some parts of the methodology—particularly the more practical aspects of how one undertakes such interventions, and the later steps of monitoring and adjustment—have been well tried. Other parts—like the substantive issue of how one can best formally analyse and then synthesise multiply perceived and multi-dimensional systems in ways that make both scholarly and 'common' sense—are still at the early stages of being worked out. Nevertheless we have found that, by emphasising both participatory methodologies and systems understandings, the process itself promotes the ends—health, sustainability, and ecological integrity—that it seeks to achieve.

References

Allen, T.F. and Starr, T.B. (1982). *Hierarchy: Perspectives for Ecological Complexity*. University of Chicago Press, Chicago.

Allen, T.F.H. and Hoekstra, T.W. (1992). *Toward a Unified Ecology*. Columbia University Press, New York.

Almendares, J., Sierra, M., Anderson, P., and Epstein, P. (1993). Critical regions: A profile of Honduras. *The Lancet*, **342**, 1400–1402.

Boyle, M., Kay, J.J., and Pond, B. (2001). Monitoring in support of policy: An adaptive ecosystem approach. In. *Encyclopedia of Global Environmental Change*, vol. 4, pp. 116–137. Munn, R. (ed. in chief). Wiley, London.

Bertalanffy, L. von (1968). *General Systems Theory*. George Brazillier, New York.

Bohm, D. (1993). In "Last words of a quantum heretic", interview with John Morgan. *New Scientist*, **137**, 1962, 27 February, 42.

Broad, C.D. (1923). *The Mind and its Place in Nature*. London, Kegan Paul, Trech and Trubner (Reissue 1980).

Bunch, M. (2000). An adaptive ecosystem approach to rehabilitation and management of the Cooum River environmental system in Chennai, India. PhD Thesis, Environmental Studies, University of Waterloo, Waterloo. URL: ersserver.uwaterloo.ca/jjkay/grad/mbunch/.

Casti, J.L. (1986). On system complexity: Identification, measurement and management. In: *Complexity, Language and Life: Mathematical Approaches*, 281pp. Casti, J.L. and Karlquist, A. (eds.). Springer-Verlag, Berlin.

Chambers, R. and Conway, G. (1992). *Sustainable Rural Livelihoods: Practical Concepts for the 21st Century*. Institute of Development Studies, Sussex.

Checkland, P. (1981). *Systems Thinking, Systems Practice*. Wiley, Chichester.

Checkland, P. and Scholes, J. (1990). *Soft Systems Methodology in Action*. Wiley, Chichester.

Engel, P. (1997). *The Social Organization of Innovation: A Focus on Stakeholder Interaction*. Royal Tropical Institute, Amsterdam.

Funtowicz, S. and Ravetz, J. (1994). Emergent complex systems. *Futures*, **26**(6), 568–582.

Gitau, T. *et al.* (2000). Agro-ecosystem health: principles and methods used in high-potential tropical agro-ecosystem. In: *Agro-Ecosystems, Natural Resources Management and Human Health Related Research in East Africa*. Jabbar, M.A., Peden, D.G., Saleem, M.A.M., and Li Pun, H. (eds.). International Livestock Research Institute, Nairobi.

Gunderson, L.H., Holling, C.S., and Light, S.S. (1995). *Barriers and Bridges to the Renewal of Ecosystems and Institutions*. Columbia University Press, New York.

Holland, J.H. (1995). *Hidden Order: How Adaptation Builds Complexity*. Addison-Wesley, Reading, MA.

Holling, C.S. (1986). The resilience of terrestrial ecosystems: Local surprise and global change. In: *Sustainable Development in the Biosphere*, pp. 292–320. Clark, W.M. and Munn, R.E. (eds.). Cambridge University Press, Cambridge.

Holling, C.S. (1995). What barriers? What bridges? pp. 3–34. In: *Barriers and Bridges to the Renewal of Ecosystems and Institutions*. Gunderson, L.H., Holling, C.S., and Light, S.S. (eds.). Columbia University Press, USA.

Kay, J. and Schneider, E.D. (1994). Embracing complexity, the challenge of the ecosystem approach. *Alternatives*, **20**(3) 32–38.

Kay, J., Regier, H., Boyle, M., and Francis, G. (1999). An ecosystem approach for sustainability: Addressing the challenge of complexity. *Futures*, **31**, 721–742.

Koestler, A. (1978). *Janus: A Summing Up*. Hutchinson, London.

McMichael, A.J. (1999). Prisoners of the proximate: Loosening the constraints on epidemiology in an age of change. *American Journal of Epidemiology*, **149**, 887–897.

Murray, T. (2000). Investigating health, biodiversity and natural resource use on the Amazon frontier: An ecosystem approach. *IDRC Reports*, April 28. (www.idrc.ca/reports/).

Murray, T.P., Kay, J.J., Waltner-Toews, D., and Raez-Luna, E. (2001). Linking human and ecosystem health on the Amazon frontier: An adaptive ecosystem approach. In: *Conservation Medicine: Ecological Health in Practice*. Aguirre, A.A., Ostfeld, R.S., House, C.A., Tabor, G.M., and Pearl, M.C. (eds.). Oxford University Press, New York.

Nicolis, G. and Prigogine, I. (1977). *Self-Organization in Non-equilibrium Systems*. Wiley, Chichester.

Nicolis, G. and Prigogine, I. (1989). *Exploring Complexity*. W.H. Freeman, New York.

Pretty, J.N. (1995). Participatory learning for sustainable agriculture. *World Development*, **23**(8), 1247–1263.

Pretty, J.N. (1998). Supporting policies and practice for scaling up sustainable agriculture. In: *Facilitating Sustainable Agriculture*. Roling, N.G. and Wagemakers, M.A. (eds.). Cambridge University Press, Cambridge.

Puccia, J.C. and Levins, R. (1985). *Qualitative Modeling of Complex Systems*. Harvard University Press, Cambridge, MA.

Ramirez, R. (1999). Stakeholder analysis and conflict management. In: *Cultivating Peace: Conflict and Collaboration in Natural Resource Management*, pp. 101–126. Buckles, D. (ed.). International Development Research Centre, Ottawa.

Roling, N.G. and Wagemakers, M.A. (1998). A new practice: Facilitating sustainable agriculture. In: *Facilitating Sustainable Agriculture*. Roling, N.G. and Wagemakers, M.A. (eds.). Cambridge University Press, Cambridge.

Rosen (1977). *Life Itself: A Comprehensive Inquiry into the Nature, Origin and Fabrication of Life*. Columbia University Press, New York.

Rowley, T., Gallopin, G., Waltner-Toews, D., and Raez-Luna, E. (1997). Development and application of an integrated conceptual framework to tropical agroecosystems based on complex systems theories: Centro Internacional de Agricultura Tropical-University of Guelph Project. *Ecosystem Health*, **3**, 155–161.

Scheffer, M. (1998). *Ecology of Shallow Lakes*. Chapman and Hall, London.

Schneider, E.D. and Kay, J.J. (1994). Complexity and thermodynamics: Towards a new ecology. *Futures*, **24**(6), 626–647.

Singh, N. and Wanmali, S. (1998). The sustainable livelihood approach: A concept paper. Framework paper for the sustainable livelihoods workshop, University of Florida in Gainesville, 1–3 October.

Waltner-Toews, D. (1996). Dog worms, buffalo cysts, and the home of the gods. Presented at the International Society for Ecological Economics in Boston. Posted on the author's web-site at http://www.ovcnet.uoguelph.ca/popmed/ecosys/research.htm#Nepal.

Waltner-Toews, D. (1999). *Mad Cows and Bad Berries.* Gatekeeper Series 84. International Institute for Environment and Development, London.

Waltner-Toews, D. and Wall, E. (1997). Emergent perplexity: In search of post-normal questions for agroecosystem health. *Social Science and Medicine,* **45**, 1741–1749.

Waltner-Toews, D. and Lang, T. (2000). A new conceptual base for food and agricultural policy: The emerging links between agriculture, food, health, environment and society. *Global Change and Human Health,* **1**, 116–130.

Woodhill, J. and Rolings, N. (1998). The second wing of the eagle: The human dimension in learning our way to more sustainable futures. In: *Facilitating Sustainable Agriculture.* Roling, N.G. and Wagemakers, M.A. (eds.) Cambridge University Press, Cambridge.

Index